Modern RF and Microwave
Filter Design

For a listing of recent titles in the
Artech House Microwave Library,
turn to the back of this book.

Modern RF and Microwave Filter Design

Protap Pramanick
Prakash Bhartia

ARTECH HOUSE
BOSTON | LONDON
artechhouse.com

Library of Congress Cataloging-in-Publication Data
A catalog record for this book is available from the U.S. Library of Congress.

British Library Cataloguing in Publication Data
A catalog record for this book is available from the British Library.

ISBN-13: 978-1-63081-157-0

Cover design by John Gomes

© 2016 Artech House
685 Canton Street
Norwood, MA 02760

All rights reserved. Printed and bound in the United States of America. No part of this book may be reproduced or utilized in any form or by any means, electronic or mechanical, including photocopying, recording, or by any information storage and retrieval system, without permission in writing from the publisher.

All terms mentioned in this book that are known to be trademarks or service marks have been appropriately capitalized. Artech House cannot attest to the accuracy of this information. Use of a term in this book should not be regarded as affecting the validity of any trademark or service mark.

10 9 8 7 6 5 4 3 2 1

Contents

Preface xi

CHAPTER 1

Introduction 1

1.1 Applications of RF and Microwave Filters 1
1.2 Impedance Matching Networks 4
1.3 The Concept of Complex Frequency 5
1.4 Useful Definitions 6
1.5 Realizable Driving-Point Impedances 10
      References 12

CHAPTER 2

Microwave Network Theory 13

2.1 Introduction 13
2.2 Concepts of Equivalent Voltage and Current 13
2.3 Impedance and Admittance Matrices 17
2.4 Scattering Matrix 19
    2.4.1 Definition of Scattering Matrix 19
    2.4.2 Transformation of Scattering Matrix Due to Shift in Reference Plane 24
    2.4.3 Scattering Matrix of a Lossless Three-Port Network 28
    2.4.4 Usefulness of Scattering Matrix 28
    2.4.5 Scattering Matrix and the Concept of Insertion Loss 28
2.5 Measurement of Scattering Matrix 30
2.6 Chain or ABCD Matrix 34
    2.6.1 Use of ABCD Matrix in Computing Network Properties 36
    2.6.2 Normalized ABCD Matrix 37
        References 40

CHAPTER 3

Properties of Microwave Transmission Lines 41

3.1 Introduction 41
3.2 Transmission Line Equations 41
3.3 Transmission Line with Electrical Discontinuity 44

3.4	Two-Conductor Transmission Lines	45
	3.4.1 Two Round Conductors of Equal Diameter	45
	3.4.2 Coaxial Line	46
	3.4.3 Other Forms of Coaxial Line	50
	3.4.4 General Equation for Attenuation in Two-Conductor Transmission Lines	50
	3.4.5 Maximum Power-Handling Capability of a Coaxial Line	52
	3.4.6 Coaxial Line Discontinuities	54
3.5	Rectangular Coaxial Line	59
	3.5.1 Higher-Order Modes in a Rectangular Coaxial Line	60
	3.5.2 Square Coaxial Line with Circular Inner Conductor	60
3.6	Strip Transmission Line	61
	3.6.1 Basic Configuration	61
	3.6.2 Modes in a Stripline	62
	3.6.3 Characteristic Impedance of a Balanced Stripline	62
	3.6.4 Unbalanced Stripline	64
	3.6.5 Propagation Constant in a Stripline	65
	3.6.6 Synthesis of a Stripline	65
	3.6.7 Attenuation Constant in a Stripline	66
	3.6.8 Power-Handling Capability of a Stripline	67
	3.6.9 Stripline Discontinuities	67
3.7	Parallel-Coupled Lines	74
	3.7.1 Edge-Coupled Striplines	74
	3.7.2 Synthesis Equations	77
	3.7.3 Attenuation in Coupled Striplines	78
	3.7.4 Broadside Coupled Striplines	79
	3.7.5 Coupled-Slab Lines	85
3.8	Inhomogeneous Transmission Lines	87
	3.8.1 Shielded Microstrip Line	88
	3.8.2 Coupled Microstrip Line	92
	3.8.3 Suspended Microstrip Line	96
	3.8.4 Shielded Suspended Microstrip Line	98
	3.8.5 Edge-Coupled Suspended Microstrip Lines	100
	3.8.6 Broadside-Coupled Suspended Striplines	102
	3.8.7 Microstrip and Suspended Stripline Discontinuities	103
3.9	Single-Conductor Closed Transmission Lines	105
	3.9.1 Hollow Metallic Waveguides	105
	3.9.2 Characteristic Impedance of a Circular Waveguide	112
	3.9.3 Attenuation in a Circular Waveguide	112
	3.9.4 Maximum Power-Handling Capability of a Waveguide	116
	3.9.5 Power-Handling Capability of a Circular Waveguide	119
	3.9.6 Discontinuities in Waveguides and the Circuit Parameters	120
	3.9.7 Waveguide Asymmetric H-Plane Step	122
	3.9.8 Mode Matching Method and Waveguide Discontinuities	124

3.9.9	Waveguide Discontinuity Analysis in General	132
3.9.10	Finite Element Modal Expansion Method	133
3.9.11	Typical Three-Dimensional Discontinuity in a Rectangular Waveguide	137
	References	141

CHAPTER 4

Low-Pass Filter Design — 145

4.1	Insertion Loss Method of Filter Design	145
4.2	Belevitch Matrix and Transfer Function Synthesis	148
	4.2.1 Butterworth Approximation	150
	4.2.2 Cauer Synthesis	153
	4.2.3 General Solution for Butterworth Low-Pass Filter Response	157
	4.2.4 Chebyshev Approximation	161
	4.2.5 Elliptic Function Approximation	168
	4.2.6 Generalized Chebyshev Low-Pass Filters	175
4.3	Concept of Impedance Inverter	178
	4.3.1 Physical Realization of Impedance and Admittance Inverters	186
4.4	Low-Pass Prototype Using Cross-Coupled Networks	189
4.5	Design of Low-Pass Prototypes Using Optimization and the Method of Least Squares	193
	References	198
Appendix 4A		200
	4A.1.1 Darlington Synthesis	201

CHAPTER 5

Theory of Distributed Circuits — 207

5.1	Distributed Element Equivalence of Lumped Elements	207
5.2	TLE or UE and Kuroda Identity	210
5.3	Effects of Line Length and Impedance on Commensurate Line Filters	212
	5.3.1 Design Example	217
	5.3.2 Low-Pass Filter Prototype with a Mixture of UEs and Impedance Inverters	219
	5.3.3 Quasi-Distributed Element TEM Filters	230
5.4	High-Pass Filter Design	233
	5.4.1 Low-Pass to High-Pass Transformation	233
	5.4.2 Quasi-Lumped Element High-Pass Filter	234
	5.4.3 Levy's Procedure for Planar High-Pass Filter Design	237
	5.4.4 Waveguide High-Pass Filter Design	242
5.5	Band-Stop Filter Design	243
	5.5.1 Low-Pass to Band-Stop Transformation	243
	5.5.2 Determination of Fractional 3-dB Bandwidth of a Single Branch	251
	5.5.3 Effects of Dissipation Loss in Band-Stop Filters	252

	5.5.4	Stub-Loaded Band-Stop Filter	253
	5.5.4	Design Steps for Waveguide Band-Stop Filters	258
		References	263

CHAPTER 6

Band-Pass Filters — 265

6.1 Theory of Band-Pass Filters — 265
6.2 Distributed Transmission Line Form of Capacitively Coupled Band-Pass Filter — 275
  6.2.1 Gap-Coupled Transmission Line Band-Pass Filters — 275
  6.2.2 Edge Parallel-Coupled Band-Pass Filters — 277
  6.2.3 Hairpin-Line Filter — 284
  6.2.4 Interdigital Filters — 289
  6.2.5 Capacitively Loaded Interdigital Filters — 292
  6.2.6 Band-Pass Filter Design Based on Coupling Matrix — 295
  6.2.7 Coaxial Cavity Band-Pass Filter Design — 303
  6.2.8 Combline Filters — 303
  6.2.9 Waveguide Band-Pass Filter Design — 310
  6.2.10 Evanescent-Mode Waveguide Band-Pass Filter Design — 314
  6.2.11 Cross-Coupled Resonator Filter Design — 319
  6.2.12 Design of Cross-Coupled Filters Using Dual-Mode Resonators — 330
  6.2.13 Folded Resonator Cross-Coupled Filters — 334
  6.2.14 Cross-Coupled Filters Using Planar Transmission Lines — 338
6.3 Cross-Coupled Band-Pass Filters with Independently Controlled Transmission Zeros — 345
6.4 Unified Approach to Tuning Coupled Resonator Filters — 346
6.5 Dielectric Resonator Filters — 349
  6.5.1 Introduction — 349
  6.5.2 Modes in a Dielectric Resonator — 351
    References — 368
Appendix 6A Slot Coupled Coaxial Combline Filter Design — 373
Appendix 6B A Step-By-Step Procedure for Waveguide Folded and Elliptic Filter Design — 375
  Step 1: Designing the Basic Filter — 375
  Step 2: Bringing the Nonadjacent Resonators to Be Cross Coupled Close to Each Other — 375
  Step 3: Generating the Cross Coupling — 375
Appendix 6C Design of Dielectric Resonator Filters — 380

CHAPTER 7

Design of Multiplexers — 385

7.1 Definition of a Multiplexer — 385
7.2 Common Junction Multiplexer with Susceptance Annulling Network — 386
7.3 Cascaded Directional Filter — 388

7.4	Circulator Based Multiplexer	392
7.5	Manifold Multiplexer	394
	7.5.1 Diplexer Design	398
	7.5.2 Multiplexer Design	401
	References	402

About the Authors   405

Index   407

Preface

This book was started about 6 years ago and is only now complete, mainly due to work and time pressures. It has been quite a challenge as new concepts were developed over the years, others were refined, and yet others became obsolete—thereby requiring regular upgrading of the information and rewriting of certain material. Over this period, other books on filters have been written and published as well, perhaps leaving the reader wondering about what this book adds to the literature. As filter designers, we realized that we were constantly searching through the literature for the specific information we needed. No book on its own covered all the required coupling equations, power-handling details, and so on, for a specific transmission line configuration. Hence we decided to put together this edition with all the practical information any designer would need for his or her configuration. This text presents the design of filters in a continuous and comprehensive manner, thereby eliminating the need for the student or filter designer to extract the relevant information from published papers or other texts. In comparison to another recent book on filter design, this book is concise, does not try to cover topics that are of academic interest only or present obsolete ideas that have been published but were impractical and never implemented.

Unlike most other microwave components, microwave filters can be designed and manufactured with predictable performance. Therefore, it is cost effective to design the microwave components in a circuit with broadband characteristics and use filters to achieve the system performance desired at a lower cost than would otherwise be possible.

Filters can be fabricated from lumped elements (inductors, capacitors, and resistors) or from distributed elements, or both. Basically, there are four types of filters: low-pass, band-pass, band-stop, and high-pass. However, once the theory and practical implementation of a low-pass filter are clearly understood, this information can be used to advantage in designing and fabricating the high-pass, band-pass, or band-stop version.

This book is therefore organized to provide the student, engineer, or designer with the fundamentals of microwave theory, microwave transmission lines, and low-pass filter design, which is then expanded and forms the basis or building blocks for the other filter types in later chapters. Thus, Chapter 1 presents some basic information on impedance-matching networks, concept of complex frequency, useful definitions that are essential to understanding the rest of the text, and realizable driving-point functions.

In Chapter 2, the critical aspects of microwave network theory, admittance and impedance matrices, scattering matrices, and ABCD matrix are discussed. This chapter contains information that is available in most microwave engineering texts and known by most microwave engineers, but is repeated here for completeness in an abbreviated form to provide the reader a ready reference instead of having to search for the material elsewhere.

All filters, unless they are lumped element designs, are built of transmission lines of some sort. They may be fabricated in coaxial line form, waveguides, or planar transmission line form. Typical forms of planar transmission lines are microstrip lines, striplines, and variations of these lines. While waveguides and coaxial line structures are more common where real estate or power requirement so dictate or permit, filters using planar transmission lines are more commonly found in most microwave integrated circuits, where size is a very limiting factor. In addition to having a clear understanding of the design and characteristics of transmission lines, all filters require a clear comprehension of coupling mechanisms involving these transmission lines. This alone adds considerable complexity as a simple stripline can be edge-coupled, broadside-coupled, balanced or unbalanced, shielded or unshielded, and so forth. In Chapter 3, all required information for understanding the nature of these transmission lines is clearly presented, as this information forms the building blocks for the design of filters that follows in the rest of the text.

After Chapter 3, the material on low-pass filter design contained in Chapter 4 is perhaps the most important part of the book. As explained in the text and mentioned above, this is because the low-pass filter design can be used to build all other types of filters. This chapter includes the most commonly used methods of low-pass filter design, including the insertion loss method, the Belevitch matrix, transfer function synthesis, and impedance inverter concepts while explaining and systematically demonstrating the design process. The design procedure is clearly enunciated through some examples.

With the background developed in Chapter 4 on low-pass filters, and using the theory of distributed network synthesis explained in the beginning of Chapter 5, one can easily extend this theory to the design of high-pass filters using a low-pass to high-pass transformation. This is covered in the chapter, together with Levy's method for planar high-pass filter design and the design of high-pass waveguide filters. The technique can be further extended and applied to band-stop filters (also called band rejects or notch filters) and this forms the topic of discussion for the balance of the chapter.

Chapter 6 covers the design of band-pass filters in its many forms. These are perhaps the most commonly used types of filters and the reader is probably familiar with most of the names, such as edge-coupled band-pass filters, hairpin line filters, interdigital filters, combline filters, and so forth. The design of the most common waveguide types of these filters is also discussed, including evanescent-mode band-pass filters and cross-coupled filters. Dielectric resonator filters are discussed with design examples, as these filters have the advantage of compact size with very narrow bandwidths. The reader and designer are led through a step-by-step procedure for designing the different filter types and design examples are presented to bring about a clear understanding.

Finally, in Chapter 7 we discuss the design of multiplexers as these are commonly used in microwave systems and are essentially an extension of filter applications.

The complexities involved in the design of multiplexers are discussed, together with the most commonly implemented configurations. These include common junction with susceptance annulling networks, cascaded directional filters, channel filters separated by circulators, and manifold multiplexers. These configurations and their designs are covered in the final sections.

Throughout the text, computer programs and computer-assisted design software available and suitable for specific filter design applications are mentioned. As this is an all-inclusive text, designers can easily develop their own computer codes for their requirements from the material here without having to use other sources.

We wish to thank our families for their patience over the years and acknowledge the support of Mr. Douglas F. Carlberg and Mr. Archie Wohlfahrt of M2G Global Inc., Kevin Asplen, Ken Sears of K & L Microwave, Mrs. Suzanne Wright, Dr. Rudolf Cheung, Chris Holman, Hong Chau, Maurice Aghion and Sandeep Palreddy of Microwave Engineering, Dr. Paul Smith of Microcommunications Inc., and Prof. Fritz Arndt and Dr. Jill Arndt of Microwave Innovation Group (WaspNet), and Arun Ray for their help and encouragement during this writing. Finally, we wish to acknowledge those who helped with the typing and artwork and granted permissions to use their intellectual property.

CHAPTER 1

Introduction

1.1 Applications of RF and Microwave Filters

A filter in an electrical network is a two-port (input port and output port) device that allows signals of a certain desired band of frequencies to pass through it with very little attenuation or loss, while stopping the passage of other undesired bands of frequencies with very high attenuation or rejection. At microwave frequencies, filters are extensively used in communication and radar systems and instrumentation. In addition, the study of theory and realization of microwave filters is very important because any passive microwave network can be decomposed into a combination of several filterlike networks for ease of analysis and design. In this book we will describe not only how to realize microwave filters but also several other passive components that can be analyzed and designed using filter theory and realization methods. Figure 1.1 shows a typical block diagram of a simplified satellite downlink using several microwave filters [1].

In general, filters are of four types: low-pass, high-pass, band-pass, and band-stop. Figure 1.2 shows the ideal and theoretical characteristics of a high-pass filter, where A is the attenuation in decibels (dB) and f is the frequency in hertz (Hz) [2].

In Figure 1.2, f_c is the cutoff frequency of the filter, A_s the stop-band attenuation, and the filter offers 0-dB attenuation in the passband. In addition, there is no frequency band separating the passband and the stopband. However, it is impossible to physically realize such a filter. According to the Paley-Wiener causality criterion [3], the transfer function $H(j\omega)$ of a network must satisfy the following equations

$$\int_{-\infty}^{\infty} \frac{\left|\log|H(j\omega)|\right|}{1+\omega^2} d\omega < \infty \tag{1.1}$$

and

$$\int_{-\infty}^{\infty} |H(j\omega)|^2 d\omega < \infty \tag{1.2}$$

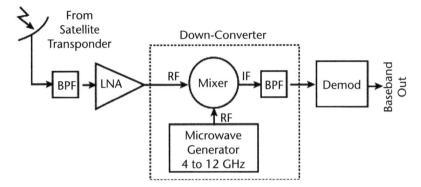

Figure 1.1 Block diagram of a simplified satellite downlink. (From [1].)

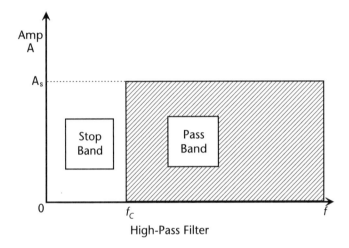

Figure 1.2 Characteristic of an ideal high-pass filter. (From [2].)

where $|H(j\omega)|$ is the modulus of the frequency domain transfer function of the network and ω is the radian frequency. The physical implication of the Paley-Wiener criterion is that the amplitude function cannot fall off to zero faster than an exponential order. The ideal high-pass response in Figure 1.2 cannot be realized because the transition from passband to stopband follows an abrupt step function violating the Paley-Wiener criterion. Therefore, in reality, every filter has a transition band between the passband and the stopband. Figure 1.3 shows a few realizable shapes of frequency responses for different types of filters.

More than one filter can be connected through a common junction network to form what is known as a *multiplexer*. A multiplexer is used when one needs to separate or combine signals belonging to different frequency bands. Figure 1.4 shows the schematic of a typical two-channel multiplexer network. It consists of two band-pass filters. Such a network is known as a diplexer. However, there can be more than two channel filters in a multiplexer. As a result, we get networks known as *triplexers* and *quadruplexers*.

The diplexer shown in Figure 1.4 consists of two bandpass filters connected at the input through a common junction network. Channel 1 covers the frequency

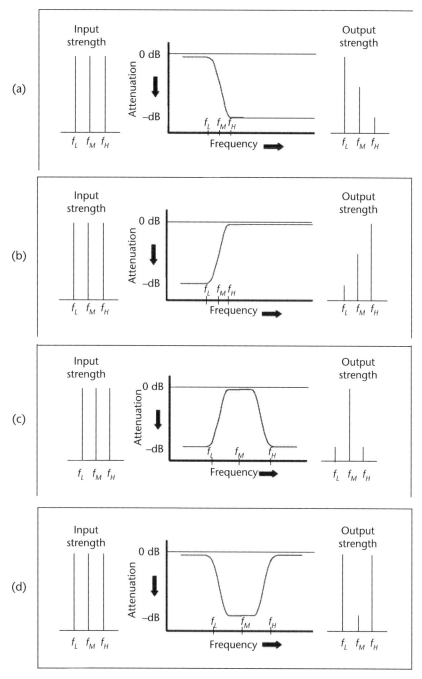

Figure 1.3 (a) Realizable low-pass filter response, (b) realizable high-pass filter response, (c) realizable band-pass filter response, and (d) realizable bandstop filter response. (From [4].)

band from 2 to 4 GHz and channel 2 covers the frequency band from 3 to 5 GHz. There is a 1-GHz guardband between the two channels. The purpose of the common junction is to reduce the power reflected at the input port of the diplexer to a negligibly small value over the frequency bands of the channel filters. In addition, the constituent filters are designed using special measures to match the impedance of the common junction network.

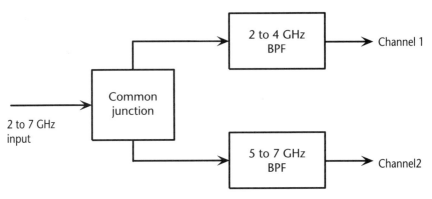

Figure 1.4 Block diagram of a diplexer.

If we reverse the directions of the arrows in Figure 1.4, we obtain a *channel combiner*. A channel combiner serves the opposite function of a multiplexer. It superimposes the signals coming out of different channel filters on a single output line. The matching network, together with the specially designed filters, assure impedance matching at all ports and minimum signal loss due to unwanted reflections at the common junction. Besides the common junction approach, the channel filters can also be connected to each other using another approach known as the manifold method. The design and analysis of multiplexers are treated in much greater detail in Chapter 7.

1.2 Impedance Matching Networks

An impedance matching network is the most widely used circuit component in any communication system. Impedance matching ensures optimum power transfer efficiency, good noise performance, and optimum gain. All impedance matching networks can be categorized under two groups, impedance transformers and filters. The purpose of a matching network is to minimize the undesired power reflection at a port over a desired passband of a system or a device.

The objective of an impedance matching network design is to obtain the minimum reflection over the widest bandwidth, unless a narrowband or single frequency matching is required. However, in network design, fulfilling that objective is always a challenge because according to Fano [5] and Bode [6], low reflection and wide bandwidth are always a trade-off when a matching network is required to match a load with a reactive part, as shown in Figure 1.5.

Bode's [5] gain-bandwidth integral restriction is given by

$$\int_0^\infty \ln\left|\frac{1}{\Gamma}\right| d\omega \leq \frac{\pi}{RC} \qquad (1.3)$$

where $\omega(=2\pi f)$ is the angular frequency, and Γ is the input reflection coefficient of the matching networks. Since for perfect matching, the designer looks for zero reflection or $\Gamma=0$, $1/\Gamma$ should become infinite. However, the above equation indicates

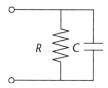

Figure 1.5 Parallel RC circuit.

that the integral always remains bounded by the values of R and C. If the reflection coefficient is a constant over a desired frequency band ranging from ω_1 to ω_2 and has the value ρ and unity at any other frequency, then (1.3) gives

$$\int_0^\infty \ln\left|\frac{1}{\Gamma}\right| d\omega \leq \frac{\pi}{RC}(\omega_1 - \omega_2)\ln|\rho| \qquad (1.4)$$

It is obvious from (1.4) that for large RC, wideband matching becomes a challenging task. Almost all matching networks behave as bandpass filters. Ideally it is desired that the reflection coefficient within the desired passband should be zero and that it should be unity at any other frequency. However, this is a brick-wall type response that, according to the Paley-Weiner criterion, cannot be physically realized. A physically realizable matching network should have a maximum but low reflection coefficient ρ_m in the passband and unity in the stopband while there must be a transition band between them. The transition band can be made narrow by the use of a larger number of matching sections, theoretically reducing it to zero with an infinite number of sections. This topic is treated in greater depth in Chapter 4.

1.3 The Concept of Complex Frequency

The generalized sinusoidal voltage can be written as

$$v(t) = |V|e^{-\sigma t}\cos(\omega t + \varphi) \qquad (1.5)$$

Then $v(t)$ is said to have a complex frequency, where

$$s = \sigma + j\omega \qquad (1.6)$$

In the above equations, t is the time in seconds, ω is the angular frequency in radians per second, and σ is the attenuation constant in nepers per unit length. Also φ is the initial phase of the voltage at $t = 0$. Figure 1.6(a) shows the dependence of $v(t)$ on t for $|V| = 38$ volts, $\sigma = -0.05$, $\omega = 0.2$ rad/sec, and $\varphi = \pi/4$ radians. It is a damped sinusoid. Figure 1.6(b) shows the voltage waveform when $\sigma = 0.05$ and Figure 1.6(c) shows the voltage waveform when $\sigma = 0.00$.

We can also write (1.5) as

$$v(t) = \mathrm{Re}\{Ve^{st}\} \qquad (1.7)$$

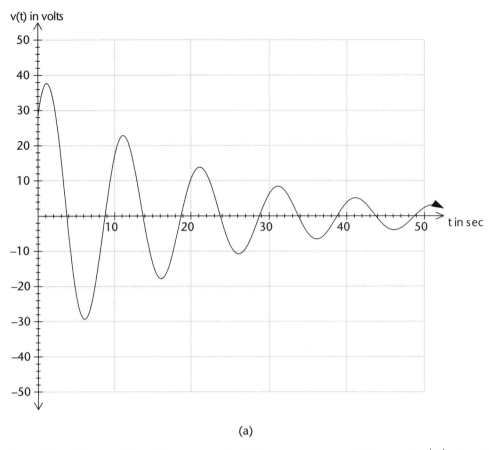

(a)

Figure 1.6 (a) Damped sinusoidal voltage with (a) damping factor $\sigma = -0.05$, $w = 0.2\pi$, $|V|=40$ and $\varphi=\pi/4$, (b) damping factor $\sigma=0.05$, $w=0.2\pi$, $|V|=40$, and $\varphi=\pi/4$, (c) damping factor $\sigma=0.00$, and $w=0.2\pi$ rad/sec, $|V|=40$ volts and $\varphi = \pi/4$. (d) Nonsinusoidal and exponentially varying voltages when $s = \pm\sigma + j0$.

and

$$V = |V|e^{j\Phi} \tag{1.8}$$

Once again, V is a complex quantity. Therefore, the real part σ of the complex frequency describes the decay or growth of the amplitude of the voltage and the imaginary part ω describes the angular frequency of the sinusoidal voltage in the usual sense.

When $\varphi = \omega = 0$, the voltage has an exponential time variation, as shown in Figure 1.6(d).

1.4 Useful Definitions

This book deals primarily with linear time-invariant passive networks. Such networks have the following properties.

The principle of superposition holds for a *linear time-invariant network*. By that principle, if, for a given network $v_{1o}(t)$ and $v_{2o}(t)$ are the voltage outputs

1.4 Useful Definitions

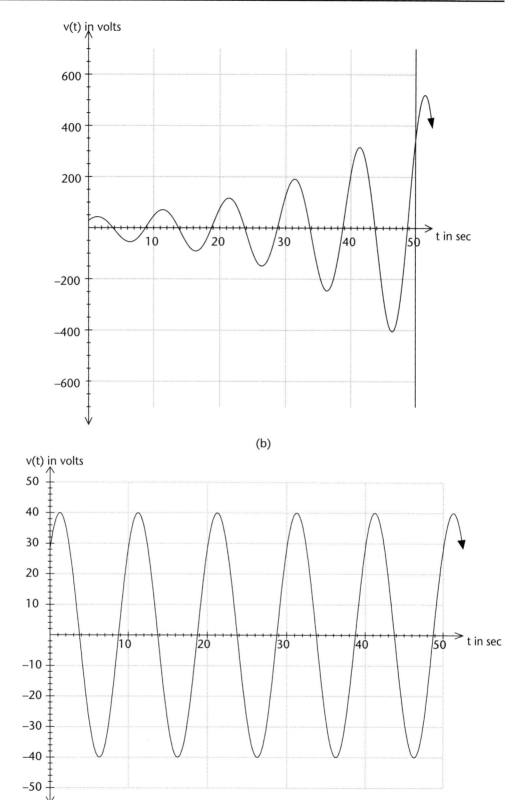

Figure 1.6 (continued)

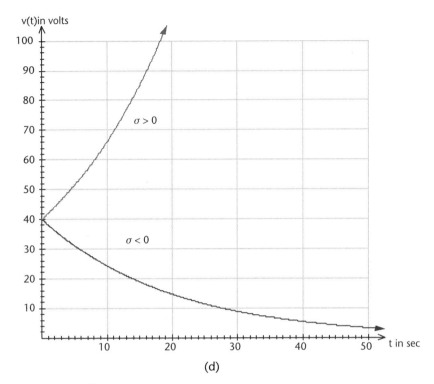

Figure 1.6 (continued)

corresponding to the voltage inputs $v_{1i}(t)$ and $v_{2i}(t)$, respectively, then corresponding to an input $v_{1i}(t) + v_{2i}(t)$ the output of the network should be $v_{1o}(t) + v_{2o}(t)$. From this it automatically follows that if the input is scaled by a factor ξ, the output is also scaled by the same factor ξ.

A passive network is *time-invariant*, meaning that its properties remain unchanged with the variation of time. For example, if $v_o(t)$ is the output of a linear time-invariant network corresponding to an input $v_i(t)$, then $v_o(t+\tau)$ will be the output corresponding to an input $v_i(t+\tau)$. This means that a time-invariant system is composed of elements that do not vary with time. As a result, if the input is delayed by any amount of time τ, the output is also delayed by the same amount of time τ. However, every linear network is not necessarily time-invariant.

A linear network is *passive* if no voltages or currents appear between any two terminals in it before an excitation is applied to the network. Also if the input signal (voltage or current) has the complex frequency $s = \sigma + j\omega$, then the output will have the same frequency. However, the phase angle φ and the amplitude $|V|$ may be different at the output.

A linear network is *reciprocal* if the response remains the same when the terminal pairs (ports) of measurement and excitation are interchanged.

A linear network is *causal* if its response is nonanticipatory. For example, if $v_o(t)$ is the output corresponding to an input $v_i(t)$ and for $t < T$

$$v_i(t) = 0 \tag{1.9}$$

then

1.4 Useful Definitions

$$v_o(t) = 0; \text{ for } t < T \tag{1.10}$$

A network function is the ratio of the response function to the excitation function in the complex frequency (s) domain. The excitation and the response can be voltage or current. A network function assumes various names depending on whether the excitation and response functions are voltage or current. Table 1.1 summarizes the relationships among various excitation, response, and network functions in a one-port network. Table 1.2 shows the relationships among various parameters in a two-port network.

Although various voltages and currents used in the network are actually in the time domain, the parameters in Table 1.1 are in the complex frequency or s domain. Therefore, the driving-point or transfer function are defined in the complex frequency or s domain. Whether it is a driving-point function or a transfer function, a network function $T(s)$ can always be described as a rational function of s. A rational function in s is the ratio of two polynomials in s. Therefore, it has the form

$$T(s) = \frac{a_0 + a_1 s + \ldots + a_n s^n}{b_0 + b_1 s + \ldots + b_m s^m} \tag{1.11}$$

From (1.11) we find that since the right-hand side is a ratio of two polynomials, the roots of the numerator polynomial determine the complex frequencies when $Z(s)$ goes to zero, while the roots of the denominator polynomial are the complex frequencies when $Z(s)$ goes to infinity. Therefore, if we write the numerator and the denominator of (1.11) in factored forms as

$$T(s) = \frac{a_n(s - z_1)(s - z_2) \ldots (s - z_n)}{b_m(s - p_1)(s - p_2) \ldots (s - p_m)} \tag{1.12}$$

then $z_1, z_2, \ldots, z_n$ are defined as the zeros of $T(s)$; while $p_1, p_2, \ldots, p_m$ are defined as the poles of $T(s)$. The locations of these poles and zeros of a driving-point or transfer function $T(s)$ in the complex frequency plane determine the behavior of a network in the time or frequency domain. Consequently, such locations of poles

Table 1.1 One-Port Network Parameters

Ratio	Symbol Used	Function Name
$\dfrac{V(s)}{I(s)}$	$Z(s)$	Driving point
$\dfrac{I(s)}{V(s)}$	$Y(s)$	Driving point

and zeros of a network in the complex frequency plane serve as a valuable aid to network synthesis.

From a physical standpoint, the poles and zeros of a network are directly related to its various natural modes of electrical resonance. If $T(s)$ is a driving-point impedance function, then the zeros of $T(s)$ denote the frequencies at which the driving point becomes a short circuit, while the poles of $T(s)$ represent the frequencies at which the driving point becomes an open circuit. Corresponding conclusions can be drawn about the poles and zeros of a driving point admittance function. For instance, the zeros of a driving point admittance function represent the frequencies at which the driving point becomes an open circuit and the poles denote the frequencies at which it is a short circuit.

If $T(s)$ represents one of the transfer functions in Table 1.2, then the poles of $T(s)$ represent the frequencies at which no transmission takes place through the network. Such frequencies are also known as *attenuation poles* or *transmission zeros*. Similarly, the zeros of $T(s)$ represent the frequencies at which attenuation is zero through the network. We will show in the following sections that not every function of the complex variable s is a realizable driving point or transfer function.

1.5 Realizable Driving-Point Impedances

Only those driving point functions that are positive real functions are physically realizable. A positive real function $Z(s)$ is defined as [6] one having the following properties [7]:

$T(s)$ is a rational analytic function in the right half complex plane and it has simple poles with real positive residues on the imaginary axis; and if the real part of $[T(j\omega)]$ is greater than zero for all real values of ω. Mathematically speaking

$$\mathrm{Re}\left[T(j\omega)\right] \geq 0; \quad \text{For } \omega \geq 0 \tag{1.13}$$

A number of other properties of $T(s)$ can be derived from the above conditions. The most significant among them is that $T(s)$ has no poles or zeros in the right half of the complex plane. Therefore, the roots of the numerator and the denominator polynomials always lie either in the left-half plane or on the imaginary axis of the complex plane.

According to Otto Brune's theorem [8], any rational function that satisfies the above conditions and are necessary and sufficient is a positive real (p.r) function and represents a physically realizable impedance function. Brune proved his theorem by developing the method of synthesizing a prescribed impedance function as a two-terminal (one-port) pair network. Darlington [9] showed how a dissipative two-terminal network could be realized as a four-terminal lossless network terminated by a resistance. Darlington's procedure later became the basis for the synthesis of many passive multiports and mainly two-port filter networks. One has to keep in mind that whether it is the synthesis of a driving-point function or a transfer function, it eventually boils down to a synthesis of a driving-point impedance function. For example, in case of a filter synthesis the initial prescribed function is always a voltage transfer function. In the next step an equivalent compatible input reflection function is derived from the transfer function, assuming that the filter is

Table 1.2 Two-Port Network Parameters

[Diagram: Two-port network with $I_1(s)$ entering, $V_1(s)$ across input port with + and − polarity, Network block, $V_2(s)$ across output port, $I_2(s)$ exiting, terminated in Z_L.]

Ratio	Symbol Used	Function Name
$\dfrac{V_1(s)}{I_1(s)}$	$Z_{11}(s)$	Driving point
$\dfrac{V_2(s)}{I_2(s)}$	$Z_{22}(s)$	Driving point
$\dfrac{-I_2(s)}{-V_1(s)}$	$Y_{21}(s)$	Transfer
$\dfrac{V_2(s)}{I_1(s)}$	$Z_{21}(s)$	Transfer
$\dfrac{-I_2(s)}{I_1(s)}$	$\alpha_{21}(s)$	Transfer
$\dfrac{V_2(s)}{V_1(s)}$	$G_{21}(s)$	Transfer

*$V_1(s)$ and $I_1(s)$ are at the excitation port and $V_2(s)$ and $I_2(s)$ are at the response port.

terminated by a 1Ω resistance at the output. The derived input reflection function is then transformed into an equivalent input driving-point impedance function assuming the source impedance to be 1Ω. The locations of the poles and zeros of the voltage transfer function in the complex frequency plane play a significant role in the above two transformations. The driving-point impedance is realized using one of the standard procedures of Brune [8] and Darlington [9].

A large number of multiport passive components can be shown to be combinations of several two-port subnetworks. Each such two-port network has its own frequency characteristics. Hence by definition, it is a filter. For example, a two-way power divider, which is a three-port device, can be divided into two two-port networks using a suitable magnetic or electric wall at a symmetry plane. The same thing can be done for parallel transmission line couplers having four ports. Therefore, the principle, briefly described in the preceding sections for two-port filter synthesis, can also be applied to those three- and four-port devices.

In conclusion, we emphasize that the term "electrical filter" is a very general one. Electrical filters have applications in every aspect of electrical and communication engineering. An excellent historical perspective of RF and microwave filters is available in Levy and Cohn's article published by the Institute of Electrical and Electronics Engineers (IEEE) in 1984 [10].

The scope of filters is virtually limitless. We should also keep in mind that since any transfer function is basically a complex quantity, filtering in general not only means energy filtering but also phase filtering. The latter is directly related to time-delay networks.

References

[1] Hindle, P., "Mil SatCom Capacity Crunch: The BUC Stops Here," *Microw. J.*, Vol. 52, No. 8, August 1, 2009.

[2] Frequency Devices, Analog and Digital Products Design/Selection Guide, Ottawa, IL, http://www.freqdev.com.

[3] Paley, R. E. A. C, and N. Wiener, *Fourier Transforms in Complex Domain,* Vol. 19, New York: American Mathematical Society Colloquium Publications, 1934, pp. 16–17.

[4] EW and Radar Systems Engineering Handbook, Naval Air Systems Command, Avionics Division, Washington D.C.)

[5] Fano, R. M., "Theoretical Limitations on the Broadband Matching of Arbitrary Impedances," *J. Franklin Inst.*, Vol. 249, January/February, 1950, pp. 57–84 and 139–154.

[6] Bode, H. W., *Network Analysis and Feedback Amplifier Design*, New York: Van Nostrand Co., 1945, pp. 360–371.

[7] Guillemin, E. A., *Mathematics of Circuit Analysis*, John Wiley and Sons, New York, 1957.

[8] Brune, O., "Synthesis of Finite Two Terminal Network Whose Driving Point Impedance is a Prescribed Function of Frequency," *J. Math. Phys.*, Vol. 10, 1931, pp. 191–236.

[9] Darlington, S. "Synthesis of Reactance Four-Poles with Prescribed Insertion Loss Characteristics," *J. Math. Phys.*, Vol.18, 1939, pp. 257–353.

[10] Levy, R., and S. B. Cohn, "A History of Microwave Filter Research, Design, and Development", *IEEE Trans. Microw. Theory Tech.*, Vol. 32, No. 9, September 1984, pp. 1055–1067.

CHAPTER 2
Microwave Network Theory

2.1 Introduction

Microwave passive and active networks can be classified as multiport networks. Such networks are also known as N-port networks. Assuming the input and the output ports of an N-port microwave network are known, its frequency and time domain responses to a known excitation can be determined. For instance, if one of the ports of a transistor is terminated in a short circuit, the frequency response of the remaining two-port network can be obtained from knowledge of the original three-port network. Also, in any microwave circuit, system or subsystem, there are many components connected in a desired fashion. The frequency response of this system can be obtained from knowledge of the individual components.

There are many equivalent ways in which the frequency response of a linear microwave network can be calculated [1]. This chapter deals with the representation of linear microwave networks as multiport black boxes.

2.2 Concepts of Equivalent Voltage and Current

The determination of network characteristics at microwave frequencies involves concepts that are substantially different from those used for low-frequency RF circuits for which voltages and currents that determine impedance can be uniquely defined. At microwave frequencies, use of either high-frequency probes or low-impedance current measurements are not possible because the parasitic impedance and capacitance cannot be made small enough. In addition, the physical dimensions of a microwave circuit are no longer small compared to the wavelength. Therefore, in most cases, the concepts of equivalent voltage and current are used. The equivalent voltage and current are so chosen that the power transmitted along a transmission line is computed correctly using the equation

$$P = \frac{1}{2}\text{Re}\int_S (E \times H).ds = \frac{1}{2}\text{Re}(VI^*) \qquad (2.1)$$

where the power P is assumed to be flowing in the z direction. V and I are the equivalent voltage and current respectively, and I^* is the complex conjugate of I. E and H are the electric and magnetic fields, respectively, in the x-y plane. S is the cross-sectional area of the transmission line. We can use the familiar definitions of voltage and current only in electrostatics. As soon as the electric field becomes time-dependent, it generates a time-varying magnetic field, which in turn gives rise to a dynamic electric field. The electric voltage corresponding to this new dynamic electric field depends on the path of the line integral chosen to calculate the voltage.

Consider the transverse part of the electromagnetic field in a transmission line (e.g., a coaxial line or a waveguide), shown in Figure 2.1.

$$E_T^+ = \xi e_T(x, y) e^{-j\beta z} \qquad (2.2a)$$

$$H_T^+ = \xi h_T(x, y) e^{-j\beta z} \qquad (2.2b)$$

where ξ is a constant of proportionality, $e_T(x,y)$ and $h_T(x,y)$ are the normalized modal functions such that

$$\int_S (e_T(x,y) \times h_T(x,y)) \cdot ds = 1 \qquad (2.2c)$$

and β is the propagation constant in the z direction. According to Collin [1] the above fields are proportional to the equivalent voltage and current. Therefore

$$E_T^+ = K_V V^+ e_T(x,y) e^{-j\beta z} \qquad (2.3a)$$

$$H_T^+ = K_I I^+ h_T(x,y) e^{-j\beta z} \qquad (2.3b)$$

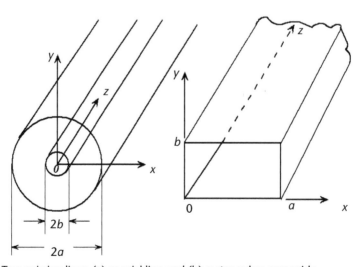

Figure 2.1 Transmission lines: (a) coaxial line and (b) rectangular waveguide.

2.2 Concepts of Equivalent Voltage and Current

where K_V and K_I are constants of proportionality and V^+ and I^+ are the forward-going equivalent voltage and current, respectively. The transmitted power associated with the forward wave is given by

$$P^+ = \frac{1}{2}\text{Re}\int_S \left(E_T^+ \times H_T^+\right) \cdot ds \quad (2.4)$$

Combining (2.2c), (2.3a), (2.3b), and (2.4) gives

$$P^+ = \frac{1}{2}K_V K_I \,\text{Re}(V^+ I^{+*}) \quad (2.5)$$

Comparing (2.1) and (2.5) gives

$$K_V K_I = 1 \quad (2.6)$$

Using transmission line theory, we can write

$$\frac{V^+}{I^+} = Z_0 \quad (2.7)$$

where Z_0 is the characteristic impedance of the line. Comparing (2.2), (2.3), and (2.7), we get

$$\frac{K_I}{K_V} = Z_0 \quad (2.8)$$

Solving equations (2.6) and (2.8) gives

$$K_V = \frac{1}{\sqrt{Z_0}} \quad (2.9a)$$

$$K_I = \sqrt{Z_0} \quad (2.9b)$$

The characteristic impedance Z_0 of a coaxial or two-conductor transmission line can be uniquely defined for the fundamental TEM mode. Therefore, the equivalent forward voltage and current can be uniquely expressed using (2.2), (2.3), and (2.9) as

$$V^+ = \zeta\sqrt{Z_0} \quad (2.10a)$$

$$I^+ = \frac{\zeta}{\sqrt{Z_0}} \tag{2.10b}$$

Similarly, the backward voltage and current can be written as

$$V^- = \kappa\sqrt{Z_0} \tag{2.11a}$$

$$I^- = \frac{\kappa}{\sqrt{Z_0}} \tag{2.11b}$$

where ζ and κ are constants of proportionality.

If the transmission line supports a non-TEM mode (i.e., TE or TM mode in a waveguide), the definition of the characteristic impedance cannot be unique. For example, if one considers the ratio of the fundamental mode average power flowing through a waveguide and the voltage at the center of the broad wall, the characteristic impedance assumes the form

$$Z_0(f) = \frac{2b}{a}\frac{\eta_0}{\sqrt{1-\left(\frac{f_c}{f}\right)^2}} \tag{2.12a}$$

where a and b are the width and the height of the waveguide cross section, respectively, η_0 is the free space impedance (120π), f is the operating frequency, and f_c is the cutoff frequency of the fundamental mode.

If one considers the ratio of the same voltage and the total longitudinal current then the characteristic impedance assumes the form

$$Z_0(f) = \left(\frac{\pi b}{2a}\right)\frac{\eta_0}{\sqrt{1-\left(\frac{f_c}{f}\right)^2}} \tag{2.12b}$$

Also, if one considers the ratio of the fundamental mode average longitudinal power and the total longitudinal current, one gets

$$Z_0(f) = \left(\frac{\pi^2 b}{8a}\right)\frac{\eta_0}{\sqrt{1-\left(\frac{f_c}{f}\right)^2}} \tag{2.12c}$$

2.3 Impedance and Admittance Matrices

Therefore, a unique definition of characteristic impedance is possible only for those transmission lines, which support the TEM mode of propagation. However, for a transmission line that supports a non-TEM mode the definition of characteristic impedance depends on the application. For example, for the purpose of impedance matching between different waveguide cross sections, any one of the above three definitions may be chosen, but in order to match a two-terminal device to a ridged waveguide the voltage current definition is preferred.

2.3 Impedance and Admittance Matrices

Consider the N-port network shown in Figure 2.2. The accessible ports are denoted by 1, 2, 3, ..., N. In addition, there is a ground terminal at each port. Let us assume that the I_i ($i = 1,2,3, ..., N$) denote the port currents and the V_i ($i=1,2,3, ..., N$) denote the port voltages, respectively at ports 1 through N. The admittance matrix of the network is defined as

$$[I] = [Y][V] \tag{2.13}$$

where

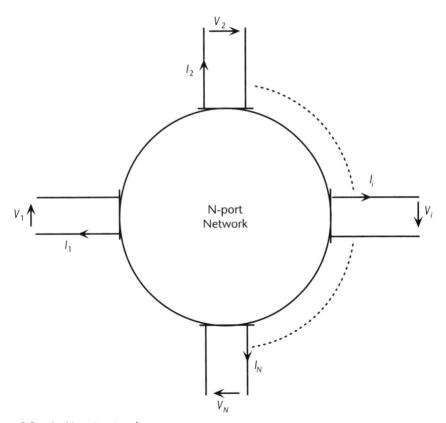

Figure 2.2 An N-port network.

$$[I] = \begin{bmatrix} I_1 \\ I_2 \\ . \\ . \\ I_N \end{bmatrix}; [V] = \begin{bmatrix} V_1 \\ V_2 \\ . \\ . \\ V_N \end{bmatrix} \text{ and } [Y] = Y \begin{bmatrix} Y_{11} & Y_{12} & Y_{13} & \cdots & Y_{1N} \\ Y_{21} & Y_{22} & Y_{23} & \cdots & Y_{2N} \\ Y_{31} & Y_{32} & Y_{33} & \cdots & Y_{3N} \\ \vdots & \vdots & \vdots & \ddots & \vdots \\ Y_{N1} & Y_{N2} & Y_{N3} & \cdots & Y_{NN} \end{bmatrix} \quad (2.14)$$

The admittance matrix [Y] is obtained by nodal analysis of the network and then solving for the port currents or voltages. The corresponding impedance matrix is defined as

$$[Z] = [Y]^{-1} \quad (2.15)$$

The above admittance and impedance matrices are also known as unnormalized admittance and impedance matrices, respectively.

A microwave component or a subsystem is connected to a larger system. As a result, the impedance offered by the larger system to which it is connected terminates each port of a component. Let us assume that the impedance vector gives the set of terminating impedances offered by the embedding system

$$[Z_I] = [Z_{I1} Z_{I2} - - - - - - Z_{IN}] \quad (2.16)$$

We define a new set of port voltages and currents, known as the normalized port voltages and currents, as follows:

$$v_i = \frac{V_i}{\sqrt{\text{Re}\, Z_{Ii}}} \text{ and } i_i = I_i \sqrt{\text{Re}\, Z_{Ii}} \quad (2.17)$$

Note that normalized voltages and currents have the same dimension, which is √(watt). Hence, the relationship between the normalized and the unnormalized parameters can be expressed as

$$[v] = [z][i] \quad (2.18)$$

where

$$[v] = [z_c]^{-\frac{1}{2}}[V] \text{ and } [i] = [z_c]^{\frac{1}{2}}[I] \quad (2.19)$$

and

$$[z_c] = \mathrm{Re} \begin{bmatrix} Z_{I1} & 0 & 0 & \cdots & 0 \\ 0 & Z_{I2} & 0 & \cdots & 0 \\ 0 & 0 & Z_{I3} & \cdots & 0 \\ \vdots & \vdots & \vdots & \ddots & \vdots \\ 0 & 0 & 0 & \cdots & Z_{IN} \end{bmatrix} \qquad (2.20)$$

$[z]$ is known as the normalized port impedance matrix and

$$[y] = [z]^{-1} \qquad (2.21)$$

is the corresponding normalized port admittance matrix. Table 2.1 shows the port admittance matrices of most commonly used microwave two-port passive elements.

Below we will show how normalized voltage and current matrices account for the interaction of the network with the system that embeds it.

2.4 Scattering Matrix

2.4.1 Definition of Scattering Matrix

The scattering matrix concept with respect to positive and real terminating impedances was introduced by Penfield [2]. Kurokawa introduced the concept of power wave variables in 1965 [3] and generalized the Penfield concept.

Following the method due to Kurokawa [3], let us consider a one-port network, as shown in Figure 2.3.

We define the following variables

$$a_1 = \frac{V_1 + Z_{I1} I_1}{2\sqrt{\mathrm{Re}\, Z_{I1}}} \qquad (2.22a)$$

$$b_1 = \frac{V_1 - Z_{I1}^* I_1}{2\sqrt{\mathrm{Re}\, Z_{I1}}} \qquad (2.22b)$$

where * denotes the complex conjugate. Solving (2.22a) and (2.22b), we obtain

$$V_1 = \frac{Z_{I1}^* a_1 + Z_{I1} b_1}{2\sqrt{\mathrm{Re}\, Z_{I1}}} \qquad (2.23a)$$

$$I_1 = \frac{a_1 - b_1}{2\sqrt{\mathrm{Re}\, Z_{I1}}} \qquad (2.23b)$$

Now, what are the parameters a_1 and b_1?

Table 2.1 Admittance Matrices of Some Common Two-Port Networks

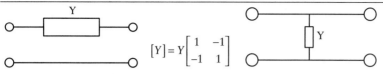

$$[Y] = Y \begin{bmatrix} 1 & -1 \\ -1 & 1 \end{bmatrix}$$

Admittance matrix does not exist

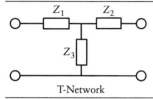

$$[Y] = \begin{bmatrix} \dfrac{Z_2 + Z_3}{Z_1 Z_2 + Z_2 Z_3 + Z_3 Z_1} & \dfrac{-Z_3}{Z_1 Z_2 + Z_2 Z_3 + Z_3 Z_1} \\ \dfrac{-Z_3}{Z_1 Z_2 + Z_2 Z_3 + Z_3 Z_1} & \dfrac{Z_1 + Z_3}{Z_1 Z_2 + Z_2 Z_3 + Z_3 Z_1} \end{bmatrix}$$

T-Network

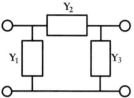

$$[Y] = \begin{bmatrix} Y_1 + Y_2 & -Y_2 \\ -Y_2 & Y_1 + Y_3 \end{bmatrix}$$

Π-network

Transmission line section

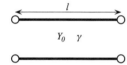

$$\gamma = \alpha + j\beta, \beta = \frac{2\pi}{\lambda_g}$$

$$[Y] = Y_o \begin{bmatrix} \coth(\gamma l) & -\cosech(\gamma l) \\ -\cosech(\gamma l) & \coth(\gamma l) \end{bmatrix}$$

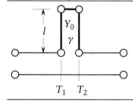

$$\gamma = \alpha + j\beta, \beta = \frac{2\pi}{\lambda_g}$$

$$[Y] = \begin{bmatrix} Y_0 \coth(\gamma l) & -Y_0 \coth(\gamma l) \\ -Y_0 \coth(\gamma l) & Y_0 \coth(\gamma l) \end{bmatrix}$$

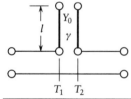

$$[Y] = \begin{bmatrix} Y_0 \tanh(\gamma l) & -Y_0 \tanh(\gamma l) \\ -Y_0 \tanh(\gamma l) & Y_0 \tanh(\gamma l) \end{bmatrix}$$

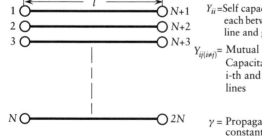

Y_{ii} = Self capacitance each between line and ground

$Y_{ij(i \neq j)}$ = Mutual Capacitance i-th and j-th lines

γ = Propagation constant

$$[Y] = \coth(\gamma l) \begin{bmatrix} [\eta] & -\cosech(\gamma l)[\eta] \\ -\cosech(\gamma l)[\eta] & [\eta] \end{bmatrix}$$

$$[\eta] = \begin{bmatrix} Y_{11} & -Y_{12} & -Y_{13} & \cdots & Y_{1N} \\ -Y_{21} & Y_{22} & -Y_{23} & \cdots & Y_{2N} \\ -Y_{31} & -Y_{32} & Y_{33} & \cdots & Y_{3N} \\ \vdots & \vdots & \vdots & \ddots & \vdots \\ -Y_{N1} & -Y_{N2} & -Y_{N3} & \cdots & Y_{NN} \end{bmatrix}$$

From: [4].

2.4 Scattering Matrix

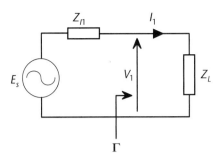

Figure 2.3 One-port network.

From Figure 2.3

$$V_1 = E_s - Z_{I1}I_1 \tag{2.24}$$

Combining (2.23a), (2.23b), and (2.24); and subsequently multiplying a_1 by its complex conjugate, we obtain

$$|a_1|^2 = \frac{|E_s|^2}{4\operatorname{Re} Z_{I1}} = P_{S\max} \tag{2.25}$$

The right-hand side of (2.25) is easily recognized as the maximum power available from the source. Therefore, we can say that $|a_1|^2$ is the incident power from the source into the load Z_L. Using (2.22a) and (2.22b), it can be shown that

$$|a_1|^2 - |b_1|^2 = \operatorname{Re}\{V_1 I_1^*\} \tag{2.26}$$

The right hand side of equation (2.26) is the total real power absorbed by the load. Therefore, we can come to the conclusion that $|b_1|^2$ is the power reflected from the load to the source. At this point we define the reflection coefficient

$$\Gamma = \frac{b_1}{a_1} = \frac{V_1 - Z_{I1}^* I_1}{V_1 + Z_{I1} I_1} = \frac{Z_L - Z_{I1}}{Z_L + Z_{I1}} \tag{2.27}$$

Using (2.26), the difference between the incident and reflected power, or in other words the power absorbed by the load, can be written as

$$P_L = P_{S\max} - P_{refl} = |a_1|^2 (1 - \Gamma^2) \tag{2.28}$$

From (2.27), under the matching condition when $Z_L = Z_{I1}^*$ we obtain $\Gamma = 0$.

Let us consider the multiport network shown in Figure 2.4.

The scattering matrix of a multiport network can be defined by the equation

$$[b] = [S][a] \tag{2.29}$$

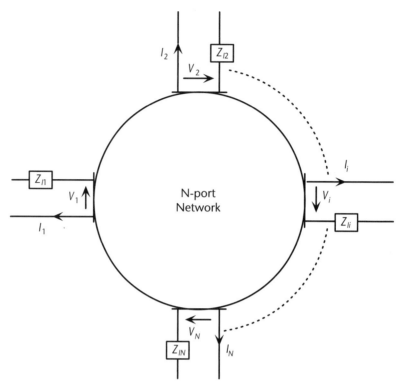

Figure 2.4 A multiport network terminated by different impedances at the ports.

where [a] and [b] are column matrices whose elements are the power wave amplitudes of the incident and reflected waves, respectively, at various ports. If the network has N-ports, then [S] is a square matrix of order N.

When the output impedances of the generators connected to the ports of the multiport network become purely real and equal to that of the transmission lines connected to the respective ports, a_i and b_i ($i=1,2,3, \ldots, N$) become the same as the complex incident and reflected voltages in the transmission lines. Then we can write

$$V_i = V_i^+ + V_i^- \qquad (2.30a)$$

$$I_i = I_i^+ + I_i^- \qquad (2.30b)$$

where

$$\frac{V_i^+}{I_i^+} = \frac{V_i^-}{I_i^-} = Z_{li} \qquad (2.31)$$

Combining (2.30) and (2.31) gives

2.4 Scattering Matrix

$$a_i = \frac{V_i^+}{\sqrt{Z_{Ii}}} = \sqrt{Z_{Ii}} I_i^+ \qquad (2.32a)$$

$$b_i = \frac{V_i^-}{\sqrt{Z_{Ii}}} = \sqrt{Z_{Ii}} I_i^- \qquad (2.32b)$$

and the reflection coefficient

$$\Gamma = \frac{b_i}{a_i} = \frac{Z_L - Z_{Ii}}{Z_L + Z_{Ii}} \qquad (2.33)$$

In (2.30) to (2.32), the + and − superscripts denote the ingoing and outgoing parameters, respectively.

The total power absorbed by all the ports is given by the difference between the sum total of the incident and reflected powers of all the ports. Mathematically speaking, if P_{total} is the total power and P_k is the power associated with each port, then

$$P_{total} = \sum_{k=1}^{N} P_k = \sum_{k=1}^{N} |a_k|^2 - \sum_{k=1}^{N} |b_k|^2 = [a^*]^T [a] - [b^*]^T [b] \qquad (2.34)$$

Combining (2.29) and (2.34), the total power absorbed by the network for a loss less condition is

$$P_{total} = [a^*]^T \big[[U] - [S^*][S]\big][a] = 0 \qquad (2.35)$$

where [U] is the unity matrix of order N. From (2.35), we get

$$[S^*]^T [S] = [U] \qquad (2.36)$$

For a reciprocal and symmetrical multiport network, (2.36) reduces to

$$[S^*][S] = [U] \qquad (2.37)$$

which indicates that the scattering matrix of a symmetrical, lossless, and reciprocal network is unitary.

Let us write down the expanded form of (2.37):

$$\begin{bmatrix} b_1 \\ b_2 \\ \vdots \\ \vdots \\ b_N \end{bmatrix} = \begin{bmatrix} S_{11} & S_{12} & \cdots & \cdots & S_{1N} \\ S_{21} & S_{22} & \cdots & \cdots & S_{2N} \\ \vdots & \vdots & \ddots & \cdots & \vdots \\ \vdots & \vdots & \vdots & \ddots & \vdots \\ S_{N1} & S_{N2} & \cdots & \cdots & S_{NN} \end{bmatrix} \begin{bmatrix} a_1 \\ a_2 \\ \vdots \\ \vdots \\ a_3 \end{bmatrix} \qquad (2.38)$$

From (2.38) it can be seen that, in general

$$S_{ij} = \frac{b_i}{a_j}\bigg|_{a_k=0, k=1,2,3,\ldots,N; k \neq j} \tag{2.39}$$

The condition

$$a_k = 0 \tag{2.40}$$

for tall k's except $k=j$ is created by perfectly matching all but the jth port. The transmitted signal at the ith port and the incident signal at the jth port is appropriately monitored and S_{ij} is computed using (2.39). For a detailed description of the procedure, the reader is referred to [1].

For a given network terminated by a set of real impedances given by $[Z_l]$ in (2.16), the scattering matrix can be computed by using the following procedure.

- Obtain the port admittance matrix $[Y]$ using nodal analysis and then solving for the node currents;
- Invert the admittance matrix $[Y]$ to obtain the corresponding impedance matrix $[Z]$;
- Normalize $[Z]$ using

$$[z] = [z_c]^{-\frac{1}{2}}[Z][z_c]^{-\frac{1}{2}} \tag{2.41}$$

where the diagonal matrix $[z_c]$ is obtained from (2.20).
Obtain the $[S]$ matrix from

$$[S] = [[z]-[U]][[z]+[U]]^{-1} \tag{2.42}$$

Solving (2.42) gives

$$[z] = [[U]-[S]]^{-1}[[U]+[S]] \tag{2.43}$$

For a reciprocal network, all the matrices associated with the network are symmetrical matrices, which means

$$[Z] = [Z]^T, \ [S] = [S]^T, \ \text{etc.} \tag{2.44}$$

2.4.2 Transformation of Scattering Matrix Due to Shift in Reference Plane

Consider Figure 2.5. The unprimed reference planes are the original reference planes with respect to which the scattering matrix $[S]$ of the N-port network is defined. Now, let us assume that the reference planes are moved away from the

2.4 Scattering Matrix

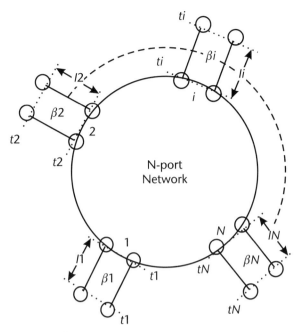

Figure 2.5 Shift in reference planes.

network to their new positions marked by the primed letters: $t'_j : j = 1,2,\ldots, N$. The new scattering matrix of the network is given by

$$[S'] = [S_L][S][S_L] \tag{2.45}$$

where the matrix

$$[S_L] = \begin{bmatrix} e^{-j\beta_1 l_1} & & \cdots & 0 \\ & e^{-j\beta_2 l_2} & \cdots & 0 \\ & & \cdots & \\ & & \vdots & \\ 0 & 0 & \cdots & e^{-j\beta_N l_N} \end{bmatrix} \tag{2.46}$$

and β_k is the propagation constant of the wave at the kth port. Keep in mind that what is defined as the reflection coefficient for a one-port network becomes the scattering matrix for a multiport network. Therefore, in a sense it is a generalized reflection coefficient. Consequently, if the set of impedances terminating the ports is altered, the scattering matrix of the network is also changed. For example, since the reflection coefficient of a one-port network is defined with respect to a reference impedance, the scattering matrix of a multiport network is also defined with respect to a reference set of impedances that terminate the ports. Consider Figure 2.5. If the set of port impedances are changed from Z_{li} to Z'_{li}, $i = 1, 2, \ldots, N$, then according to (2.45), the definitions of normalized voltage and current wave vectors will also change. As a result, the new scattering matrix is defined as [4,5]

$$[S'] = \left[[U]-[\Gamma]^2\right]^{-1}\left[[S]-[\Gamma]\right]\left[[U]-[\Gamma][S]\right]^{-1}\left[[U]-[\Gamma]^2\right] \tag{2.47}$$

where

$$[\Gamma] = \left[[z'_c] - [z_c]\right]\left[[z'_c] + [z_c]\right]^{-1} \tag{2.48}$$

and the matrix $[z_c]$ has been defined in (2.20) and the matrix $[z'_c]$ is obtained by replacing the diagonal terms Z_{Ii} by $\{Z_{Ii}^{-1}\}$, $i = 1, 2,..., N$ in (2.20). Table 2.2 shows the scattering matrices of the most commonly used microwave circuit elements.

Table 2.2 Scattering Matrices of Common Networks

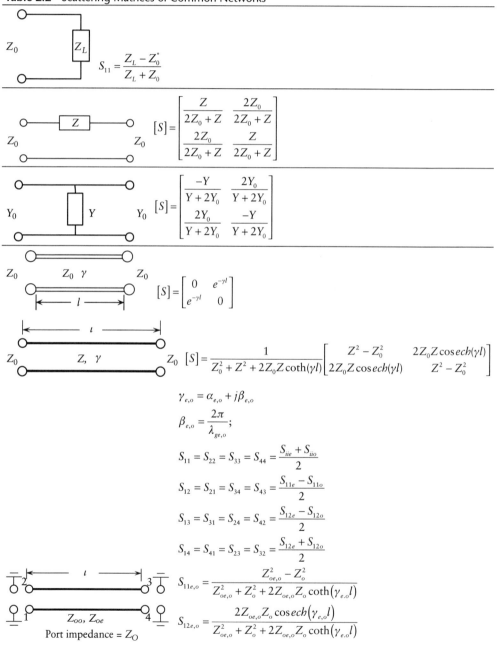

From: [4].

2.4 Scattering Matrix

All the elements of the scattering matrix of a passive, lossless, and reciprocal network cannot be chosen independently. Let us assume that Figure 2.6 represents one such two-port network whose scattering matrix is given by

$$[S] = \begin{bmatrix} S_{11} & S_{12} \\ S_{21} & S_{22} \end{bmatrix} \quad (2.49)$$

From reciprocity, [S] must be symmetrical. Therefore

$$[S] = \begin{bmatrix} S_{11} & S_{12} \\ S_{12} & S_{11} \end{bmatrix} = \begin{bmatrix} |S_{11}|e^{j\theta_{11}} & |S_{12}|e^{j\theta_{12}} \\ |S_{12}|e^{j\theta_{12}} & |S_{11}|e^{j\theta_{11}} \end{bmatrix} \quad (2.50)$$

Combining (2.37) and (2.50) gives

$$|S_{11}|^2 + |S_{12}|^2 = 1 \quad (2.51)$$

$$\theta_{11} = \theta_{12} \pm \frac{\pi}{2} + 2n\pi; \quad n = 0, 1, 2\ldots \quad (2.52)$$

This means that if S_{11} is known in complex form, S_{12} can be obtained from (2.51) and (2.52). Equation (2.51) represents conservation of power in a lossless two-port network.

For a lossless nonreciprocal network, the above relations become

$$|S_{11}|^2 + |S_{21}|^2 = |S_{22}|^2 + |S_{12}|^2 = 1 \quad (2.53)$$

$$S_{11}^* S_{12} + S_{21}^* S_{22} = 0 \quad (2.54)$$

$$\theta_{11} + \theta_{22} = \theta_{12} + \theta_{21} \mp \frac{\pi}{2} + 2n\pi; \, n = 0, 1, 2\ldots \quad (2.55)$$

where θ_{21} and θ_{22} are the phase angles associated with the elements S_{21} and S_{22}, respectively.

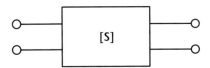

Figure 2.6 Two-port network.

2.4.3 Scattering Matrix of a Lossless Three-Port Network

An application of a unitary condition given by (2.37) shows that it is impossible to simultaneously match all the ports of a three-port lossless reciprocal network. However, it is possible to match all three ports if the circuit is lossy or nonreciprocal. Examples are a Wilkinson power divider or a three-port circulator.

2.4.4 Usefulness of Scattering Matrix

At this point it is worthwhile to discuss the usefulness of a scattering matrix. The concept of a scattering matrix is more general than admittance and impedance matrices. Many circuits may not possess admittance or impedance matrix. Typical examples are ideal transformers having nonfinite elements. On the contrary such transformers have scattering matrices. According to H. J. Carlin [6], all passive networks possess scattering matrices.

In microwave engineering, power flow is of primary consideration. Therefore, a scattering matrix is extremely useful. Let us consider the network in Figure 2.7. Let P_a represent the available power from the generator and P_L the power dissipated in the load Z_L. Then it can be shown that the magnitude of the forward transmission coefficient is given by

$$|S_{21}|^2 = \frac{P_L}{P_a} \tag{2.56}$$

2.4.5 Scattering Matrix and the Concept of Insertion Loss

Let us consider the two-port network shown in Figure 2.8(a). The transfer coefficient of the network is given by

$$S_{21} = \frac{2V_L}{E_1}\left[\frac{R_g}{R_L}\right]^{\frac{1}{2}} \tag{2.57}$$

where R_g is the real part of the generator impedance Z_{I1} and R_L is that of the load impedance Z_L
Now

$$|S_{21}|^2 = S_{21}S_{21}^* \tag{2.58}$$

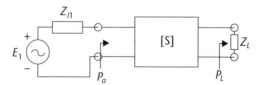

Figure 2.7 Circuit representation of a two-port network.

2.4 Scattering Matrix

Therefore, from (2.58) and (2.25), we obtain

$$|S_{21}|^2 = \frac{\frac{|V_L|^2}{2R_L}}{\frac{E_1^2}{8R_g}} = \frac{P_L}{P_a} \qquad (2.59)$$

where P_a is the available power from the generator and P_L is the power dissipated in the load R_L.

Let us consider the network in Figure 2.8(a). Suppose we have interposed a two-port network between the reference planes t_1 and t_2 as shown in Figure 2.8(b). Let the voltages across the load resistance R_L before and after the interposition of the two-port network be V_L and V_L^p, respectively. Also, let the powers dissipated in R_L before and after the interposition of the two-port network be P_L and P_L^p, respectively. The insertion power ratio of the two-port network is defined as

$$IPL = \frac{P_L^p}{P_L} \qquad (2.60)$$

Analyzing the circuit in Figure 2.8(b), we obtain

$$\frac{P_L^p}{P_l} = \frac{|E_1|^2}{|V_L|^2} \frac{R_L^2}{\left(R_g + R_L\right)^2} \qquad (2.61)$$

Combining (2.60) and (2.61) gives

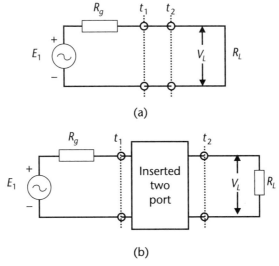

Figure 2.8 (a) Two-port network and (b) two-port network inserted.

$$\frac{P_L^p}{P_L} = \frac{4R_L R_g}{\left(R_g + R_L\right)^2} \frac{1}{|S_{21}|^2} \tag{2.62}$$

Under a perfectly matched condition

$$\frac{P_L^p}{P_L} = \frac{1}{|S_{21}|^2} \tag{2.63}$$

Equation (2.63) is a very useful relationship in network synthesis.

2.5 Measurement of Scattering Matrix

As mentioned in Section 2.4, scattering or S-matrix is a generalized reflection coefficient matrix. Such a matrix supplies not only the information on reflection coefficient but also the information on transmission coefficient. The instrument that is used for quantitative measurement of an S-matrix is known as a reflectometer. A frequency domain reflectometer is an instrument or special system that allows fast and accurate measurement of S-parameters of a two-port device over a frequency band. The heart of a reflectometer is a passive device called a directional coupler. A frequency domain reflectometer consists of two precision directional couplers that sample the transmitted and reflected signals and feed the sampled signals to two separate detectors so that the ratio between the two sampled signals can be determined.

Figure 2.9 shows the schematic diagram of a directional coupler. The *primary line* is the section of the transmission line that carries the power of the signal to be sampled. The secondary line is the section of transmission line carrying a fraction of the power that travels in the forward direction in the primary line and appears at port 3 or the desired coupled port. A reflectionless load terminates the remaining port or the undesired port of the secondary line and thereby prevents any power coupled to that port from being reflected back to the system. The primary line

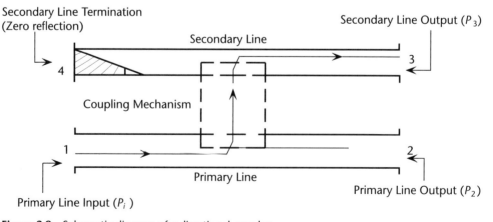

Figure 2.9 Schematic diagram of a directional coupler.

return loss is the return loss introduced in a matched transmission line when the directional coupler is inserted. The secondary line is terminated in a matched load. Since a directional coupler is a symmetrical and reciprocal device, the primary line return loss will be the same for either orientation of the coupler if the matchings offered by the connectors and terminations at all ports are identical. The secondary line return loss is the return loss measured looking into Port 3 while all other ports are terminated by matched loads and the detector at Port 3 offers a matched termination.

Coupling is the ratio of the power injected into the main line through port 1 to the power coming out of port 3 when all ports are matched. This is coupling to the wave in the forward direction. The coupling C expressed in decibels (dB) is defined as

$$C = 10\log\frac{P_i}{P_3} \tag{2.64}$$

while the insertion loss of the couplers is defined as

$$I = 10\log\frac{P_i}{P_2} \tag{2.65}$$

which basically accounts for power loss from the input signal due to coupling and the ohmic loss in the forward path (i.e., the transmission line connecting port 1 and port 3). Besides coupling and insertion loss, the other parameter of major importance for a directional coupler is directivity, which is the ratio of the power from the secondary line P_3 to the secondary line P_4 when a certain amount of power is injected into the input port. For example, suppose the power impinging on the matched load is P_4 and the power coming out of port 3 is P_3. Then directivity of the coupler is defined as

$$D = 10\log\frac{P_3}{P_4} \tag{2.66}$$

Directivity is a measure of isolation of port 4 or the undesired port from the input port. Therefore, one has to keep in mind that a directional coupler is basically a four-port device. However, it is reduced to a three-port one by terminating one of the ports in a matched load. Figure 2.10 shows how a directional coupler is employed in a reflection meter [7]. Let the forward transmission coefficient of the directional coupler be C_T and the coupling coefficient be C_P so that

$$S_{21} = 20\log|C_T| \tag{2.67}$$

and

$$S_{31} = 20\log|C_P| \tag{2.68}$$

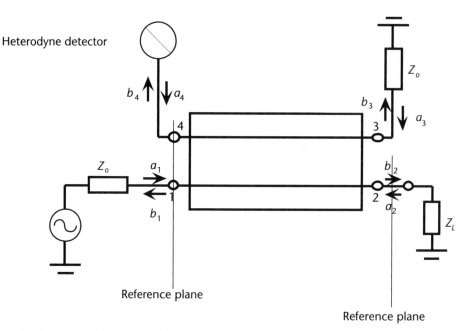

Figure 2.10 Schematic diagram of reflection meter employing a directional coupler. (From [7]. Reprinted with permission from John Wiley and Sons.)

The directional coupler being symmetrical and reciprocal, the scattering matrix is given by

$$[S_D] = \begin{bmatrix} 0 & C_T & jC_P & 0 \\ C_T & 0 & 0 & jC_P \\ jC_P & 0 & 0 & C_T \\ 0 & jC_P & C_T & 0 \end{bmatrix} \quad (2.69a)$$

with

$$C_T = \sqrt{1 - C_P^2} \quad (2.69b)$$

Using the above equations, we can write

$$b_2 = C_T a_1 \quad (2.70)$$

$$a_2 = C_T a_1 \Gamma_L \quad (2.71)$$

$$b_3 = C_P a_1 \quad (2.72)$$

$$b_4 = C_T C_P a_1 \Gamma_L \quad (2.73)$$

2.5 Measurement of Scattering Matrix

Assuming the heterodyne receiver or a detector, in the case of a scalar network analyzer, is matched to the directional coupler and port 3 is terminated into a matched load, we can write

$$a_4 = a_3 = 0 \tag{2.74a}$$

The magnitude of the reflection coefficient of the one-port device is then given by

$$|\Gamma_L| = |S_{11}| = \frac{b_4}{C_T C_P a_1} \tag{2.74b}$$

Figure 2.11 shows the basic configuration of a scalar network analyzer for full two-port scattering parameter measurements [7].

In Figure 2.11, $m_i (i = 1,2,3)$ are the measured quantities proportional to a_1, b_1 and b_2, respectively. The scattering parameters of the device under test (DUT) can be obtained from the following equations

$$|S_{11}| = c_{11} \frac{m_2}{m_1} \quad |S_{21}| = c_{21} \frac{m_3}{m_1} \tag{2.75a}$$

The proportionality constants c_{11} and c_{21}, functions of the directional coupler, are obtained by using the equations

$$c_{11} = -1 \frac{m_1^{Short}}{m_2^{Short}} \quad c_{21} = 1 \frac{m_1^{Through}}{m_3^{Through}} \tag{2.75b}$$

The quantities with a Short superscript are obtained by connecting a short circuit at the reference plane of port 1. The quantities with a Through superscript are obtained by connecting both ports. Repeating the above steps with the DUT ports reversed gives $|S_{22}|$ and $|S_{12}|$.

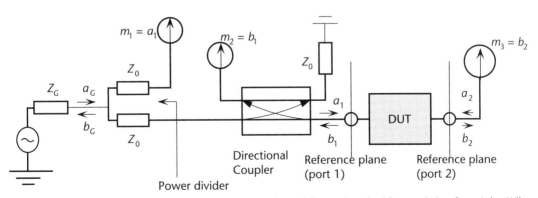

Figure 2.11 Unidirectional scalar network analyzer. (From [7]. Reprinted with permission from John Wiley and Sons.)

The simplest network analyzers are the scalar network analyzers that use power detectors. For most applications, only the magnitudes of the scattering parameters are needed. However, when the phases of the scattering parameters are required, the power detectors are replaced by heterodyne receivers. For an excellent account on the subject of vector network analyzers (VNAs) the reader is referred to reference [7].

2.6 Chain or ABCD Matrix

The chain matrix or ABCD matrix is particularly useful in cascading or chain connecting microwave networks. The networks may be purely two-port or may have multiports. For a single isolated network, the ABCD matrix relates the output voltages and currents to the input voltages and currents. Let us consider the multiport network shown in Figure 2.12. The input voltages and currents are $V_1, ..., V_N$ and $I_1, ..., I_N$. The output voltages and currents are $V_{N+1}, ..., V_{2N}$ and $I_{N+1}, ..., I_{2N}$, respectively. The input and the output parameters are related using the following matrix equation

$$\begin{bmatrix} V_1 \\ V_2 \\ \vdots \\ V_N \\ I_1 \\ I_2 \\ \vdots \\ I_N \end{bmatrix} = \begin{bmatrix} [A] & [B] \\ [C] & [D] \end{bmatrix} \begin{bmatrix} V_{N+1} \\ V_{N+2} \\ \vdots \\ V_{2N} \\ -I_{N+1} \\ -I_{N+2} \\ \vdots \\ -I_{2N} \end{bmatrix} \quad (2.76)$$

where $[A], [B], [C], [D]$ are $N \times N$ square matrices. For a two-port network, (2.76) reduces to

$$\begin{bmatrix} V_1 \\ I_1 \end{bmatrix} = \begin{bmatrix} A & B \\ C & D \end{bmatrix} \begin{bmatrix} V_2 \\ -I_2 \end{bmatrix} \quad (2.77)$$

For a reciprocal two-port network, shown in Figure 2.13, it can be seen that

$$AD - BC = 1 \quad (2.78)$$

Or in general

$$[A][D] - [B][C] = [U] \quad (2.79)$$

where $[U]$ is an identity matrix of order N.

2.6 Chain or ABCD Matrix

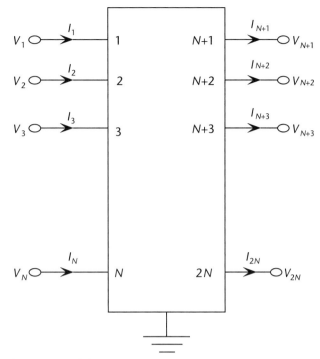

Figure 2.12 Multiport network for ABCD matrix representation.

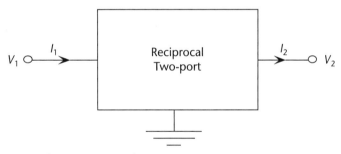

Figure 2.13 Reciprocal two-port network.

As mentioned above, the $ABCD$ matrix is very useful in obtaining the overall response of a chain connection of a number of two-port networks. Let us consider the chain connection of two two-port networks, as shown in Figure 2.14
Let the $ABCD$ matrices of the individual networks with respect to the reference planes $[t_1, t_2]$ and $[t_2, t_3]$ be $[T_1]$ and $[T_2]$, respectively. Then it can be shown that the ABCD matrix of the cascaded network is given by

$$[T] = [T_1][T_2] \qquad (2.80)$$

Equation (2.80) can be generalized for a cascade of more than two networks using the same principle.

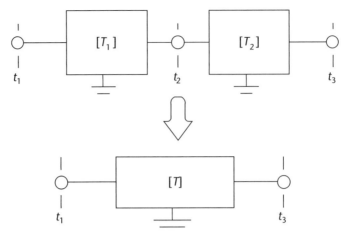

Figure 2.14 Two-port networks in tandem.

2.6.1 Use of ABCD Matrix in Computing Network Properties

In the previous section we showed that a chain of cascaded two-port networks can be reduced to an equivalent two-port network by finding the overall *ABCD* matrix. Let such an equivalent two-port network be fed at the input port by a voltage source of output impedance Z_g and terminated at the output-port by a load Z_L, as shown in Figure 2.15.

Using (2.77) and the relationship

$$V_2 = V_L = Z_L I_2 \tag{2.81}$$

it can be shown that the input impedance of the network is given by

$$Z_{in} = \frac{Z_L A + B}{Z_L C + D} \tag{2.82}$$

and the output impedance is

$$Z_{out} = \frac{Z_g D + C}{Z_g B + A} \tag{2.83}$$

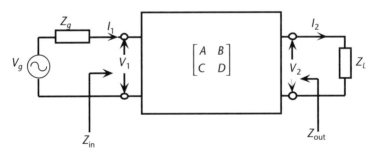

Figure 2.15 ABCD matrix representation of a two-port network.

2.6 Chain or ABCD Matrix

The reflection coefficients looking into the input and the output are given by

$$\Gamma_{in} = \frac{Z_{in} - Z_g^*}{Z_{in} + Z_g} \tag{2.84}$$

and

$$\Gamma_{out} = \frac{Z_{out} - Z_L}{Z_{out} + Z_L} \tag{2.85}$$

respectively.

The voltage gain is given by

$$A_v = \frac{V_L}{V_g} = \frac{V_2}{V_g} = \frac{Z_L}{Z_L A + B + Z_L Z_g C + Z_g D} \tag{2.86}$$

The power gain is given by

$$G = \frac{P_L}{P_1} = \frac{\operatorname{Re} Z_{in}}{\operatorname{Re} Z_L} \left| \frac{Z_L}{Z_L A + B} \right|^2 \tag{2.87}$$

where

$$P_1 = V_1 I_1 \tag{2.88}$$

is the power delivered to the input port of the network and

$$P_L = V_L I_2 \tag{2.89}$$

is the power delivered to the load.

The transducer power gain is given by

$$G_T = \frac{P_L}{P_{AG}} = 4 \frac{\operatorname{Re} Z_g}{\operatorname{Re} Z_L} \left| \frac{Z_L}{Z_L A + B + Z_L Z_g C + Z_g D} \right|^2 \tag{2.90}$$

where P_{AG} is available power from the source.

2.6.2 Normalized ABCD Matrix

Using the normalized voltage and current definitions in (2.17), we can define the normalized *ABCD* matrix as follows [5, 8]. Suppose the multiport network in

Table 2.3 Unnormalized ABCD Matrices of Common Networks

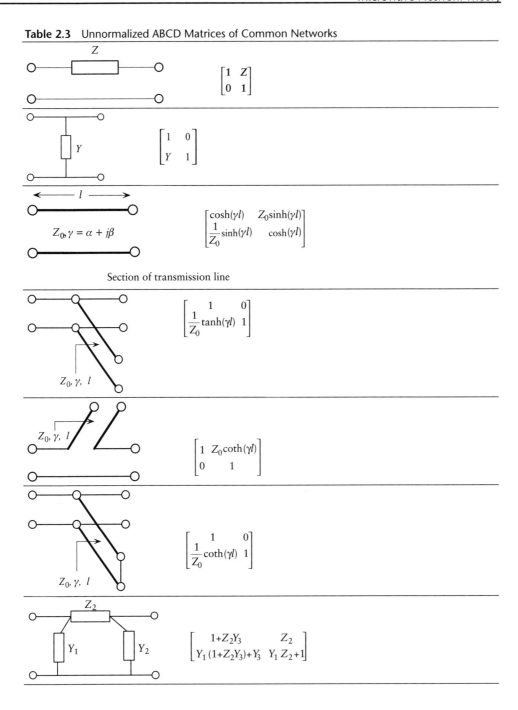

Figure 2.4 is terminated by a set of impedances Z_i ($i = 1,2,..., 2N$). Then we can define a set of normalized voltages and currents v_i and i_i, respectively ($i = 1,2,..., 2N$) according to (2.76). These normalized voltages and currents can be related using the normalized *ABCD* matrix as follows:

2.6 Chain or ABCD Matrix

Table 2.3 (continued)

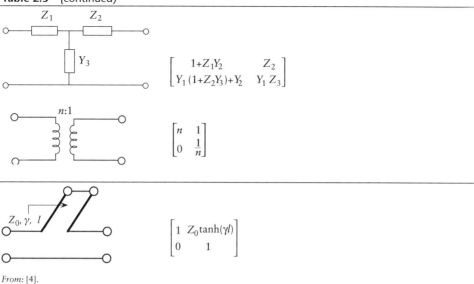

From: [4].

$$\begin{bmatrix} v_1 \\ v_2 \\ \vdots \\ v_N \\ i_1 \\ i_2 \\ \vdots \\ i_N \end{bmatrix} = \begin{bmatrix} [A_n] & [B_n] \\ [C_n] & [D_n] \end{bmatrix} \begin{bmatrix} v_{N+1} \\ v_{N+2} \\ \vdots \\ v_{2N} \\ -i_{N+1} \\ -i_{N+2} \\ \vdots \\ -i_{2N} \end{bmatrix} \quad (2.91)$$

where $[A_n], [B_n], [C_n], [D_n]$, are $N \times N$ square matrices. For a two-port network, (2.91) reduces to

$$\begin{bmatrix} v_1 \\ i_1 \end{bmatrix} = \begin{bmatrix} A_n & B_n \\ C_n & D_n \end{bmatrix} \begin{bmatrix} v_2 \\ -i_2 \end{bmatrix} \quad (2.92)$$

The unnormalized and normalized *ABCD* matrices are related through

$$\begin{bmatrix} [A] & [B] \\ [C] & [D] \end{bmatrix} = \begin{bmatrix} [Z_{IN}]^{-\frac{1}{2}} & [0] \\ [0] & [Z_{IN}]^{\frac{1}{2}} \end{bmatrix}^{-1} \begin{bmatrix} [A_n] & [B_n] \\ [C_n] & [D_n] \end{bmatrix} \begin{bmatrix} [Z_{IN}]^{-\frac{1}{2}} & [0] \\ [0] & [Z_{IN}]^{\frac{1}{2}} \end{bmatrix} \quad (2.93)$$

$[Z_{IN}]$ is a diagonal matrix of real elements $Z_{I1}, Z_{I2},..., Z_{IN}$, and $[Z_{I2N}]$ is a diagonal matrix of real elements $Z_I(N+1), Z_I(N+2),..., Z_I(2N)$. In the case of a simple two-port network, (2.93) reduces to

$$\begin{bmatrix} A & B \\ C & D \end{bmatrix} = \begin{bmatrix} \dfrac{1}{\sqrt{Z_{I1}}} & 0 \\ 0 & \sqrt{Z_{I1}} \end{bmatrix}^{-1} \begin{bmatrix} A_n & B_n \\ C_n & D_n \end{bmatrix} \begin{bmatrix} \dfrac{1}{\sqrt{Z_{I2}}} & 0 \\ 0 & \sqrt{Z_{I2}} \end{bmatrix} \qquad (2.94)$$

Obviously the normalized $ABCD$ matrix takes into consideration the effects of the impedances terminating the ports of a network. It can be shown that the scattering matrix of a two-port network is related to its normalized $ABCD$ matrix via

$$[S] = \frac{1}{A_n + B_n + C_n + D_n} \begin{bmatrix} A_n + B_n - C_n - D_n & 2(A_n D_n - C_n B_n) \\ 2 & D_n + B_n - A_n - C_n \end{bmatrix} \qquad (2.95)$$

The unnormalized $ABCD$ matrices of some useful networks are shown in Table 2.3.

References

[1] Collin, R. E., *Foundations for Microwave Engineering* New York: McGraw Hill Publications, 1965.

[2] Penfield, P., "Noise in Negative Resistance Amplifiers," *IRE Trans. Circuit Theory*, Vol. CT-7, 1960, pp. 160–170.

[3] Kurokawa, K., Power Waves and the Scattering Matrix," *IEEE Trans. Microwave Theory and Techniques*, Vol. MTT-13, No. 2, March 1965, pp. 194–202.

[4] Dobrowolski, A., and W. Ostrowski, *Computer-Aided Analysis, Modeling and Design of Microwave Networks*, Norwood, MA: Artech House, 1966.

[5] Grivet, P., *Microwave Circuits and Applications*, Vol. 2, New York: Academic Press, 1964.

[6] Carlin, H. J., and A. B. Giorando, *Network Theory*, Englewood Cliffs, NJ: Prentice Hall, 1964.

[7] Rolfes, I., A. Gronfield, and B. Schiek, "Standing Wave Meters and Network Analyzers," in *Wiley Encyclopedia of Electrical and Electronics Engineering*, Vol. 20, 1999, pp. 403–423.

[8] Kajfez, D., *Notes on Microwave Circuits*, Vol. 1, Kajfez Consulting, Oxford, MS, 1986.

CHAPTER 3

Properties of Microwave Transmission Lines

3.1 Introduction

The purpose of a uniform transmission line is to transfer energy from a generator to a load. Transmission lines can be of various types depending on the application and the microwave and millimeter-wave frequency band used. The length of a transmission line may vary from a fraction of a wavelength in lumped element applications, to several wavelengths in distributed element circuitry. Open-wire lines are used at frequencies far below the microwave band and are replaced by coaxial lines above 1 GHz. Above around 3 GHz, hollow metal tubes or waveguides and their various derivatives, such as finlines, are used. Besides the metallic waveguides, surface waveguides and dielectric waveguides are also used at microwave and millimeter-wave frequencies. Other types of planar transmission lines, which can be realized using printed circuit board (PCB) techniques, such as microstrip, stripline, suspended stripline, slotline, and coplanar lines are used over the entire band of microwave and millimeter-wave frequencies. In this chapter, we present a brief description of transmission line theory and the properties of various transmission lines used in microwave passive circuit design.

3.2 Transmission Line Equations

Transmission lines are used at microwave frequencies to support modes, which can be broadly divided into two types, the transverse electromagnetic (TEM) and the nontransverse electromagnetic (non-TEM) modes of propagation. The four basic parameters that characterize a transmission line are

1. The characteristic impedance, Z_0;
2. The propagation constant, β;
3. The attenuation constant α;
4. The peak power handling capability, P_{max}.

For the transmission line supporting a non-TEM mode, the above four parameters are dependent on the type of the supported mode, and the definition of Z_0 is nonunique. However, in all cases the parameters are functions of the geometrical cross section and the material properties of the transmission line. Figure 3.1 shows the lumped element equivalent circuit of a transmission line supporting the TEM mode, where λ is the operating wavelength, L and R are the per unit lengths of series inductance and resistance, respectively, and C and G are the per unit lengths of the shunt capacitance and conductance, respectively. Table 3.1 shows the functional relationships among various parameters of a TEM transmission line.

The voltage and current waves in a two-conductor transmission line, including all planar transmission line supporting TEM and quasi-TEM modes, can be represented as

$$V(z) = V_0^+ e^{-\gamma z} + V_0^- e^{+\gamma z} \tag{3.1}$$

$$I(z) = I_0^+ e^{-\gamma z} + I_0^- e^{+\gamma z} = \frac{1}{Z_0}(V_0^+ e^{-\gamma z} - V_0^- e^{+\gamma z}) \tag{3.2}$$

respectively, where the terms associated with $e^{-\gamma z}$ represent the forward-traveling wave and the term associated with $e^{+\gamma z}$ represent the backward-traveling waves. In the above equations the characteristic impedance Z_0 and the propagation constant γ of the transmission line are given by

$$Z_0 = \sqrt{\frac{R + j\omega L}{G + j\omega C}} \tag{3.3}$$

$$\gamma = \sqrt{(R + j\omega L)(G + j\omega C)} = \alpha + j\beta \tag{3.4}$$

where α is the attenuation constant (nepers/unit length) and β is the phase or propagation constant (radians/unit length).

For a lossless line with $R = G = 0$ we see that

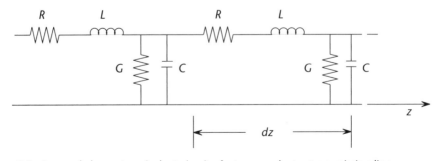

Figure 3.1 Lumped element equivalent circuit of a two-conductor transmission line.

3.2 Transmission Line Equations

Table 3.1 Transmission Line Parameters

Quantity	General Line	Ideal Line	Approximate Result or Low-Loss Line
$\alpha + j\beta$	$\sqrt{(R+j\omega L)(G+j\omega C)}$	$j\omega\sqrt{LC}$	See α and β below
β	$\text{Im}(\gamma)$	$\omega\sqrt{LC} = \dfrac{\omega}{v_p} = \dfrac{2\pi}{\lambda}$	$\omega\sqrt{LC\left(1 - \dfrac{RG}{4\omega^2 LC} + \dfrac{G^2}{8\omega^2 C^2} + \dfrac{R^2}{8\omega^2 L^2}\right)}$
α	$\text{Re}(\gamma)$	0	$\dfrac{R}{2Z_0} + \dfrac{GZ_0}{2}$
Z_0	$\sqrt{\dfrac{R+j\omega L}{G+j\omega C}}$	$\sqrt{\dfrac{L}{C}}$	$\sqrt{\dfrac{L}{C}}\left[1 + j\left(\dfrac{R}{2\omega L} + \dfrac{G}{2\omega C}\right)\right]$
Z_{in}	$Z_0\left[\dfrac{Z_L \cosh\gamma l + Z_0 \sinh\gamma l}{Z_0 \cosh\gamma l + Z_L \sinh\gamma l}\right]$	$Z_0\left[\dfrac{Z_L \cos\beta l + Z_0 \sin\beta l}{Z_0 \cos\beta l + Z_L \sin\beta l}\right]$	
For $Z_L = 0$			
Z_{in}	$Z_0 \tanh\gamma l$	$Z_0 \tan\beta l$	(See Figure 3.2)

$$Z_0 = \sqrt{\frac{L}{C}} \tag{3.5}$$

$$\gamma = j\omega\sqrt{LC} = j\beta \quad \text{radians/unit length} \tag{3.6}$$

In reality, a transmission line will have some finite amount of attenuation and the overall attenuation constant is given by

$$\alpha_t = \alpha_c + \alpha_d \tag{3.7}$$

where α_c and α_d are the attenuation constants due to conductor and dielectric losses, respectively. For small losses

$$\alpha_c = \frac{\beta}{2Q_c} = \frac{R}{2Z_0} \quad \text{Nepers/unit length} \tag{3.8}$$

$$\alpha_d = \frac{\beta}{2Q_d} = \frac{G}{2Y_0} \quad \text{Nepers/unit length} \tag{3.9}$$

where $Q_c = \omega L/R$ and $Q_d = \omega C/G$ are the conductor and the dielectric Q-factors of the line. The overall Q-factor of the line is given by

$$Q = \frac{Q_c Q_d}{Q_c + Q_d} \tag{3.10}$$

For moderately lossy transmission lines, the propagation constant γ and the characteristic impedance Z_0 are given by

$$\gamma = j\beta\sqrt{1 - j\left(\frac{1}{Q_c} + \frac{1}{Q_d}\right)} \qquad (3.11)$$

$$Z_0 = \sqrt{\frac{L}{C}}\sqrt{1 + j\left(\frac{1}{Q_c} + \frac{1}{Q_d}\right)} \qquad (3.12)$$

The concept of transmission line loss and Q-factor (also known as unloaded Q-factor) is extremely useful when the lines are used as resonators. Q-factor determines the passband insertion loss in a band-pass filter and the rejection depth of a band reject filter composed of such resonators.

3.3 Transmission Line with Electrical Discontinuity

Figure 3.2 shows a transmission line that has been abruptly terminated by a known complex load Z_L. When such a termination occurs, the forward- and backward-traveling waves on the line interfere and give rise to a standing-wave pattern, as shown in Figure 3.2.

The fraction of the incident voltage that is reflected by the load Z_L, when it is not equal to the characteristic impedance Z_0, of the line is given by

$$\Gamma = \frac{V_0^-}{V_0^+} = \frac{Z_L - Z_0}{Z_L + Z_0} \qquad (3.13)$$

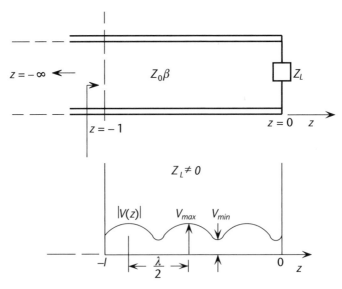

Figure 3.2 A terminated transmission line and the voltage standing wave.

In general Z_0 is a complex quantity. However, for a lossless line it is a real number for all practical purposes. Therefore, if Z_L is complex then Γ is also complex. Equation (3.13) gives the value of the reflection coefficient at $z = 0$. At any other point, $z = -l$ on the line it assumes the general form

$$\Gamma(z) = \frac{Z_{in}(-l) - Z_0}{Z_{in}(-l) - Z_0} = \frac{V_0^- e^{+j\gamma(-l)}}{V_0^+ e^{-j\gamma(-l)}} = \Gamma e^{-j2\gamma l} \tag{3.14}$$

Knowing the reflection coefficient, we may find the standing-wave ratio

$$S = \frac{1 + |\Gamma|}{1 + |\Gamma|} \tag{3.15}$$

Combining (3.13) and (3.14), we obtain

$$Z_{in} = Z_0 \frac{Z_L + jZ_0 \tan(\beta l)}{Z_0 + jZ_L \tan(\beta l)} \tag{3.16}$$

For a short-circuited transmission line ($Z_L = 0$)

$$Z_{in} = jZ_0 \tan(\beta l) \tag{3.17}$$

For an open-circuited line ($Z_L = \infty$)

$$Z_{in} = -jZ_0 \cot(\beta l) \tag{3.18}$$

For a matched load ($Z_L = Z_0$)

$$Z_{in} = Z_0 \tag{3.19}$$

The above equations are useful in realization of reactive elements in microwave circuits. In the following few sections we will discuss the properties of a number of most commonly used two-conductor transmission lines

3.4 Two-Conductor Transmission Lines

3.4.1 Two Round Conductors of Equal Diameter

Figure 3.3 illustrates a two-conductor transmission line with round conductors. The per unit length inductance (L), capacitance (C), and resistance (R) for the line are, respectively, given by

$$L = 0.4 \ln\left\{\frac{2H}{d}\right\} \text{ (mH/m)} \tag{3.20}$$

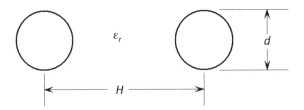

Figure 3.3 Cross section of a two-round conductor transmission line.

$$C = \frac{27.669\varepsilon_r}{\ln\frac{2H}{d}} \text{ (F/m)} \tag{3.21}$$

$$R = \sqrt{\frac{f\mu}{\pi\sigma}} \ln\frac{2H}{d} \text{ (}\Omega/m\text{)} \tag{3.22}$$

where f is the operating frequency, μ is the permeability, σ is the conductivity of the rods, and ε_r the dielectric constant of the medium. A two-conductor line is used up to frequencies at which the radiation loss is negligible.

3.4.2 Coaxial Line

Coaxial lines find extensive use over the widest frequency spectrum, starting from extra-low frequency (ELF) to the millimeter-wave band. Although a coaxial line can support non-TEM modes, the dominant mode is TEM. As a result, the characteristics of a coaxial line can be obtained from static field analysis. Figure 3.4(a) shows the most basic form of a coaxial line. Figure 3.4(b) shows the dependence of characteristic impedance on the ratio of the diameters of the inner and outer conductors. The parameters of the coaxial line, shown in Figure 3.4(a), are shown in Table 3.2 as functions of the geometrical dimensions. For values of the attenuation constant at temperatures other than 20°C, one should multiply the value of α in Table 3.2 by $[1 + 3.9 \times 10^{-3}(T - 20)]^{1/2}$, where T is in degrees celsius. Figure 3.4(b) shows the variations of coaxial line parameters with respect to the ratio b/a.

3.4.2.1 Q-Factor of a Coaxial Line

The Q-factor of a coaxial line is an important parameter in realizing microwave filters using coaxial line resonators. It is obtained from (3.23) as

$$Q = \frac{Q_c Q_d}{Q_c + Q_d} \tag{3.23}$$

where the conductor Q-factor, Q_c, is given by

3.4 Two-Conductor Transmission Lines

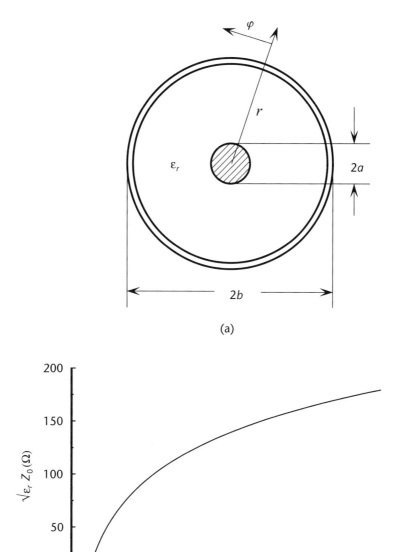

Figure 3.4 (a) Cross section of a coaxial line, and (b) variation of characteristic impedance of a coaxial line as a function of b/a of the coaxial line shown in (a).

$$Q_c = \frac{379.1\sqrt{f_c}\ln\left(\dfrac{b}{a}\right)}{\left(\dfrac{1}{2a}+\dfrac{1}{2b}\right)} \quad (3.24)$$

and a and b are in inches, f_c is in gigahertz. The dielectric Q-factor Qd is given by

Table 3.2 Coaxial Line Characteristics

Parameter	Expression	Unit
Capacitance	$C = \dfrac{55.556\varepsilon_r}{\ln\left(\dfrac{b}{a}\right)}$	pF/m
Inductance	$L = 200\ln\left(\dfrac{b}{a}\right)$	nH/m
Characteristic Impedance	$Z_0 = \dfrac{60}{\sqrt{\varepsilon_r}}\ln\left(\dfrac{b}{a}\right)$	Ω
Phase velocity	$v_p = \dfrac{3\times 10^8}{\sqrt{\varepsilon_r}}$	m/s
Delay	$3.33\sqrt{\varepsilon_r}$	ns/m
Conductor attenuation constant	$\alpha_c = \dfrac{9.5\times 10^{-5}\sqrt{f(a+b)}\sqrt{\varepsilon_r}}{ab\ln\left(\dfrac{b}{a}\right)}$	db/unit length
Dielectric attenuation constant	$\alpha_d = \pi\sqrt{\varepsilon_r}\,\dfrac{\tan\delta}{\lambda_0}$	db/unit length
Cutoff wavelength for the next higher-order mode (TE)	$\lambda_c = \pi\sqrt{\varepsilon_r}(a+b)$	Unit of a & b
Maximum peak power	$P_{max} = 44\lvert E_{max}\rvert^2 a^2\sqrt{\varepsilon_r}\ln\left(\dfrac{b}{a}\right)$	kW

λ_0 = Free-space wavelength; f = operating frequency in GHz; $\tan\delta$ = loss tangent of the dielectric; E_{max} = maximum breakdown electric field in the dielectric uniformly filling the space between the conductors; α_c = copper at 20°C.

$$Q_d = \frac{1}{\tan\delta} \tag{3.25}$$

where $\tan\delta$ is the loss tangent of the material filling the space between the conductors. The above equation is valid for copper as the conductor. For other conductors, (3.24) should be multiplied by a factor ζ, where

$$\zeta = \frac{\sigma_{Cu}}{\sigma_m} \tag{3.26}$$

where σ_{Cu} and σ_m are the bulk conductivities of copper and the metal in question, respectively.

The electric and magnetic fields of the dominant TEM mode of a coaxial line are

$$E_r = \frac{V}{r \ln\left(\frac{b}{a}\right)} \tag{3.27}$$

$$H_\varphi = \frac{I}{2\pi r} \tag{3.28}$$

$$E_\varphi = H_r = E_z = H_z = 0 \tag{3.29}$$

where V is the voltage between the two conductors, I is the current flowing through the conductors, and z is the direction of wave propagation. The cutoff wavelength of the dominant TEM mode in a coaxial line is infinite. In other words, the cutoff frequency of the dominant mode of a coaxial line is zero. Figure 3.5 shows the electromagnetic field pattern of the dominant (TEM) mode in a coaxial line.

A coaxial line can also support higher-order modes like TE and TM modes. Figure 3.6 shows the field patterns of a first few non-TEM higher-order modes in a

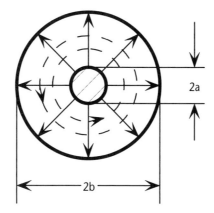

Figure 3.5 Field pattern of TEM mode in a coaxial line (--- H, – E).

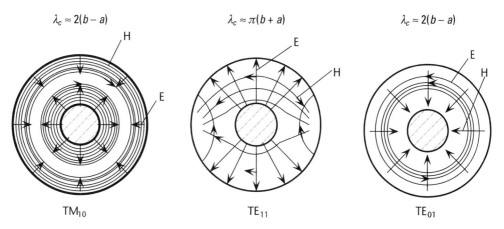

Figure 3.6 Higher-order mode transverse field patterns of a coaxial line (see Figure 3.5).

coaxial line. The higher-order modes are referred to as TE_{mn} and TM_{mn} modes. For TE modes there is no axial electric field (E_z) component and for TM mode there is no axial magnetic field (H_z) component. The subscript m denotes the number of half-cycle variations of the radial component of the field in the angular (Φ) direction and the subscript n denotes the total number of half-cycle variations of angular field component in the radial (r) direction. Figure 3.7 shows the order in which the first few higher-order modes appear in a coaxial line. TE_{11} is the first higher-order mode above the fundamental mode. Here c is the velocity of light in the dielectric.

From the above figures, it is recommended that in order to ensure the propagation of the fundamental TEM mode, the operation frequency in coaxial line should not exceed $f_u = c/\{\pi(a+b)\}$. In other words, the operating wavelength λ should exceed $\pi(a+b)$.

3.4.3 Other Forms of Coaxial Line

3.4.4 General Equation for Attenuation in Two-Conductor Transmission Lines

Consider a two-conductor transmission line with an arbitrary cross section, as shown in Figure 3.8. The attenuation in the line at a frequency f in gigahertz is given by [1]

$$\alpha = \frac{\pi\sqrt{\varepsilon_r}f}{0.2998}\left[1 - \frac{Z_0'}{Z_0}\right] \quad \text{(Nepers/meter)} \tag{3.30}$$

where Z_0 is the characteristic impedance of the line and Z_0' is that of the line when the dimensions are changed by the amount of half the skin depth, δ_s, as shown in Figure 3.8; δ_s is given by

$$\delta_s = 0.0822\sqrt{\frac{\rho_{rcu}}{f}} \quad \text{(mils)} \tag{3.31}$$

where ρ_{rcu} is the resistivity of the conductor with respect to copper. Equation (3.30) is applicable to any of the transmission lines described so far. The conductor attenu-

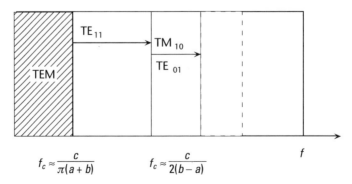

Figure 3.7 Order of cutoff frequencies of various modes of a coaxial line.

3.4 Two-Conductor Transmission Lines

Table 3.3 Formulas for Z_0 (Ω) of Various Derivatives of Coaxial Line

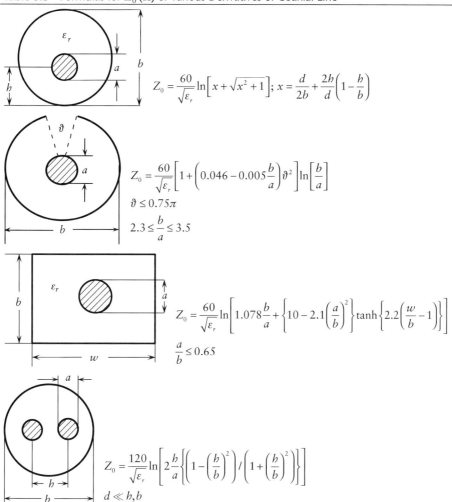

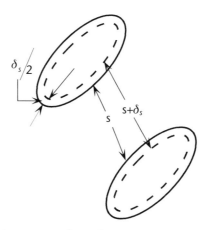

Figure 3.8 Loss calculation in a two-conductor line.

ation constant of a coaxial line based on the above approach is shown in Figure 3.9. The corresponding conductor Q-factor is shown in Figure 3.10.

The overall attenuation in a coaxial line is obtained by adding the dielectric attenuation to the conductor attenuation (see Table 3.2). Figure 3.10 shows the variation of conductor Q-factor of a coaxial line with conductor diameter ratio. Table 3.4 shows the attenuation in standard commercially available coaxial lines.

3.4.5 Maximum Power-Handling Capability of a Coaxial Line

The maximum power-handling capability of a coaxial line is determined by the breakdown electric field E_m between the two conductors. Under normal atmospheric pressure in an air-filled coaxial line the breakdown field is approximately 2.80×10^4 volts/cm. The maximum average power that can be handled by a coaxial line is given by

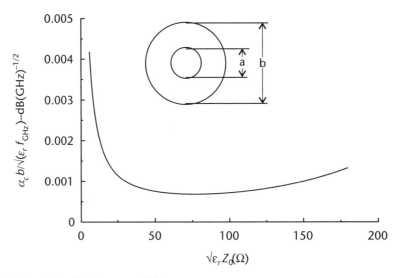

Figure 3.9 Conductor loss in a coaxial line.

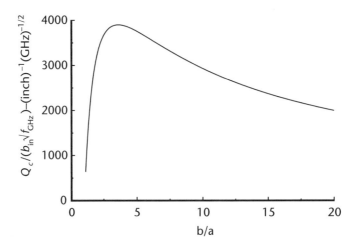

Figure 3.10 Conductor Q of a coaxial line.

3.4 Two-Conductor Transmission Lines

Table 3.4 Attenuation in Standard Commercially Available Coaxial Lines (db/100 ft)

RG/U Type/ Freq (MHz	1.0	10	50	100	400	1000
8, 8A, 10A, 213	0.15	0.55	1.3	1.9	4.1	8.0
9, 9A, 9B, 214	0.21	0.66	1.5	2.3	5.0	8.8
14, 14A, 217	0.12	0.41	1.0	1.4	3.1	5.5
17, 17A, 18, 18A, 218, 219	0.06	0.24	0.62	0.95	2.4	4.4
55B, 223	0.30	1.2	3.2	4.8	10.0	16.5
58	0.33	1.2	3.1	4.6	10.5	17.5
59, 59B	0.33	1.1	2.4	3.4	7.0	12.0
141, 141A, 400, 142, 142A	0.30	0.90	2.1	3.3	6.9	13.0
LDF4-50A	0.06	0.21	0.47	0.68	1.4	2.3
LDF5-50A	0.03	0.11	0.25	0.36	0.78	1.4
7/8" Air Heliflex	0.035	0.110	0.251	0.359	0.75	1.24
1-5/8" Air Heliflex	0.020	0.064	0.142	0.202	0.411	0.662
3-1/8" Air Heliflex	0.11	0.037	0.077	0.109	0.230	0.384
4-1/2" Air Heliflex	–	–	0.048	0.076	0.132	0.214
6-1/8" Air Heliflex	–	–	0.037	0.051	0.103	0.0164

$$P_{\max(av)} = \frac{E_m^2}{480} \left\{ \frac{b^2 \ln \frac{b}{a}}{\left(\frac{b}{a}\right)^2} \right\} \text{ Watts} \quad (3.32)$$

Figure 3.11 shows the variation of maximum average power-handling capability of an air-filled coaxial line as a function of the ratio of diameters of the conductors for a fixed outer diameter. The maximum power is transmitted when the diameter ratio is 1.65. The corresponding characteristic impedance is 30 Ω.

The average power-handling capability of a coaxial line is determined by the insulator dielectric strength and the thermal properties of the line that are related

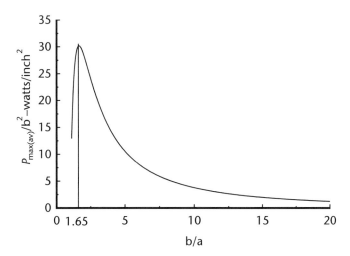

Figure 3.11 Maximum power handling of a coaxial line.

to the breakdown voltage and the insertion loss, respectively. High temperature and a good heat transfer mechanism increases the average power handling while higher dielectric strength and good dielectric strength with high breakdown field increases the peak power-handling capability. Breakdown due to large peak power is independent of frequency. However, it varies with the density of the pressuring gas and other factors in a coaxial line. High altitude and high ambient temperature reduce the maximum average power-handling capability of a line by impeding the heat transfer out of the transmission line. In most cases, the information carrying microwave or RF power in coaxial line is modulated. The peak power-handling capability is therefore determined to a large extent by the degree and type of modulation. The continuous wave (CW) power handling of a coaxial line is often treated as the average power-handling capability. However, it is always necessary to check both the average and the peak power-handling capabilities. Table 3.5 shows the peak power-handling capability of standard commercially available coaxial lines.

3.4.6 Coaxial Line Discontinuities

No transmission line is useful in microwave circuits without discontinuities in them. The three most important discontinuities in a coaxial line are (1) step in the inner or outer conductor, (2) T-junction, and (3) bend. Of these three, the most important and the most frequently encountered one is the step discontinuity. Figure 3.12 shows the two types of coaxial line step discontinuity in a coaxial line. They are formed by abrupt changes in either the inner or the outer coaxial line conductor. Step discontinuities in coaxial lines were analyzed using the variational technique by Marcuvitz [2]. The discontinuity susceptances in the two cases are given by

$$\frac{B}{Y_0} = \frac{2b_0 A_1}{\lambda} \left[\begin{array}{l} 2\ln\left(\dfrac{1-\alpha^2}{4\alpha}\right)^{\frac{1+\alpha}{2\alpha}} \sqrt{\dfrac{1+\alpha}{1-\alpha}} + 4\dfrac{A+A'+2C}{AA'-C^2} \\ +\dfrac{1}{2}\left(\dfrac{b}{\lambda}\right)^2 \left(\dfrac{1-\alpha}{1+\alpha}\right)^{4\alpha} \left\{ \dfrac{5\alpha^2-1}{1-\alpha^2} + \dfrac{4}{3}\dfrac{C\alpha^2}{A} \right\} + \dfrac{A_2}{2} \end{array} \right] \qquad (3.33)$$

where

$$\frac{Y_0'}{Y_0} = \frac{\ln\left(\dfrac{c}{a}\right)}{\ln\left(\dfrac{c}{b}\right)} \qquad (3.34)$$

Y_0 and Y_0' are the characteristic admittances on the two sides of the step discontinuity.

3.4 Two-Conductor Transmission Lines

Table 3.5 Maximum Power Handling Capability of Coaxial Cable (in Watts)

	Dielectric Diameter	Overall Diameter	@ Frequency				
			400 MHz	1 GHz	3 GHz	5 GHz	10 GHz
M17/RG178, CN178SC, CN178TC, RG196 A/U	.033"	.071"	123	78	41	28	14
M17/131-RG403	0.33"	.116"	123	78	41	28	14
M17/113-RG316, CN316SC, CN316TC, RG188 A/U, SB316	.060"	.098"	240	160	80	57	30
M17/152-00001, CN316SCSC, CN316TCTC	.060"	.114"	240	160	80	57	30
M17/94-RG179, CN176SC, CN179TC, RG187 A/U	.063"	.100"	310	200	110	76	41
SS405	.064"	.104"	240	160	80	57	30
SS75086	.064"	.100"	240	160	80	57	30
LL120	.080"	.120"	720	460	250	190	140
M17/95-RG180, RG195 A/U	.102"	.141"	400	250	135	93	50
M17/60-RG142, CN142SCSC, CN142TCTC	.116"	.195"	1100	550	350	245	140
M17/111-RG303	.116"	.170"	1100	550	350	245	140
M17/128-RG400, SB400	.116"	.195"	1100	550	350	245	140
SS402	.117"	.163"	1100	550	350	245	140
SB142	.117"	.195"	1100	550	350	245	140
LL142	.145"	.195"	1200	720	400	310	220
LL235	.160"	.235"	1500	900	540	410	300
LL393-2	.185"	.270"	1900	1100	680	510	380
M17/112-RG304	.185"	.280"	1900	1100	680	510	380
SB304	.185"	.280"	1450	870	460	330	190
LL335	.250"	.335"	2900	1800	1050	850	600
M17/127-RG393, SB393	.285"	.390"	2800	1700	880	620	350
LL450	.360"	.450"	7250	4200	2200	1600	1015

$$A = \left[\frac{1+\alpha}{1-\alpha}\right]^{2\alpha} \frac{1+\sqrt{1-\left(\frac{2b_0}{\lambda}\right)^2}}{1-\sqrt{1-\left(\frac{2b_0}{\lambda}\right)^2}} - K \quad (3.35)$$

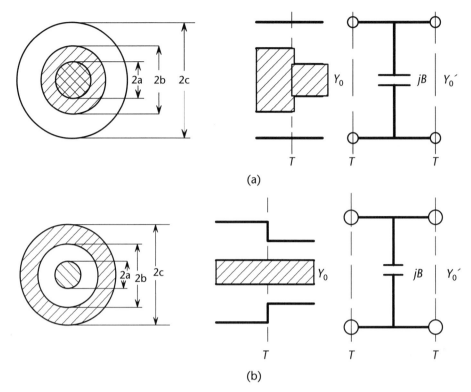

Figure 3.12 Step discontinuity in a coaxial line: (a) change in inner diameter, and (b) change in outer diameter.

$$A' = \left[\frac{1+\alpha}{1-\alpha}\right]^{2\alpha} \frac{1+\sqrt{1-\left(\frac{2b'_0}{\lambda}\right)^2}}{1-\sqrt{1-\left(\frac{2b'_0}{\lambda}\right)^2}} + K \quad (3.36)$$

$$K = \frac{3+\alpha^2}{1-\alpha^2} \quad (3.37)$$

$$C = \left(\frac{4\alpha}{1-\alpha}\right)^2 \quad (3.38)$$

where

$$\alpha = \frac{c-b}{c-a}, \; b_0 = c-a, \; b'_0 = c-b \quad (3.39)$$

$$A_1 = \frac{b}{a} \frac{\ln \frac{c}{a}}{\frac{c}{a} - 1} \left[\frac{\frac{c}{b} - 1}{\ln \frac{c}{b}} - 1 \right]^2 \tag{3.40}$$

$$A_2 = \frac{\pi^2 \frac{a}{b}}{\gamma_1 \sqrt{1 - \left(\frac{2b_0}{\gamma_1 \lambda}\right)^2}} \frac{\frac{c}{a} - 1}{\frac{J_o^2(\chi)}{J_o^2(\chi c/a)} - 1} \left[\frac{J_o(\chi) N_o\left(\frac{\chi b}{a}\right) - N_o(\chi) J_o\left(\frac{\chi b}{a}\right)}{\frac{c}{b} - 1} \right]$$

$$- \frac{1}{\sqrt{1 - \left(\frac{2b_0}{\lambda}\right)^2}} \left\{ \frac{2b_0}{\pi b_0'} \sin \frac{\pi b_0'}{b_0} \right\} \tag{3.41}$$

$$\gamma_1 = \frac{\chi}{\pi} \left(\frac{c}{a} - 1 \right) \tag{3.42}$$

χ is the first nonvanishing root of

$$J_o(\chi) N_o\left(\frac{\chi c}{a}\right) - N_o(\chi) J_o\left(\frac{\chi c}{a}\right) = 0$$

The above equations are applicable to both types of step discontinuities shown in Figure 3.12. However, for the type shown in Figure 3.12(b), the following changes are to be made:

$$\alpha = \frac{c-b}{c-a}, \; b_0 = c - a \text{ and } b_0' = c - b \tag{3.43}$$

The equivalent circuit is valid in the wavelength range $\lambda \geq 2(c-a)/\gamma_1$, provided the field is rotationally symmetrical. Inaccuracy in (3.33) is within a few percent for $c/a < 5$ and the operating wave length is not too close to $2(c-a)/\gamma 1$. For a rigorous, general, and accurate description of a coaxial line discontinuity one must use numerical electromagnetic (EM) analysis of the junction. A suitable method is mode matching analysis [3]. An approximate scattering matrix of the junction can be obtained from the above equivalent circuit using the following equation

$$[s] = \begin{bmatrix} \dfrac{-jB}{Y_0 + jB} & \dfrac{2}{Y_0 + jB} \sqrt{\dfrac{Y_0}{Y_0'}} \\ \dfrac{2}{Y_0 + jB} \sqrt{\dfrac{Y_0'}{Y_0}} & \dfrac{-jB}{Y_0 + jB} \end{bmatrix} \tag{3.44}$$

The coaxial line step discontinuity is a very important circuit component in coaxial line-harmonic reject low-pass filters and matching transformers, as will be shown in Chapter 5.

The above section is an example of how the effect of a discontinuity in a coaxial line can be described using an equivalent circuit. Such equivalent circuits were indispensable [4] until the advent of modern digital computers and highly accurate numerical techniques [5,6]. Today, virtually all types of discontinuities in microwave transmission lines can be described by their scattering matrices using numerical techniques and digital computers. In fact, any complex microwave structure involving several different discontinuities and straight sections of uniform transmission lines can be analyzed and optimized as a whole using a single analysis without considering each discontinuity separately. Therefore, it is sufficient to know the approximate equivalent circuit of a discontinuity in order to have a qualitative understanding of its behavior. Figure 3.13 shows the equivalent circuits of other types of coaxial line discontinuities. Such equivalent circuits can be derived from the scattering matrix obtained through full EM analysis.

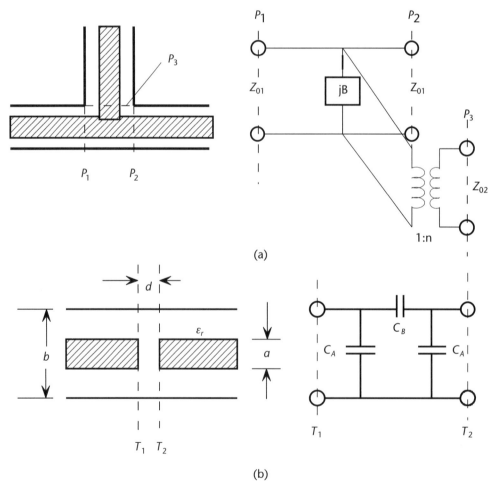

Figure 3.13 Coaxial line discontinuity: (a) T-junction, and (b) gap,

T-junctions are commonly used in stub tuners and band-pass and low-pass filters. Gap discontinuities are used in band-pass and high-pass filters. There are various other types of discontinuities that are formed between a coaxial line and another type of transmission line; for example, a transition from a coaxial line to a waveguide or a coaxial line and a microstrip line. Such discontinuities will be treated at an appropriate point, once the properties of waveguides and other planar transmission lines such as stripline, suspended stripline, and microstripline are described.

3.5 Rectangular Coaxial Line

Rectangular coaxial lines are often used at the input and output of interdigital and combline filters using rectangular bar resonators, hybrids, power dividers, and couplers [7]. Figure 3.14 shows the basic configuration of the cross section of a rectangular coaxial line. No accurate closed-form equation exists for the characteristic impedance of a rectangular coaxial line except for the exactly square cross-section case. For a square cross section ($a_1 = a_2 = a$ and $b_1 = b_2 = b$), the characteristic impedance is given by [8]

$$Z_0 = \frac{47.086\left[1 - \frac{b}{a}\right]}{0.279 + 0.721\frac{b}{a}} \Omega \tag{3.45}$$

where a and b are the dimensions of the outer and the inner conductors of the coaxial line. Figure 3.15 shows the variation of the characteristic impedance Z_0 of a square coaxial line with b/a ratio. The attenuation constant of a square coaxial line is given by

$$\alpha_c = 0.0231 \frac{Z_0(a+b)}{15\pi(a-b)^2} R_s \tag{3.46}$$

where R_s is the surface resistance of the walls of the outer conductors. The overall attenuation constant α can be calculated using (3.46) and the equation for dielectric loss α_d in Table 3.2. The total attenuation constant is the sum of α_c and α_d.

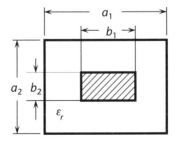

Figure 3.14 Cross section of a rectangular coaxial line.

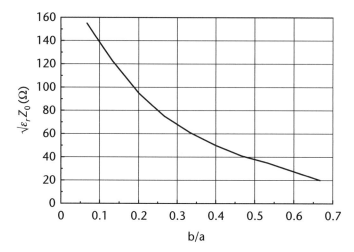

Figure 3.15 Characteristic impedance of square coaxial line as a function of b/a ratio ($b_1 = b_2 = b$) and $a_1 = a_2 = a$).

3.5.1 Higher-Order Modes in a Rectangular Coaxial Line

Like circular coaxial lines, rectangular coaxial lines support higher-order modes. The propagation characteristics of higher-order modes in a rectangular coaxial line can be calculated using numerical techniques based on full electromagnetic analysis. However, for conventional circuit and transmission line theory based passive component design, only the cutoff frequency of the first higher-order mode is significant because it sets the highest frequency range for the fundamental TEM mode operation. Figure 3.16 shows the variation of first higher-order mode cutoff frequency as a function of the outer conductor size for $b/a = 0.4$. For more detailed information on square coaxial lines refer to [9,10].

3.5.2 Square Coaxial Line with Circular Inner Conductor

The most commonly used coaxial line used as a resonator in microwave filters is the coaxial line with square outer conductor and a circular inner conductor. Figure 3.17 shows the cross section of such a coaxial line.

The characteristic impedance of a coaxial line with square outer conductor and circular inner conductor is given by Lin [11] (see Figure 3.17).

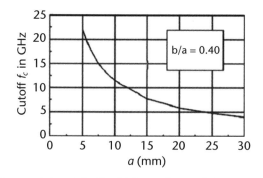

Figure 3.16 Cutoff characteristics of the first higher-order mode in a square coaxial line.

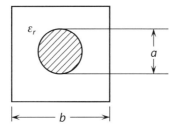

Figure 3.17 Cross section of coaxial line with square outer conductor and circular inner conductor.

$$Z_0 = 59.952\left\{\ln\left(\frac{b}{a}\right) + 0.06962\right\} \quad (3.47)$$

The attenuation in the line is minimum when Z_0 equals 77Ω. Substitution of that value for Z_0 in (3.47) gives $b/a = 3.369$. Choice of such impedance in cavity resonator design offers the maximum unloaded Q-factor.

3.6 Strip Transmission Line

Striplines are one of the most common types of transmission lines used in microwave circuits. This chapter presents the fundamental characteristics of this type of transmission line together with its governing equations. Striplines are useful for designing virtually all microwave circuit components and these are described in the rest of the chapter.

3.6.1 Basic Configuration

The stripline, shown in Figure 3.18 is obtained by theoretically moving the side walls of a rectangular coaxial line, shown in Figure 3.14, to infinity. It is the oldest planar transmission line that has been in use in microwave integrated circuits since its creation by R. M. Barrett in 1950 [12]. In its simplest form, it consists of a conducting strip, of width W and thickness t, separated from a pair of common conducting ground planes of theoretically infinite extent compared to the width W of the strip conductor, $W \ll a$; where a is the width of the ground plane. The ground planes are separated by a thickness b and the entire space is homogeneously filled with a dielectric material of complex dielectric constant $\varepsilon_r(1 - j \tan \delta)$. The

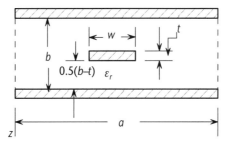

Figure 3.18 Basic balanced stripline configuration.

ground planes are kept at the same potential. In a balanced stripline, the strip conductor is equidistant from the ground planes. In an unbalanced stripline, there is an offset and the strip is not equidistant from the two ground planes as shown in Figure 3.22. The first significant theoretical investigation on striplines was carried out by S. Cohn in the mid-1950s [13]. While Sanders Associates used the trade name Triplate [14], the term stripline was first introduced by Airborne Instruments Laboratories (AIL).

3.6.2 Modes in a Stripline

Although striplines can support waveguide type modes (TE or TM), the fundamental mode of propagation in a stripline is the TEM mode having no cutoff frequency. The field configuration for the fundamental mode is shown in Figure 3.19. The usable single mode bandwidth of a stripline is determined by the cutoff frequency of the lowest-order waveguide mode. For that mode, the two ground planes have the same potential, the electric field is normal to the strip and the ground planes, and the longitudinal electric field is zero with cutoff frequency [15] given by

$$f_c = \frac{c}{\sqrt{\varepsilon_r}\left(2\frac{W}{b} + 4\frac{d}{b}\right)b} \quad (3.48)$$

where c is the velocity of light in free space (3×10^8 meters/sec), $4d/b$ is a function of the cross section of the stripline. For a balanced stripline when $t/b = 0$ and $W/b > 0.35$, then $4d/b$ is a function of b/λ_c alone and is given in Figure 3.20.

3.6.3 Characteristic Impedance of a Balanced Stripline

The characteristic impedance of a balanced strip transmission line can be accurately calculated from [16]:

$$Z_0 = \frac{60}{\sqrt{\varepsilon_r}} \ln\left[\frac{4b}{\pi d}\right] \quad (3.49)$$

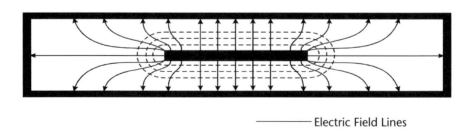

Figure 3.19 Field configuration in a stripline.

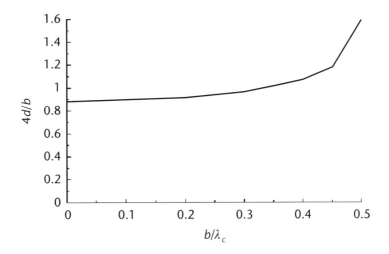

Figure 3.20 Cutoff wavelength in a stripline as a function of d/b ratio.

for $W/b < 0.35$, where

$$d = \frac{W}{2}\left[1 + \frac{t}{\pi W}\left(1 + \ln\frac{4\pi W}{t} + 0.51\pi\left(\frac{t}{W}\right)^2\right)\right] \quad (3.50)$$

and

$$Z_0 = \frac{94.15}{\sqrt{\varepsilon_r}} \frac{1}{\dfrac{C_f}{\varepsilon} + \dfrac{W}{b\left(1 - \dfrac{t}{b}\right)}} \, \Omega \quad (3.51)$$

for $W/b \geq 0.35$
where

$$\frac{C_f}{\varepsilon} = \frac{1}{\pi}\left\{\frac{2}{1-\dfrac{t}{b}}\ln\left(\dfrac{1}{1-\dfrac{t}{b}}+1\right)\right\} - \frac{1}{\pi}\left\{\left(\dfrac{1}{1-\dfrac{t}{b}}-1\right)\ln\left(\dfrac{1}{\left(1-\dfrac{t}{b}\right)^2}-1\right)\right\} \quad (3.52)$$

$2C_f/\varepsilon$ is the per-unit length fringing field capacitance between the strip and each ground plane and $\varepsilon = \varepsilon_0\varepsilon_r$; $\varepsilon_0 = 8.854e^{-12}$. Farads/meter (permittivity of free space). Figure 3.21 shows the dependence of characteristic impedance on W/b with strip thickness t/b as a parameter.

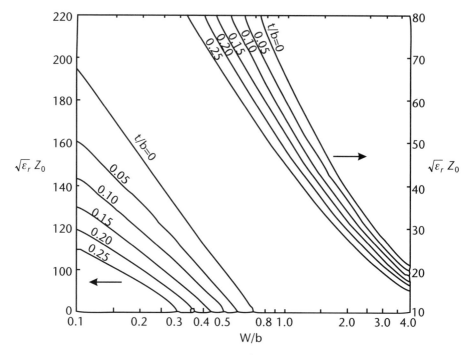

Figure 3.21 Variation of a balanced stripline characteristic impedance with t/b and w/b.

3.6.4 Unbalanced Stripline

In an unbalanced stripline the strip is nonequidistant from the top and the bottom ground planes, as shown in Figure 3.22. Such striplines find many applications in microwave circuit design.

The characteristic impedance of an unbalanced stripline is given by [16]:

$$Z_0 = \frac{120\pi}{\sqrt{\varepsilon_r}\dfrac{C}{\varepsilon}} \tag{3.53}$$

where C/ε is the per unit length static capacitance between the strip and the two ground planes, normalized by the permittivity ε of the medium.

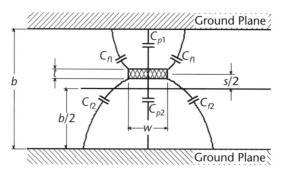

Figure 3.22 Unbalanced (offset) stripline.

$$\frac{C}{\varepsilon} = \frac{C_{p1}}{\varepsilon} + \frac{C_{p2}}{\varepsilon} + \frac{2C_{f1}}{\varepsilon} + \frac{2C_{f2}}{\varepsilon} \qquad (3.54)$$

$$\frac{C_{p1}}{\varepsilon} = \frac{2\dfrac{W}{b-s}}{1 - \dfrac{t}{b-s}} \qquad (3.55)$$

$$\frac{C_{p2}}{\varepsilon} = \frac{2\dfrac{W}{b+s}}{1 - \dfrac{t}{b+s}} \qquad (3.56)$$

The per-unit length fringe-field capacitances C_{f1} and C_{f2} are obtained from (3.52) by replacing b with $(b-s)$ and $(b+s)$, respectively. The above formulas are applicable to single strips only. Figure 3.21 shows the variations of Z_0 with W/b for various values of t/b for a balanced stripline. Note that for $s = 0$ an unbalanced or offset stripline becomes a balanced or symmetric stripline.

3.6.5 Propagation Constant in a Stripline

The propagation constant β of a stripline is given by

$$\beta = \frac{2\pi}{\lambda_g} = \frac{2\pi\sqrt{\varepsilon_r}}{\lambda_0} \text{ radians/unit length} \qquad (3.57)$$

where λ_g and λ_0 are the wavelengths in the stripline and free space, respectively.

3.6.6 Synthesis of a Stripline

In order to obtain the structural dimensions for a stripline to be designed, when the characteristic impedance Z_0 and the substrate dielectric constant ε_r are given, the following formula is used [17]:

$$W = W_0 - \Delta W_0 \qquad (3.58)$$

$$W_0 = \frac{8(b-t)\sqrt{B+0.568}}{\pi(B-1)} \qquad (3.59a)$$

$$\Delta W_0 = \frac{t}{\pi}\left\{1 - 0.5\ln\left[\left(\frac{t}{2b-t}\right)^2 + \left(\frac{0.0796t}{W_0 - 0.26t}\right)^m\right]\right\} \qquad (3.59b)$$

$$B = e^{\left(\frac{Z_0\sqrt{\varepsilon_r}}{30}\right)} \tag{3.60}$$

and

$$m = 6\left[\frac{b-t}{3b-t}\right] \tag{3.61}$$

There is no closed-form design equation for the unbalanced stripline shown in Figure 3.22. An iterative procedure based on analysis and optimization is used to synthesize an unbalanced stripline for a given characteristic impedance Z_0 and substrate dielectric constant ε_r. The above equations have been programmed in many commercial software packages [18–20] for prompt analysis and synthesis of an uncoupled single stripline. Some of those software packages use simple optimization routines in order to synthesize unbalanced striplines.

3.6.7 Attenuation Constant in a Stripline

The attenuation constant of a stripline, balanced or unbalanced, is given by [21]

$$\alpha = \alpha_c + \alpha_d \text{ Nepers/unit length} \tag{3.62}$$

The attenuation constant α_c due to conductor loss in the line at a frequency f in gigahertz is obtained from [22]:

$$\alpha_c = \frac{\pi\sqrt{\varepsilon_r}f}{0.2998}\left[1 - \frac{Z_0'}{Z_0}\right] \text{ Nepers/meter} \tag{3.63}$$

where Z_0 is the characteristic impedance of the line and Z_0' is the characteristic impedance of the line when W, t and b are replaced by, $W = W + \delta_s$, $t + \delta_s$, and $b - \delta_s$, respectively, in (3.49) to (3.56).

$$\delta_s = 0.0822\sqrt{\frac{\delta_{rcu}}{f}} \text{ mils} \tag{3.64}$$

is the skin depth at frequency f in GHz and δ_{rcu} is the conductivity relative to copper. The attenuation constant α_d due to dielectric loss is given by

$$\alpha_d = \frac{\beta \tan \delta}{2} \text{ Nepers/meter} \tag{3.65}$$

The unloaded Q-factor of a stripline is given by

$$Q = \frac{8.686\pi\sqrt{\varepsilon_r}}{\lambda_0 \alpha} \qquad (3.66)$$

3.6.8 Power-Handling Capability of a Stripline

The average power P, in kilowatts, that can be carried by a matched balanced stripline with rounded edges is shown in Figure 3.23 [23]. The ground plane to ground plane distance is measured in inches. Although the strip edges are assumed to be round, an approximate value of Z_0 can be obtained from either Figure 3.21 or from the analysis equations presented above.

3.6.9 Stripline Discontinuities

Stripline discontinuities are as essential an element of microwave circuits as their uniform line counterpart. Any arbitrary discontinuity in a stripline can be decomposed into a few basic forms of discontinuities, as shown in Figure 3.24. These include an abrupt change in width or step discontinuity, a gap, a circular hole in the strip, an open end, a cross junction, a T-junction, and an angled bend. The appearance of discontinuities causes alterations in the electromagnetic field configurations of an otherwise uniform stripline. Therefore, the modified field configuration can be taken into account by appropriate incorporation of a shunt or a series capacitance or inductance and transformers. For example, an open end can be represented

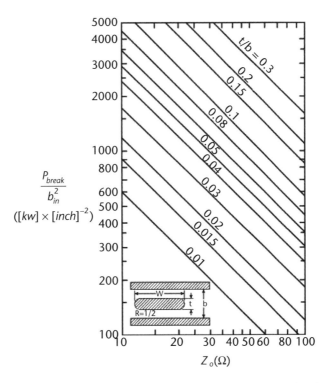

Figure 3.23 Power-handling capability of a stripline. (From [23]. Reprinted with permission from Artech House.)

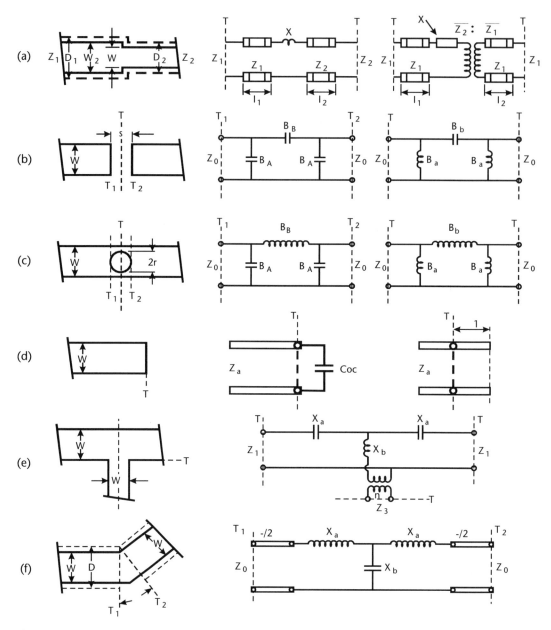

Figure 3.24 Basic stripline discontinuities and the equivalent circuits.

by a shunt capacitance. Figure 3.24 shows the configurations and the corresponding equivalent circuits of the discontinuities [24,25,26]. The equivalent width D, shown by the dashed lines in Figure 3.24 (D_1 and D_2 in Figure 3.24(a) and D in Figure 3.24(f)), is obtained by conformal mapping techniques, as shown in Figure 3.25. The parameters b, t, and W are defined in Figure 3.18 and 3.22 for a balanced and an unbalanced stripline, respectively.

$$D = b\frac{K(k)}{K(k')} + \frac{t}{\pi}\left(1 - \ln\frac{2t}{b}\right) \quad (3.67)$$

3.6 Strip Transmission Line

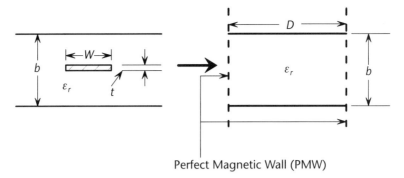

Figure 3.25 Parallel plate equivalent of a stripline.

For $W/b \leq 0.5$
where $K(k)$ is the complete elliptic integral of the first kind

$$k = \tanh\left(\frac{\pi W}{2b}\right) \tag{3.68}$$

$$K(k) = \int_0^1 \frac{dx}{\sqrt{(1-x^2)(1-k^2x^2)}} \tag{3.69}$$

and the associated complementary elliptic integral is defined as

$$K(k') = K\left(\sqrt{1-k^2}\right) \tag{3.70}$$

and

$$D = W + \frac{2b}{\pi}\ln 2 + \frac{t}{\pi}\left(1 - \ln\frac{2t}{b}\right) \tag{3.71}$$

for $W/b > 0.5$

3.6.9.1 Step Discontinuity

A change in strip width or step discontinuity is essential for the design of stripline matching transformers and low-pass filters. The equivalent circuit parameters, shown in Figure 3.24(a), are given by

$$X = Z_1 \frac{2D_1}{\lambda_g} \ln \csc \frac{\pi D_2}{D_1} \tag{3.72}$$

$$l_1 = -l_2 = \frac{b \ln 2}{\pi} \qquad (3.73)$$

The normalized scattering matrix of the discontinuity can be written as

$$[S] = \frac{1}{\Delta S}\begin{bmatrix} S_{11} & S_{12} \\ S_{21} & S_{22} \end{bmatrix} \qquad (3.74)$$

$$S_{11} = (Z_2 - Z_1 + jX)e^{-j2\beta l_1} \qquad (3.75a)$$

$$S_{12} = S_{21} = 2\sqrt{Z_1 Z_2} \qquad (3.75b)$$

$$S_{22} = (Z_1 - Z_2 + jX)e^{-j2\beta l_2} \qquad (3.75c)$$

$$\Delta S = S_{11}S_{22} - S_{12}S_{21} \qquad (3.75d)$$

The equivalent network for equal normalizations at the input and the output ports includes a transformer, as shown in Figure 3.24(a).

3.6.9.2 Gap Discontinuity

A series capacitance in a stripline is realized by a gap discontinuity, as shown in Figure 3.24(b). The equivalent circuit is, however, a pi network of one series and two shunt capacitances. The series component is due to the fringing capacitance from one strip to the other strip and the shunt components are due to the field disturbance at the edge of each strip. As the gap increases, the series capacitance decreases and the two shunt capacitances tend toward that of an open-ended stripline. The normalized susceptance parameters of the equivalent pi network are given by

$$\overline{B_A} = \frac{1 + \overline{B_a}\cot(\beta s/2)}{\cot(\beta s) - \overline{B_a}} = \frac{\omega C_1}{Y_0} \qquad (3.76)$$

$$2\overline{B_B} = \frac{1 + (2\overline{B_b} + \overline{B_a})\cot(\beta s/2)}{\cot(\beta s/2) - (2\overline{B_b} + \overline{B_a})} - \overline{B_A} = \frac{2\omega C_{12}}{Y_0} \qquad (3.77)$$

$$\lambda \overline{B_a} = 2b\ln\left\{\operatorname{sech}\left(\frac{2b}{\pi s}\right)\right\} \qquad (3.78)$$

$$\lambda \overline{B_b} = b \ln\left\{\coth\left(\frac{\pi s}{2b}\right)\right\} \tag{3.79}$$

3.6.9.3 Circular Hole Discontinuity

A hole discontinuity in a stripline is introduced to realize reactive tuning in filters and resonators. Such discontinuities are predominantly inductive in nature. The most common hole discontinuity is a circular hole discontinuity, as shown in Figure 3.24(c). The susceptance parameters of the equivalent pi-network are given by

$$\overline{B_A} = \frac{1 + \overline{B_a}\cot(\beta r)}{\cot(\beta r) - \overline{B_b}} \tag{3.80}$$

$$2\overline{B_B} = \frac{1 + 2\overline{B_b}\cot(\beta r)}{\cot(\beta r) - \overline{B_b}} - \overline{B_A} \tag{3.81}$$

where

$$\overline{B_b} = -\left(\frac{3bD}{16\beta r^3}\right), \quad \overline{B_a} = \frac{1}{3\overline{B} - b} \tag{3.82}$$

Note that the equivalent networks for gap and circular hole discontinuities depend on where the reference plane is considered to be situated.

3.6.9.4 Open-End Discontinuity

An open-end discontinuity occurs whenever an open-circuited stripline stub is used in components such as matching networks, and filters. Figure 3.24(d) shows a stripline open end and two equivalent networks. The network can be a shunt capacitance C_{oc} or an extended length Δl. The second representation assumes that a perfect magnetic wall exists at a distance Δl from the physical open circuit. The open circuit capacitance is given by

$$C_{oc} = \frac{1}{\omega Z_0}\tan^{-1}\left[\frac{\xi + 2W}{4\xi + 2W}\tan(\beta\xi)\right] \tag{3.83}$$

where

$$\lambda = \frac{\lambda_0}{\sqrt{\varepsilon_r}}, \quad \beta = \frac{2\pi}{\lambda}, \quad \xi = 0.2206b \tag{3.84}$$

The length extension Δl can be obtained from the open-end capacitance

$$\Delta l = \frac{1}{\beta}\tan^{-1}(Z_0 \omega C_{oc}) \tag{3.85}$$

The reflection coefficient from the open end discontinuity can be calculated as

$$S_{11} = \frac{1 - jZ_0 \omega C_{oc}}{1 + jZ_0 \omega C_{oc}} \tag{3.86}$$

In the above equations, Z_0 is the characteristic impedance of the stripline.

3.6.9.5 T-Junction Discontinuity

A T-junction discontinuity occurs in stripline stub matching, stub-loaded low-pass and band-pass filters, branchline couplers, hybrid rings, and in many other components. Figure 3.24(e) shows the stripline T-junction and the equivalent network. The network parameters are obtained from

$$\frac{X_a}{Z_1} = -(0.785n)^2 \frac{D_3^2}{D_1 \lambda} \tag{3.87}$$

$$\frac{X_b}{Z_1} = -\frac{X_a}{2Z_1} + \frac{1}{n^2}\left\{\frac{B_1}{2Y_1} + \frac{2D_1}{\lambda}\left[0.6931 + \frac{\pi D_3}{6D_1} + 1.5\left(\frac{D_1}{\lambda}\right)^2\right]\right\} \tag{3.88}$$

for $D_3/D_1 < 0.5$, and

$$\frac{X_b}{Z_1} = -\frac{X_a}{2Z_1} + \frac{2D_1}{\lambda n^2}\left\{\ln\frac{1.43D_1}{D_3} + 2\left(\frac{D_1}{\lambda}\right)^2\right\} \tag{3.89}$$

for $D_3/D_1 > 0.5$. The transformer turns ratio n is given by

$$n = \frac{\sin\left(\dfrac{\pi D_3}{\lambda}\right)}{\dfrac{\pi D_3}{\lambda}} \tag{3.90}$$

and

$$\frac{B_1}{2Y_1} = \frac{2D_1}{\lambda}\left[\ln\csc\frac{\pi D_1}{2D_1} + 0.5\left(\frac{d_1}{\lambda}\right)^2 \cos^4\frac{\pi D_3}{2D_1}\right] \tag{3.91}$$

In the above equations D_1 and D_3 are the widths of the equivalent parallel plate waveguides for strips of widths W and W' respectively, Z_1 and Z_3 are corresponding characteristic impedances, and Y_1 and Y_3 are the respective characteristic admittances. The normalized scattering matrix of the T-junction is obtained from

$$S_{11} = S_{22} = \frac{j2(Z_3/n^2)X_a - (Z_1^2 + 2X_aX_b + X_a^2)}{(Z_1 + jX_a)\Delta} \quad (3.92)$$

$$S_{12} = S_{21} = \frac{2Z_1(Z_3/n^2 + jX_a)}{(Z_1 + jX_b)\Delta} \quad (3.93)$$

$$S_{13} = S_{23} = S_{31} = S_{32} = \frac{2\sqrt{Z_1 Z_2}/n^2}{\Delta} \quad (3.94)$$

$$S_{33} = \frac{Z_1 - 2(Z_3/n^2) + j(X_a + 2X_b)}{\Delta} \quad (3.95)$$

and

$$\Delta = Z_1 + 2Z_3/n^2 + j(X_a + 2X_b) \quad (3.96)$$

3.6.9.6 Bend Discontinuity

A bend discontinuity occurs mainly in stripline transitions and hybrids. Figure 3.24(f) shows a stripline bend discontinuity and the equivalent network. The parameters of the network are obtained from the following equations derived from Babinet's principle and the equivalent parallel plate waveguide model.

$$\lambda \overline{X_a} = 2D\left\{\psi(x) + 1.9635 + \frac{1}{x}\right\} \quad (3.97)$$

$$\overline{X_b} = -\frac{\lambda}{2\pi D}\cot\frac{\theta}{2} \quad (3.98)$$

with θ, in degrees, x is given by

$$x = 0.5\left\{1 + \frac{\theta}{180}\right\} \tag{3.99}$$

and

$$\psi(x) = 0.5223\ln(x) + 0.394 \tag{3.100}$$

Equation (3.97) is an approximation of the T-function [26]. Accurate values of the T-function for various x are available in [27]. The reference planes T_1 and T_2 meet at an angle θ. This modifies the scattering parameters of the bend by multiplying S_{11} and S_{22} by $e^{j2\beta\zeta}$ and S_{12} and S_{21} by $e^{j\beta\zeta}$, where

$$\zeta = (D - W)\tan\frac{\theta}{2} \tag{3.101}$$

The above models for the discontinuities have been included as a library in many commercially available microwave circuit solvers [28,29,30]. However, any stripline circuit comprising several such discontinuities and uniform transmission lines can also be analyzed, as a whole, very accurately using full-wave (EM) solvers [31,32].

3.7 Parallel-Coupled Lines

Parallel-coupled lines find extensive use in microwave circuits and components such as filters, couplers, baluns, and delay lines. The coupling takes place by the capture of electromagnetic fields produced by one of the lines by the other and vice versa. Figure 3.26 shows the most commonly used and the oldest parallel-coupled line in microwave circuit design, known as the edge-coupled stripline.

3.7.1 Edge-Coupled Striplines

The electrical behavior of the coupled stripline structure shown in Figure 3.26 can be described in terms of four characteristic impedances known as the even-and the odd-mode impedances. The impedances are given by [33]

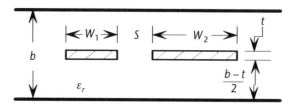

Figure 3.26 Edge-coupled stripline.

3.7 Parallel-Coupled Lines

$$Z_{0e}^i = \frac{\sqrt{\varepsilon_r}\eta_0}{C_e^i} \qquad (3.102)$$

$$Z_{0o}^i = \frac{\sqrt{\varepsilon_r}\eta_0}{C_o^i} \qquad (3.103)$$

where

$$C_e^i = 2\left(C_f^i + C_p^i + C_{fe}^i\right) \qquad (3.104)$$

$$C_e^i = 2\left(C_f^i + C_p^i + C_{fe}^i\right) \qquad (3.105)$$

$$C_{fe}^i = \Xi_1 G_1 G_2 G_3 \qquad (3.106)$$

$$G_1 = 1 + \frac{C_f^i + C_p^i - C_f^j - C_p^j}{C_f^i + C_p^i + C_f^j + C_p^j}\left(1 - \tanh\frac{\pi S}{2b}\right) \qquad (3.107)$$

$$G_2 = 1 - \frac{1}{\pi}e^{\left[-\frac{8\pi W_t}{3b}\left(1+\frac{1/2}{1+(\pi S/b)^2}\right)\right]} \qquad (3.108a)$$

$$G_3 = \varepsilon_r \frac{2}{\pi}\left(1 + \tanh\frac{\pi S}{2b}\right) \qquad (3.108b)$$

$$\begin{aligned}\Xi_1 &= \left\{-\left(0.2933 + 3.333\frac{S}{b}\right)\frac{t}{b}\right\} & \frac{S}{b} \le 0.08 \\ \Xi_1 &= -0.56\frac{t}{b} & \frac{S}{b} \ge 0.08 \end{aligned} \qquad (3.109)$$

$$C_{fo}^i = C_{fe}^i + C_{12}^i \qquad (3.110)$$

$$C_{12}^i = \Xi_2 \left[C_{12}^i\right]_{t=0} \qquad (3.111)$$

where

$$\left[C_{12}^i\right]_{t=0} = \frac{2}{\pi} \ln\left[\coth\frac{\pi s}{2b}\right]\Theta \qquad (3.112)$$

$$\Theta = \frac{1}{2}\left\{1 + \frac{C_f^i + C_p^i}{C_f^j + C_p^j}\right\}\left\{1 - \frac{1}{\pi}e^{\left[-\frac{8\pi\sqrt{W_1 W_2}}{6b}\right]}\right\} \qquad (3.113)$$

$$\Xi_2 = 1 + 2.13507\frac{t}{b}\left[\frac{s}{b}\right]^{-0.57518} \qquad \frac{S}{b} \le 0.3 \qquad (3.114)$$

$$\Xi_2 = 1 + 3.89531\frac{t}{b}\left[\frac{s}{b}\right]^{-0.11467} \qquad \frac{S}{b} \ge 0.3 \qquad (3.115)$$

$$\Delta = \frac{1}{\pi}\left[2t_b \ln(t_b + 1) - (t_b - 1)\ln(t_b^2 - 1)\right] \qquad (3.116)$$

$$C_p^i = 2\varepsilon_r \frac{W_i}{b} \qquad (3.117)$$

For $W_1 = W_2 = W$ and $t/b < 0.1$, the above equations simplify as follows:

$$Z_{0e} = \frac{30\pi(b-t)}{\sqrt{\varepsilon_r}\left(W + A_e b C_f\right)} \qquad (3.118)$$

$$Z_{0o} = \frac{30\pi(b-t)}{\sqrt{\varepsilon_r}\left(W + A_o b C_f\right)} \qquad (3.119)$$

where

$$A_e = \frac{\ln 2 + \ln\left(1 + \tanh\frac{\pi S}{2b}\right)}{2\pi \ln 2} \qquad (3.120)$$

$$A_o = \frac{\ln 2 + \ln\left(1 + \coth\frac{\pi S}{2b}\right)}{2\pi \ln 2} \qquad (3.121)$$

3.7 Parallel-Coupled Lines

$$C_f = 2\ln\left(\frac{2b-t}{b-t}\right) - \frac{t}{b}\left(\frac{t(2b-t)}{(b-t)^2}\right) \quad (3.122)$$

Figure 3.27 shows the variations of even- and odd-mode impedances of coupled symmetrical striplines with strip width and separation. Note that for $S \gg W$ and $S \gg b$ the even- and odd-mode impedances become equal and the above equations become applicable to isolated single striplines.

3.7.2 Synthesis Equations

Closed-form equations for synthesis of symmetrical coupled striplines are available for zero thickness ($t = 0$) strips [34]. For a given set of even- and odd-mode characteristic impedances Z_{0e} and Z_{0o}, ε_r, b, the strip width W and spacing S between them are given by

$$W = \frac{2b}{\pi}\tanh^{-1}\sqrt{k_e k_o} \quad (3.123)$$

$$S = \frac{2b}{\pi}\tanh^{-1}\sqrt{\frac{k_e}{k_o}} \quad (3.124)$$

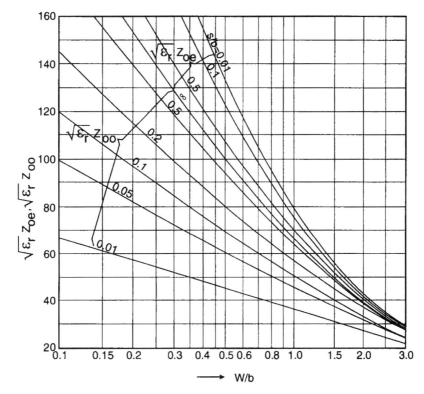

Figure 3.27 Analysis curves for coupled striplines. (Reprinted from [33] with permission from the IEEE.)

where for $\sqrt{\varepsilon_r}Z_{0i}$ ($i=e$ or o) $\geq 188.5\ \Omega$

$$k_{e,o} = 4e^{-\left(2\pi\frac{\sqrt{\varepsilon_r}Z_{0i}}{\eta_0}\right)}$$

and for $\sqrt{\varepsilon_r}Z_{0i}$ ($i=e$ or o) $\leq 188.5\ \Omega$

$$k_{e,o} = \left[\tanh\left\{\frac{\pi\eta_0}{8\sqrt{\varepsilon_r}Z_{0i}} \frac{\ln 2}{2}\right\}\right]^2$$

For synthesis of coupled striplines of nonzero strip thickness, one can use an optimization routine based on analysis equations (3.102) to (3.122), where (3.118) and (3.122) are used for the initial guesses for W and S. However, today there exist a number of analysis and synthesis software solutions for general stripline structures, which are based on either full EM analysis or quasi-static analysis [35]. A huge storehouse of information on stripline design is available at http://www.circuitsage.com.

3.7.3 Attenuation in Coupled Striplines

The generalized equation based on the incremental inductance rule described in Section 3.4.4 can be used to calculate the conductor losses in coupled striplines. The process is very simple and it is separately applied to the even and odd modes as follows. Let Z_{0e} and Z_{0o} be the even- and the odd-mode characteristic impedances of the coupled lines and Z'_{0e} and Z'_{0o} be those of the dimensions changed by the skin depth, as shown in Figure 3.28. Then the conductor losses for the even and odd modes are given by [36]

$$\alpha_{0e}^i = 28.97\sqrt{\varepsilon_r}\pi\left[1 - \frac{Z_{0e}^i}{Z_{0e}^{i'}}\right]\sqrt{f}\ \text{dB/m} \qquad (3.125)$$

$$\alpha_{0o}^i = 28.97\sqrt{\varepsilon_r}\pi\left[1 - \frac{Z_{0o}^i}{Z_{0o}^{i'}}\right]\sqrt{f}\ \text{dB/m} \qquad (3.126)$$

where f is in GHz. The dielectric loss in coupled stripline is calculated as [36]

$$\alpha_d = 28.97\pi\sqrt{\varepsilon_r}f\tan(\delta)\ \text{dB/m} \qquad (3.127)$$

The above equation applies to both the even and odd modes of both lines. It will be shown in later chapters how the above equations are used in analysis and synthesis of planar microwave passive components.

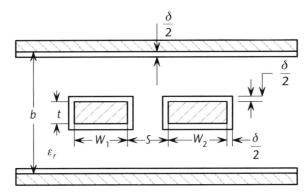

Figure 3.28 Modified coupled stripline for loss calculation.

3.7.4 Broadside Coupled Striplines

Broadside coupled striplines, shown in Figure 3.29, are extensively used in couplers of very tight coupling and stripline high-pass and wideband filters. The analysis equations for broadside coupled striplines ($t \approx 0$) are as follows [37,38]:

$$Z_{0o} = \frac{Z^a_{0\infty} - \Delta Z^a_{0\infty}}{\sqrt{\varepsilon_r}} \qquad (3.128)$$

where

$$Z^a_{0\infty} = 60 \ln \left(\frac{3S}{W} + \sqrt{\left(\frac{S}{W}\right)^2 + 1} \right) \qquad (3.129)$$

and

$$\begin{aligned} \Delta Z^a_{0\infty} &= P \text{ for } \frac{W}{S} \leq \frac{1}{2} \\ &= PQ \text{ for } \frac{W}{S} \geq \frac{1}{2} \end{aligned} \qquad (3.130)$$

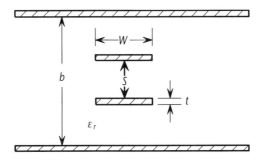

Figure 3.29 Broadside coupled stripline.

$$P = 270\left[1 - \tanh^{-1}\left(0.28 + 1.2\sqrt{\frac{b-S}{S}}\right)\right] \quad (3.131)$$

$$Q = 1 - \tanh^{-1}\left\{\frac{0.48\sqrt{\frac{2W}{S} - 1}}{\left(1 + \frac{b-S}{S}\right)^2}\right\} \quad (3.132)$$

The even mode impedance is obtained from

$$Z_{0e} = \frac{60\pi}{\sqrt{\varepsilon_r}}\frac{K(k')}{K(k)} \quad (3.133)$$

$$k = \tanh\left(\frac{294S/b}{Z_{0o}}\right) \quad (3.134)$$

$$k' = \sqrt{1 - k^2} \quad (3.135)$$

where $K(k)$ is the complete elliptic integral of first kind with k as an argument.

3.7.4.1 Synthesis of Broadside-Coupled Striplines

Given Z_{0o} and Z_{0e} the corresponding W/b and S/b are obtained as follows [37,39,40]:
From (3.133) we get

$$\frac{K(k')}{K(k)} = \frac{\sqrt{\varepsilon_r}Z_{0e}}{60\pi} \quad (3.136)$$

for

$$\frac{K(k')}{K(k)} \le 2$$

$$\frac{K(k')}{K(k)} \le 2k = 4e^{\left\{-\frac{\pi K(k')}{2 K(k)}\right\}} \quad (3.137)$$

otherwise

$$k = \left[\tanh\left\{\frac{\pi}{2}\frac{K(k')}{K(k)}\right\} - \frac{\ln 2}{2}\right]^2 \qquad (3.138)$$

$$\frac{S}{b} = \frac{\sqrt{\varepsilon_r}Z_{0o}}{294}\tanh^{-1}(k) \qquad (3.139)$$

$$\frac{W}{b} = \frac{1}{\pi}\left\{\ln\left(\frac{1+R}{1-R}\right) - \frac{S}{b}\ln\left(\frac{1+R/k}{1-R/k}\right)\right\} \qquad (3.140)$$

$$R = \sqrt{\left(k\frac{b}{S} - 1\right)\bigg/\left(\frac{1}{k}\frac{b}{S} - 1\right)} \qquad (3.141)$$

3.7.4.2 Broadside-Coupled Offset Striplines

Offset coupled striplines are a useful planar transmission line in microwave circuit design. The structure is shown in Figure 3.30. It offers more coupling than is offered by edge-coupled striplines and is more versatile than broadside-coupled striplines. However, analysis of offset coupled striplines is more cumbersome than the analysis of other coupled striplines. There exist no closed-form analysis equations for offset-coupled striplines. Consequently, the designer has to use EM analysis-based computer-aided design (CAD) programs to estimate the quantitative electrical behavior of this important transmission line. Shelton [41] has given closed-form synthesis equations for offset-coupled striplines. These are as follows.

We assume that the strip thickness is negligible (i.e., $t \approx 0$) and first normalize the dimensions $s = S/b$, $w = W/b$, $w_0 = W_0/b$, and $w_c = W_c/b$. Then we define two coupling ranges.

Loose coupling:

$$\frac{w}{1-s} \geq 0.35 \quad \text{and} \quad \frac{w_c}{s} \geq 0.70$$

Tight coupling:

$$\frac{w}{1-s} \leq 0.35 \quad \text{and} \quad \frac{2w_0}{1+s} \geq 0.85$$

and define the parameter

$$\zeta = \frac{Z_{0e}}{Z_{0o}} \qquad (3.142)$$

Synthesis equations for loose coupling case:

$$C_0 = \frac{\eta_0 \sqrt{\zeta}}{\sqrt{\varepsilon_r Z_{0e} Z_{0o}}} \qquad (3.143)$$

$$\Delta C = \frac{\eta_0 (\zeta - 1)}{\sqrt{\varepsilon_r Z_{0e} Z_{0o} \zeta}} \qquad (3.144)$$

$$K = \frac{1}{e^{\left(\frac{\pi \Delta C}{2}\right)} - 1} \qquad (3.145)$$

$$q = \frac{K}{a} \qquad (3.146)$$

$$a = \sqrt{\left(\frac{s-K}{s+1}\right)^2 + K} - \frac{s-K}{s+1} \qquad (3.147)$$

$$w_c = \frac{1}{\pi}\left[s\ln\left(\frac{q}{a}\right) + (1-s)\ln\left(\frac{1-r}{1+a}\right)\right] \qquad (3.148)$$

$$w = \frac{1-s^2}{4}\left[C_o - C_f - C_{f0}(a=\infty)\right] \qquad (3.149)$$

$$C_{f0} = \frac{2}{\pi}\left[\frac{1}{1+s}\ln\frac{1+a}{a(1-q)} - \frac{1}{1-s}\ln(q)\right] \qquad (3.150)$$

$$C_f(a=\infty) = -\frac{2}{\pi}\left[\frac{1}{1+s}\ln\left(\frac{1-s}{2}\right) + \frac{1}{1-s}\ln\left(\frac{1}{1+s}\right)\right] \qquad (3.151)$$

Synthesis equations for tight coupling case:

$$A = e^{\left[\frac{60\pi^2}{\sqrt{\varepsilon_r Z_{0e} Z_{0o}}}\left(\frac{1-\zeta s}{\sqrt{\zeta}}\right)\right]} \quad (3.152)$$

$$B = \frac{A - 2 + \sqrt{A^2 - 4A}}{2} \quad (3.153)$$

$$p = \frac{(B-1)\left(\frac{1+s}{2}\right) + \sqrt{(B-1)^2\left(\frac{1+s}{2}\right)^2 + 4Bs}}{2} \quad (3.154)$$

$$r = \frac{Bs}{p} \quad (3.155)$$

$$C_{f0} = \frac{1}{\pi}\left[-\frac{2}{1-s}\ln(s) + \frac{1}{s}\ln\left\{\frac{pr}{(p+s)(1+p)(1-r)(r-s)}\right\}\right] \quad (3.156)$$

$$w = (C_0 - C_{f0})\frac{s(1-s)}{2} \quad (3.157)$$

$$w = \frac{1}{2\pi}\left[(1+s)\ln\frac{p}{s} + (1-s)\ln\left\{\frac{(1+p)(r-s)}{(p+s)(1-r)}\right\}\right] \quad (3.158)$$

where C_0 is given by (3.143).

3.7.4.3 Broadside Slot-Coupled Striplines

The amount of coupling in broadside-coupled striplines, shown in Figure 3.30, can be controlled by controlling the strip width W and the separation S. However, an added degree of freedom can be achieved by introducing a slot between the strip, as shown in Figure 3.31. The structure is known as a broadside slot-coupled stripline. Such a structure is efficiently used in a constant group delay stripline filter. The analysis and synthesis equations are as follows [42].

Analysis:

$$Z_{0e} = \frac{\eta_0}{2\left(\frac{K'(k_1)}{K(k_1)} + \frac{K(k_2)}{K'(k_2)}\right)} \quad (3.159)$$

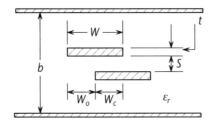

Figure 3.30 Broadside-coupled offset stripline.

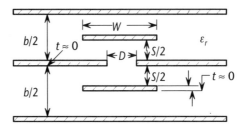

Figure 3.31 Broadside slot-coupled stripline.

$$k_1 = \sqrt{\frac{\sinh^2\left(\frac{\pi W}{2S}\right)}{\sinh^2\left(\frac{\pi W}{2S}\right) + \cosh^2\left(\frac{\pi D}{2S}\right)}} \tag{3.160}$$

$$k_2 = \tanh\left\{\frac{\pi W}{2(b-s)}\right\} \tag{3.161}$$

where $K'(k) = K(k')$ are elliptic integrals of first kind and $k_i' = \sqrt{1-k_i^2}$.

$$Z_{0e} = \frac{\eta_0}{2\left(\frac{K'(k_3)}{K(k_3)} + \frac{K(k_2)}{K'(k_2)}\right)} \tag{3.162}$$

$$k_3 = \tanh\left\{\frac{\pi W}{2S}\right\} \tag{3.163}$$

Synthesis:
Given ε_r, b, S, Z_{0e}, and Z_{0o}, W/b and D/b are obtained from [40]

$$\frac{W}{b} = \frac{1}{\pi}\ln\frac{\Theta(\alpha+a)}{\Theta(\alpha-a)} \tag{3.164}$$

$$\frac{D}{b} = \frac{1}{\pi} \ln \frac{H(\beta+a)}{H(\beta-a)} \tag{3.165}$$

where

$$\alpha = sn^{-1}\left[\frac{1}{k}\frac{\sqrt{Z(a)}}{\sqrt{Z(a)sn^2(a)+sn(a)cn(a)dn(a)}}, k\right] \tag{3.166}$$

$$\beta = sn^{-1}\left(\frac{k_e}{k_0}, k_0\right) \tag{3.167}$$

$$a = \frac{S}{b}K(k_0) \tag{3.168}$$

Use Z_{0e} and Z_{0o} in (3.136) to (3.138) to obtain the values of k_e and k_0, respectively. In the above equations, Θ is the Jacobian theta function, H is the Jacobian eta function, and $Z(a)$ is the Jacobian zeta function, while sn, cn, and dn are the Jacobian elliptic functions [43].

3.7.5 Coupled-Slab Lines

Coupled slab lines, shown in Figure 3.32, are extensively used in combline and interdigital filter realizations. Theoretical analysis of coupled slablines was performed by Edward Crystal [44] and Stracca et al. [45]. However, Rosleniec [46] presented accurate equations and computer algorithm for CAD of coupled slablines. The equations are as follows:

$$Z_{0e}(x,y) = 60\ln\left\{\frac{0.5239}{f_1 f_2 f_3}\right\} \tag{3.169}$$

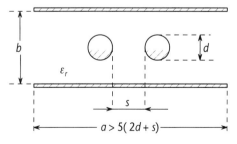

Figure 3.32 Coupled slabline.

$$Z_{0o}(x,y) = 60\ln\left\{\frac{0.5239 f_3}{f_1 f_4}\right\} \qquad (3.170)$$

With $x = \dfrac{d}{b}$ and $y = \dfrac{s}{b}$, f_1, f_2, f_3 and f_4 are defined as

$$f_1 = \frac{a(x)}{b(x)}$$
$$f_2 = c(y) - xd(y) + e(x)g(y), \text{ for } y \le 0.9$$
$$ = 1 + 0.004 e^{(0.9-y)}, \qquad \text{ for } y > 0.9$$

$$f_3 = \tanh\{\pi(x+y)/2\}$$
$$f_4 = k(y) - xl(y) + m(x)n(y) \text{ for } y < 0.9$$
$$ = 1.0, \qquad \text{ for } y \ge 0.9$$

$$a(x) = 1 + e^{(16x - 18.272)}$$
$$b(x) = \sqrt{5.905 - x^4}$$
$$c(y) = -0.8107 y^3 + 1.3401 y^2 - 0.6929 y + 1.0892$$
$$ + 0.014002 / y - 0.000636 / y^2$$
$$d(y) = 0.11 - 0.83 y + 1.64 y^2 - y^3$$
$$e(x) = -0.15 e^{(-0.13x)}$$
$$g(y) = 2.23 e^{\left(-7.01y + 10.24 y^2 - 27.58 y^3\right)}$$
$$k(y) = 1 + 0.01\begin{pmatrix} -0.0726 - 0.2145 / y \\ +0.222573 / y^2 - 0.012823 / y^3 \end{pmatrix}$$
$$l(y) = 0.01\begin{pmatrix} -0.26 + 0.6886 / y \\ +0.01831 / y^2 - 0.0076 / y^3 \end{pmatrix}$$
$$m(x) = -0.1098 + 1.2138 x - 2.2535 x^2 + 1.1313 x^3$$
$$n(y) = -0.019 - 0.016 / y + 0.0362 / y^2 - 0.00243 / y^3$$

The above analysis equations can be used in a suitable optimization routine for synthesis of a coupled pair of slablines. Rosleniec has given the following algorithm for such an optimization scheme. Let $V_1(x,y)$ and $V_2(x,y)$ be the penalty functions, defined as

$$V_1(x,y) = Z_{0e}(x,y) - Z_{0e} \qquad (3.171)$$

$$V_2(x,y) = Z_{0o}(x,y) - Z_{0o} \qquad (3.172)$$

to be minimized. A simple FORTRAN code can be written using IMSL library subroutine in DEC Visual FORTRAN, based on Powell's minimization. The initial approximation of the solution is given by [47]

$$x^{(0)} = \frac{4}{\pi} e^{\left[-\frac{Z_0}{60\sqrt{0.987-0.171k-1.723k^2}}\right]} \tag{3.173}$$

$$y^{(0)} = \frac{1}{\pi} \ln\left\{\frac{r+1}{r-1}\right\} - x^{(0)} \tag{3.174}$$

where

$$Z_0 = \sqrt{Z_{0e}Z_{0o}},\; k = \frac{Z_{0e}-Z_{0o}}{Z_{0e}+Z_{0o}} \text{ and } r = \left[\frac{4}{\pi x^{(0)}}\right]^{(0.001+1.117k-0.683k^2)}$$

3.8 Inhomogeneous Transmission Lines

In an inhomogeneous transmission line the space between the conductors is inhomogeneously filled with different dielectric materials. Figure 3.33 shows a shielded microstrip line as an example of inhomogeneous transmission line. The line can be considered as an asymmetrical strip transmission line in which the space below the strip is filled with a dielectric substrate of dielectric constant ε_r and the space above the strip is filled with air ($\varepsilon_r = 1$). Inhomogeneous transmission lines have various advantages over their homogeneous counterparts. For example, inhomogeneous transmission lines have lower loss and especially in planar inhomogeneous transmission lines, mounting surface mount devices are easier. However, almost all inhomogeneous transmission lines suffer from the existence of undesired higher-order modes and have a reduced fundamental mode band of operation. Below, we present equations for the basic electrical parameters of the most commonly used inhomogeneous transmission lines.

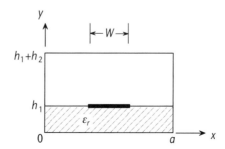

Figure 3.33 Inhomogeneous transmission line (shielded microstrip line).

3.8.1 Shielded Microstrip Line

A shielded microstrip line is a special form of inhomogeneous stripline where the space above and below the strip is filled with different dielectric materials. In most cases the space above the strip is filled with air ($\varepsilon_r = 1$). Figure 3.34 shows the homogeneous equivalent of shielded microstrip line.

The parameter ε_{eff} is known as the effective dielectric constant of the microstrip line. It is given by [48]

$$\varepsilon_{eff} = \frac{1}{2}\{\varepsilon_r + 1 + q(\varepsilon_r - 1)\} \qquad (3.175)$$

where

$$q = q_c(q_\infty - q_T) \qquad (3.176)$$

$$q_\infty - q_T = \left[1 - \frac{10h_1}{W}\right]^{-a(u_r)b(\varepsilon_r)} \qquad (3.177)$$

$$q_T = \frac{2}{\pi} \frac{\ln 2}{\sqrt{\frac{W}{h_1}}} \frac{t}{h_1} \qquad (3.178)$$

and

$$q_c = \tanh\left[1.043 + 0.121\frac{h_2}{h_1} - 1.164\frac{h_1}{h_2}\right] \qquad (3.179)$$

$$u_r = \frac{W}{h_1} + \Delta\left[\frac{W}{h_1}\right] \qquad (3.180)$$

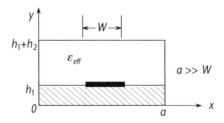

Figure 3.34 Homogeneous equivalent of shielded microstrip line.

3.8 Inhomogeneous Transmission Lines

$$\Delta\left[\frac{W}{h_1}\right] = \frac{t}{2\pi h_1}\left[1 - \frac{1}{\cosh\sqrt{\varepsilon_r - 1}}\right]\ln\left[1 + \frac{h_1}{t}\frac{4e}{\coth^2\sqrt{6.517\frac{W}{h_1}}}\right] \quad (3.181)$$

$$Z_0^\infty = (Z_{0\infty}^a - \Delta Z_0^a)/\sqrt{\varepsilon_{eff}} \quad (3.182)$$

where Z_0^∞ is the characteristic impedance of the microstrip line when $h_2 = \infty$ and the structure is uniformly filled with a material of dielectric constant unity. The reduction in characteristic impedance for a finite value of h_2 is given by the formulas below.

For $W/h_1 \leq 1$:

$$\Delta Z_0^a = 270\left\{1 - \tanh\left(0.28 + 1.2\sqrt{\frac{h_2}{h_1}}\right)\right\}$$

For $W/h_2 \geq 1$:

$$\Delta Z_0^a = 1 - \tanh^{-1}\left[\frac{0.48\sqrt{\frac{W}{h_1}}}{\left(1 + \frac{h_2}{h_1}\right)^2}\right] \quad (3.183)$$

$$Z_{0\infty}^a = 60\ln\left\{\frac{f\left(\frac{W}{h_1}\right)}{\frac{W}{h_1}} + \sqrt{1 + \left(\frac{2h}{W}\right)^2}\right\} \quad (3.184)$$

$$f\left(\frac{W}{h_1}\right) = 6 + 0.283 e^{\left\{-\left(\frac{30.666 h_1}{W}\right)^{0.7528}\right\}} \quad (3.185)$$

The above equations are accurate to within ± 0.5% for all practical purposes. However, the equations are based on curve fitting of results based on static analysis of shielded microstrip line. Such analyses are valid only in the low frequency range. As the frequency increases, dispersive effects become significant to substantially alter the values of characteristic impedance and effective dielectric constant, as shown in Figure 3.35. The asymptotic values of the frequency-dependent effective dielectric constant are $\varepsilon_{eff}(0)$ and ε_r. The corrections for dispersion are given by [49]

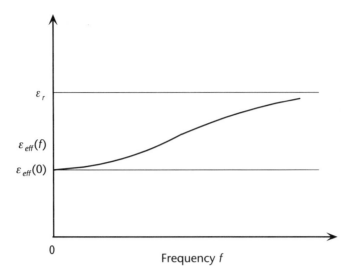

Figure 3.35 Frequency dispersion in effective dielectric constant of a microstrip.

$$\varepsilon_{eff}(f) = \varepsilon_r - \frac{\varepsilon_r - \varepsilon_{eff}(0)}{1 + Af_x + Bf_x^2} \qquad (3.186)$$

where A and B are functions of normalized shield height h_2/h_1, and given by for $W/h_1 \geq 0.41$

$$A = 0.77 + 0.31 \ln\left(\frac{h_2}{h_1}\right)$$

$$B = 0.94$$

and for $W/h_1 \leq 0.41$

$$A = \left(0.5 + \frac{W}{h_1}\right) + \left(1 + 0.5\frac{W}{h_1}\right)\ln\left(\frac{h_1}{h_2}\right)$$

$$B = 1.4 - 1.5\left(\frac{W}{h_1}\right)$$

$$f_x = \frac{f}{f_T} \qquad (3.187)$$

f_T is the cut of frequency of the lowest-order TE mode of the equivalent parallel plate waveguide model of microstrip at zero frequency [49]. f_T is given by

$$f_T = \frac{c}{2(W + \Delta W)\sqrt{\varepsilon_r}} \qquad (3.188)$$

where c is the velocity of light in free space and ΔW, which accounts for the fringing field effects at the edges of the strip is given by

$$\Delta W = \frac{h_1 \sqrt{\varepsilon_r}}{Z_0 c \varepsilon_0 \varepsilon_r} - W \qquad (3.189)$$

where Z_0 is the characteristic impedance of the microstrip at zero frequency and ε_0 is the free space permittivity. One should use $h_2/h_1 \geq 5$ for open microstrip. The above dispersion equations yield accuracy within 98% over the frequency range:

$$f_L \leq f \leq 2f_L \qquad (3.190)$$

and mostly 1% inaccuracy elsewhere. f_L is the frequency at which the onset of lowest order TE surface wave mode takes place in the corresponding open microstrip. f_L can be calculated from [50]

$$f_L \approx \frac{110}{h_1(mm)\sqrt{\varepsilon_r - 1}} \text{ (GHz)} \qquad (3.191)$$

It can be regarded as the highest frequency below which a microstrip can be used with the stated accuracy for all practical circuit design and analysis applications. However, the achievable accuracy in circuit realization depends on fabricational tolerances of the thin-film or thick-film structures and the variations of the properties of the materials used.

Using (3.188) and (3.189) it can be shown that lowest-order TE mode cutoff frequency is given by

$$f_T = \frac{Z_0}{2\mu_0 h_1} \sqrt{\frac{\varepsilon_r}{\varepsilon_{eff}}} \qquad (3.192)$$

The dispersion model presented here has been developed by Wells and Pramanick [51] using numerical data from Pozar and Das's hybrid model technique and the associated computer program PCAAMT [52]. PCAAMT is a commercially available general-purpose software for analysis of any type of multilayer transmission lines.

The frequency dependence of microstrip characteristic impedance is obtained from Pramanick and Bhartia's model [53]:

$$Z(f) = \sqrt{\frac{2(q-1)}{\varepsilon_e(f) - \sqrt{\varepsilon_e^2(f) - 4q(1-q)\varepsilon_r}}} \qquad (3.193a)$$

where

$$q = \frac{\varepsilon_{eff} - 1}{\varepsilon_r - 1} \qquad (3.193b)$$

Equation (3.193a) yields results within ±1% of spectral domain technique.

There exists no closed form synthesis equation for a shielded microstrip line. However, the analysis equations above can be used in an optimization routine in a computer program for accurate synthesis of a shielded microstrip line. Any such optimization routine begins with the initial approximation using Wheeler's [17] equation for static synthesis of a microstrip line. The equations are as follows:

$$\frac{W}{h_1} = \left[\frac{e^H}{8} - \frac{e^{-H}}{4} \right]^{-1} \qquad (3.194)$$

where

$$H = \frac{Z_0 \sqrt{2(\varepsilon_r + 1)}}{120} + \frac{1}{2}\left(\frac{\varepsilon_r - 1}{\varepsilon_r + 1}\right)\left(\ln\frac{\pi}{2} + \frac{1}{\varepsilon_r}\ln\frac{4}{\pi}\right) \qquad (3.195)$$

3.8.2 Coupled Microstrip Line

The structure of a coupled pair of unshielded microstrip lines is shown in Figure 3.36. These lines form the basic building block in numerous microwave integrated circuits and VLSI circuits. There exist reasonably accurate analysis equations for coupled pairs of unshielded or open microstrip lines only [54]. For any other coupled pair of microstrip lines one has to use computer programs based on static or dynamic electromagnetic analysis. One such example is PCAAMT [52]. Below we present closed-form analysis equations for open or unshielded coupled microstrip lines on low or moderate dielectric [54] constant substrates. First we define the pertinent parameters as follows:

$Z_{0e}(0)$ = quasistatic even-mode characteristic impedance;
$Z_{0o}(0)$ = quasistatic odd-mode characteristic impedance;
$\varepsilon_{0e}(0)$ = quasistatic even-mode effective dielectric constant;
$\varepsilon_{0o}(0)$ = quasistatic odd-mode effective dielectric constant.

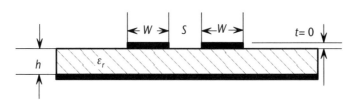

Figure 3.36 Cross section of coupled open microstrip lines.

3.8 Inhomogeneous Transmission Lines

We also define $u = W/h$ and $g = S/h$. Then

$$\varepsilon_e(0) = \frac{1}{2}\left\{\varepsilon_r + 1 + (\varepsilon_r - 1)\left(1 + \frac{10}{v}\right)^{-a_e(v)b_e(\varepsilon_r)}\right\} \tag{3.196}$$

where

$$v = u(20 + g^2)/(10 + g^2) + ge^{-g} \tag{3.197}$$

$$a_e(v) = 1 + \frac{1}{49}\ln\left\{\frac{v^4 + (v/52)^2}{v^4 + 0.432}\right\} + \frac{1}{18.7}\ln\left\{1 + \left(\frac{v}{18.1}\right)^2\right\} \tag{3.198}$$

$$b_e(\varepsilon_r) = 0.564\{(\varepsilon_r - 0.9)/(\varepsilon_r + 3)\}^{0.053} \tag{3.199}$$

The static odd-mode effective dielectric constant is given by

$$\varepsilon_o(0) = \left\{\frac{\varepsilon_r + 1}{2} + a_0(u, \varepsilon_r) - \varepsilon_{eff}(0)\right\}e^{-c_0 g^{d_0}} + \varepsilon_{eff}(0) \tag{3.200}$$

where

$$a_0(u, \varepsilon_r) = 0.7287\left[\varepsilon_{eff}(0) - \frac{\varepsilon_r + 1}{2}\right]\left[1 - e^{-0.179u}\right] \tag{3.201}$$

$$b_0(\varepsilon_r) = 0.747\varepsilon_r/(0.15 + \varepsilon_r) \tag{3.202}$$

$$c_0 = b_0(\varepsilon_r) - [b_0(\varepsilon_r) - 0.207]e^{(-0.414u)} \tag{3.203}$$

$$d_0 = 0.593 + 0.694e^{[-0.562u]} \tag{3.204}$$

$\varepsilon_{eff}(0)$ in the above equations is the effective dielectric constant of a single open microstrip of width W and zero strip thickness on a substrate height h and dielectric constant ε_r; see (3.175) with $h_1 = h$ and $t = 0$.

The frequency dependence of the even- and odd-mode open-coupled microstrips is given by

$$\varepsilon_{e,o}(f_n) = \varepsilon_r - \frac{\varepsilon_r - \varepsilon_{e,o}(0)}{1 + F_{e,o}(f_n)} \tag{3.205}$$

where
$$f_n = [f(GHz)h(mm)] \tag{3.206}$$

and
$$F_e(f_n) = P_1 P_2 \{(P_3 P_4 + 0.1844 P_7) f_n\}^{\frac{\pi}{2}} \tag{3.207}$$

$$F_o(f_n) = P_1 P_2 \{(P_3 P_4 + 0.1844) P_{15} f_n\}^{\frac{\pi}{2}} \tag{3.208}$$

$$P_1 = 0.27488 + \left[0.6315 + 0.525/(1+0.0157 f_n)^{20}\right] u - 0.065683 e^{(-8.7513u)} \tag{3.209}$$

$$P_2 = 0.33622 \left[1 - e^{(-0.03442\varepsilon_r)}\right] \tag{3.210}$$

$$P_3 = 0.0363 e^{(-4.6u)} \left\{1 - e^{-(f_n/38.7)^{4.97}}\right\} \tag{3.211}$$

$$P_4 = 1 + 2.75 \left[1 - e^{-(\varepsilon_r/15.916)^2}\right] \tag{3.212}$$

$$P_5 = 0.334 e^{\left[-3.3(\varepsilon_r/15_r)^3\right]} + 0.746 \tag{3.213}$$

$$P_6 = P_5 e^{(f_n/18)^{0.368}} \tag{3.214}$$

$$P_7 = 1 + 4.069 P_6 g^{0.479} e^{-1.347 g^{0.595 - 0.17g}} \tag{3.215}$$

$$P_8 = 0.7168 \{1 + 1.076/[1 + 0.0576(\varepsilon_r - 1)]\} \tag{3.216}$$

$$P_9 = P_8 - 0.7913 \left\{1 - e^{\left[-(f_n/20)^{1.424}\right]}\right\} \cdot \tan^{-1}\left[2.481(\varepsilon_r/8)^{0.946}\right] \tag{3.217}$$

$$P_{10} = 0.242(\varepsilon_r - 1)^{0.55} \tag{3.218}$$

$$P_{11} = 0.6366 \left[e^{(-0.3401 f_n)} - 1\right] \tan^{-1}\left[1.263(u/3)^{1.629}\right] \tag{3.219}$$

$$P_{12} = P_9 + (1 - P_9)/(1 + 1.183 u^{1.376}) \tag{3.220}$$

$$P_{13} = 1.695P_{10} / (0.414 + 1.605)P_{10}) \tag{3.221}$$

$$P_{14} = 0.8928 + 0.1027\left\{1 - e^{\left[-0.42(f_n/20)^{3.215}\right]}\right\} \tag{3.222}$$

$$P_{15} = \left|1 - 0.8928(1 + P_{11})P_{12}e^{\left(-P_{13}g^{1.092}\right)} / P_{14}\right| \tag{3.223}$$

The even-mode characteristic impedance is given by

$$Z_e(0) = \frac{Z_L(0)\sqrt{\dfrac{\varepsilon_{eff}}{\varepsilon_e(0)}}}{1 - Q_4\left[\dfrac{Z_L(0)}{120\pi}\right]\sqrt{\varepsilon_{eff}}} \tag{3.224}$$

where

$$Q_1 = 0.8695 u^{0.194} \tag{3.225}$$

$$Q_2 = 1 + 0.7519g + 0.189g^{2.31} \tag{3.226a}$$

$$Q_3 = 0.1975 + [16.6 + (8.4/g)^6]^{-0.387} + \ln\left(g^{10} / \left[1 + (g/3.4)^{10}\right]\right)/241 \tag{3.226b}$$

$$Q_4 = \frac{2Q_1}{Q_2\left[u^{Q_3}e^{-g} + (2 - e^{-g})u^{-Q_3}\right]} \tag{3.227}$$

The static odd-mode characteristic impedance is given by

$$Z_o(0) = \frac{Z_L(0)\sqrt{\dfrac{\varepsilon_{eff}}{\varepsilon_o(0)}}}{1 - Q_{10}\left\{\dfrac{Z_L(0)}{\eta_0}\right\}\sqrt{\varepsilon_{eff}}} \tag{3.228}$$

where

$$Q_5 = 1.794 + 1.14\ln\left[1 + 0.638/(g + 0.517g^{2.43})\right] \tag{3.229}$$

$$Q_6 = 0.2305 + \ln\left\{g^{10} / \left[1 + (g/5.8)^{10}\right]\right\}/281.3 + \ln(1 + 0.598g^{1.154})/5.1 \tag{3.230}$$

$$Q_7 = (10 + 190g^2)/(1 + 82.3g^3) \tag{3.231}$$

$$Q_8 = e^{\left(-6.5 - 0.95\ln(g) - (g/0.15)^5\right)} \tag{3.232a}$$

$$Q_9 = \ln(Q_7)(Q_8 + 1/16.5) \tag{3.232b}$$

$$Q_{10} = \{Q_2 Q_4 - Q_5 \exp[\ln(u) Q_6 u^{-Q_9}]\} / Q_2 \tag{3.232c}$$

The above equations are accurate within ±1% for the following most commonly encountered ranges of geometrical and material parameters:

$$0.1 \leq W/h \leq 10$$
$$1 \leq \varepsilon_r \leq 10$$
$$0.01 \leq S/h \leq \infty$$

and

$$t/h \leq 0.01$$

There exist no closed-form design equations for coupled microstrip lines outside the above ranges of geometrical and material parameters. However, a number of full-wave analysis based software programs are available [52]. Synthesis of coupled microstrip is required in many filter and coupler designs. In each case, the designer needs to find the normalized strip width W/h and the normalized edge to edge gap S/h for a desired coupling value and impedance matching. Either an iterative method or an optimization routine, based on the above equations, is used to obtain those values for a given operating frequency, substrate dielectric constant [52].

3.8.3 Suspended Microstrip Line

Since the effective dielectric constant of conventional microstrip lines described in the preceding sections is high enough to render the dimensions of millimeter-wave microstrip circuits impractically small in size, use of an air gap between the ground-plane and the substrate reduces the effective dielectric constant and the modified structure is known as a suspended microstrip line. The cross section of an open or unshielded suspended microstrip line is shown in Figure 3.37.

The characteristic impedance and the effective dielectric constant of an open suspended microstrip are obtained from [55].

$$Z_0 = \frac{60}{\sqrt{\varepsilon_{eff}}} 60 \ln\left\{\frac{f(u)}{u} + \sqrt{1 + \frac{4}{u^2}}\right\} \tag{3.233}$$

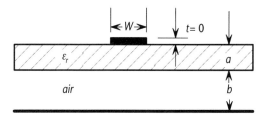

Figure 3.37 Open suspended microstrip line.

where $u = W/(a+b)$ and

$$f(u) = 6 + (2\pi - 6)e^{\left\{-\left(\frac{30.666}{u}\right)^{0.7528}\right\}} \tag{3.234}$$

$$\varepsilon_{eff} = \left[1 + \left(a_1 - b_1 \ln \frac{W}{b}\right)\left(\frac{1}{\sqrt{\varepsilon_r}} - 1\right)\right]^{-2} \tag{3.235}$$

where the coefficients a_1 and b_1 are given by

$$a_1 = 0.155 + 0.505\left(\frac{a}{b}\right) \tag{3.236}$$

$$b_1 = 0.023 + 0.1863\left(\frac{a}{b}\right) - 0.194\left(\frac{a}{b}\right)^2 \tag{3.237}$$

for $0.2 \leq \frac{a}{b} \leq 0.6$, and

$$a_1 = 0.307 + 0.239\left(\frac{a}{b}\right) \tag{3.238}$$

$$b_1 = 0.0727 - 0.0136\left(\frac{a}{b}\right) \tag{3.239}$$

for $0.6 \leq \frac{a}{b} \leq 1$

In order to fabricate the appropriate suspended microstrip, one requires the strip width W, the substrate and the air-gap thickness a and b, respectively, and the substrate dielectric constant ε_r for a specified characteristic impedance Z at a specified frequency f. Under the assumption of a quasi-static mode supported by the line, f does not play any role (dispersion is neglected), and the following synthesis equations are valid:

$$\frac{W}{b} = \left(1 + \frac{a}{b}\right) e^{(x)} \qquad (3.240)$$

where x is the solution of the cubic equation

$$x^3 + c_2 x^2 + c_1 x + c_0 = 0 \qquad (3.241)$$

$$c_2 = \frac{A}{B} - 8.6643 \qquad (3.242)$$

$$c_1 = 18.767 - 8.66433 \left(\frac{A}{B}\right) \qquad (3.243)$$

$$c_0 = 18.767 \left(\frac{A}{B}\right) - 8.9374 \left(\frac{Z}{60B}\right) \qquad (3.244)$$

$$A = 1 + \left[a_1 - b_1 \ln\left(1 + \frac{a}{b}\right)\right]\left(\frac{1}{\sqrt{\varepsilon_r}} - 1\right) \qquad (3.245)$$

$$B = b_1 \left(1 - \frac{1}{\sqrt{\varepsilon_r}}\right) \qquad (3.246)$$

3.8.4 Shielded Suspended Microstrip Line

The configuration of a shielded suspended microstrip line is shown in Figure 3.38(a). The characteristic impedance is given by

$$Z = \frac{1}{c\sqrt{CC_a}} \qquad (3.247)$$

$$\varepsilon_{ef} = \frac{C}{C_a} \qquad (3.248)$$

3.8 Inhomogeneous Transmission Lines

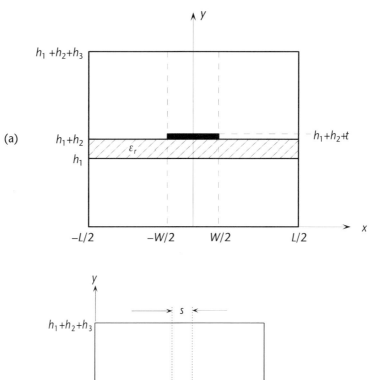

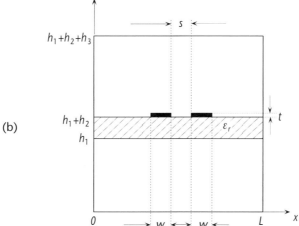

Figure 3.38 (a) Shielded suspended microstrip line, and (b) edge-coupled suspended stripline.

$$C = \frac{\left(1 + \dfrac{A}{4}\right)^2}{\displaystyle\sum_{n(odd)} \dfrac{T_n P_n}{Y}} \tag{3.249}$$

$$T_n = (L_n + AM_n)^2 \tag{3.250}$$

$$L_n = \sin\left[\frac{\beta_n W}{2}\right] \tag{3.251}$$

$$M_n = \left(\frac{2}{\beta_n W}\right)^3 \left\{ \begin{array}{l} 3\left[\left(\frac{\beta_n W}{2}\right)^2 - 2\right]\cos\left(\frac{\beta_n W}{2}\right) + \left(\frac{\beta_n W}{2}\right) \\ \left[\left(\frac{\beta_n W}{2}\right)^2 - 6\right]\sin\left(\frac{\beta_n W}{2}\right) + 6 \end{array} \right\} \quad (3.252)$$

$$P_n = \left(\frac{2}{n\pi}\right)\left(\frac{2}{\beta_n W}\right)^2 \quad (3.253)$$

$$A = -\frac{\sum_{n(odd)} \dfrac{(L_n - 4M_n)L_n P_n}{Y}}{\sum_{n(odd)} \dfrac{(L_n - 4M_n)M_n P_n}{Y}} \quad (3.254)$$

$$Y = Y_2 + Y_3 \quad (3.255)$$

$$Y_1 = \coth(\beta_n h_1) \quad (3.256)$$

$$Y_3 = \coth(\beta_n h_3) \quad (3.257)$$

$$Y_2 = \varepsilon_r \frac{Y_1 + \varepsilon_r \tanh(\beta_n h_2)}{\varepsilon_r + Y_1 \tanh(\beta_n h_2)} \quad (3.258)$$

In the above equations $\beta_n = n\pi/L$ and C^a is obtained by using (3.249) with $\varepsilon_r = 1$. The substrate thickness h_2 in the above equations is kept sufficiently small so that it only serves the purpose of mechanically supporting the metallization pattern. As a result, the dispersion in effective dielectric constant is practically negligible. The equations can be programmed very easily for rapid evaluation of single suspended stripline parameters for all practical purposes.

3.8.5 Edge-Coupled Suspended Microstrip Lines

Like edge-coupled microstrip and strip transmission lines, edge-coupled suspended strip lines are also used in various millimeter-wave filters, couplers, and hybrids. The cross-sectional shape of a pair of edge-coupled suspended striplines is shown in Figure 3.38(b). Such coupled lines support two distinctive modes called the even and the odd modes, as in coupled microstrip and striplines. The pertinent parameters of the modes are given by the following analytical equations [56]:

3.8 Inhomogeneous Transmission Lines

$$C_{\binom{0e}{0o}} = \frac{\left[1 + A_{\binom{0e}{0o}}/4\right]^2}{\sum_{n\binom{odd}{even}} g_n \left(L_n + A_{\binom{0e}{0o}} M_n\right)^2} \quad (3.259)$$

where

$$M_n = \left(\frac{2L}{n\pi w}\right)^3 \sin\left\{\frac{n\pi w}{2L}\left(\frac{L-s}{w}-1\right)\right\} \\ \left[3\left\{\left(\frac{n\pi w}{2L}\right)^2 - 2\right\}\cos\left(\frac{n\pi w}{2L}\right) + \left(\frac{n\pi w}{2L}\right)\left\{\left(\frac{n\pi w}{2L}\right)^2 - 6\right\}\sin\left(\frac{n\pi w}{2L}\right) + 6\right] \quad (3.260)$$

$$L_n = \sin\left\{\left(\frac{n\pi w}{2L}\right)\left(\frac{L-s}{w}-1\right)\right\}\sin\left(\frac{n\pi w}{2L}\right) \quad (3.261)$$

$$g_n = \frac{4}{n\pi Y}\left(\frac{2L}{n\pi w}\right)^2 \quad (3.262)$$

$$A_{\binom{0e}{0o}} = -\frac{\sum_{n\binom{odd}{even}} (4M_n - L_n) L_n g_n}{\sum_{n\binom{odd}{even}} (4M_n - L_n) M_n g_n} \quad (3.263)$$

$$Y = \varepsilon_0 \left[\coth\left(\frac{n\pi h_3}{L}\right) + \varepsilon_r \frac{\coth\left(\frac{n\pi h_1}{L}\right)\coth\left(\frac{n\pi h_2}{L}\right) + \varepsilon_r}{\varepsilon_r \coth\left(\frac{n\pi h_2}{L}\right) + \coth\left(\frac{n\pi h_1}{L}\right)}\right] \quad (3.264)$$

$$Z_{\binom{0e}{0o}} = \frac{1}{c\sqrt{C_{\binom{0e}{0o}} C^{air}_{\binom{0e}{0o}}}} \quad (3.265)$$

$$\varepsilon_{\binom{0e}{0o}} = \frac{C_{\binom{0e}{0o}}}{C^{air}_{\binom{0e}{0o}}} \quad (3.266)$$

In the above equations, the capacitance $C^{air}_{\binom{0e}{0o}}$ for the completely air-filled and homogeneous stripline is obtained by using (3.259) through (3.264) and $\varepsilon_r = 1$.

3.8.6 Broadside-Coupled Suspended Striplines

The cross-sectional shape of a broadside-coupled suspended stripline is shown in Figure 3.39. It is extensively used in the realization of quasi-planar millimeter-wave high-pass filters and very tight couplers. The even- and odd-mode characteristic impedances are given by [57]

$$Z^s_{0e} = \frac{Z^a_{0e}}{\sqrt{\varepsilon^s_{0e}}} \qquad (3.267)$$

$$Z^s_{0o} = \frac{Z^a_{0o}}{\sqrt{\varepsilon^s_{0o}}} \qquad (3.268)$$

where the characteristic impedances Z^a_{0e} and Z^a_{0o} of the corresponding homogeneous broadside-coupled air-filled stripline are obtained from (3.128) through (3.131) and $\varepsilon_r = 1$. The even- and odd-mode effective dielectric constants ε^s_{0e} and ε^s_{0o}, respectively, are given by

$$\varepsilon^s_{0o} = \frac{1}{2}(\varepsilon_r + 1) + \frac{q}{2}(\varepsilon_r - 1) \qquad (3.269)$$

where the filling factor q is given by

$$q = q_\infty q_c \qquad (3.270)$$

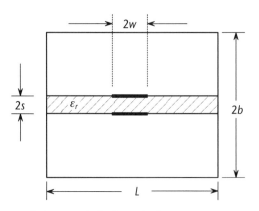

Figure 3.39 Broadside-coupled suspended microstrip line.

$$q_\infty = \left[1 + \frac{5S}{W}\right]^{-a(u)b(\varepsilon_r)} \quad (3.271)$$

$$a(u) = 1 + \frac{1}{49}\left\{\frac{u^4 + \left(\frac{u}{52}\right)^2}{u^4 + 0.432}\right\} + \frac{1}{18.7}\ln\left\{1 + \left(\frac{u}{18.1}\right)^3\right\} \quad (3.272)$$

$$u = \frac{2W}{S} \quad (3.273)$$

$$b(\varepsilon_r) = 0.564\left\{\frac{\varepsilon_r - 0.9}{\varepsilon_r + 3}\right\}^{0.053} \quad (3.274)$$

$$q_c = \tanh\left\{1.043 + 0.121\left(\frac{b-S}{S}\right) - 1.164\left(\frac{S}{b-S}\right)\right\} \quad (3.275)$$

$$\varepsilon_{0e}^s = \left\{1 + \frac{S}{b}\left[a_1 - b_1 \ln\left(\frac{W}{b}\right)\left(\sqrt{\varepsilon_r} - 1\right)\right]\right\}^2 \quad (3.276)$$

$$a_1 = \left\{0.8145 - 0.05824\frac{S}{b}\right\}^8 \quad (3.277)$$

$$b_1 = \left\{0.7581 - 0.07143\frac{S}{b}\right\}^8 \quad (3.278)$$

The above equations offer an accuracy of 1% when compared with results from full-wave analysis for $\varepsilon_r \leq 16$, $S/b \leq 0.4$, and $W/b \leq 1.2$. Figure 3.40 shows the graphical representation of the above equations.

3.8.7 Microstrip and Suspended Stripline Discontinuities

As in stripline circuits, discontinuities in microstrip and suspended striplines play significantly important roles in microwave planar integrated circuits. The types of discontinuities and their lumped element equivalent circuits are the same as those of their stripline counterparts shown in Figure 3.24. From the early 1970s to the late 1980s, closed-form equations were used for the characterization of microstrip

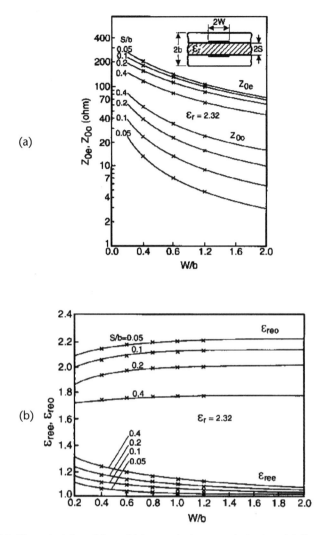

Figure 3.40 (a) Characteristics of broadside-coupled suspended microstripline, and (b) effective dielectric constant. (From [56], reprinted with permission from the IEEE.)

discontinuities. As well, to this day no closed-form model for suspended stripline discontinuities exists. However, the existing closed-form equations for microstrip discontinuities are purely empirical in nature and have very limited accuracy and range of validity with regard to frequency and structural parameters. Fortunately, the advent of highly accurate full-wave EM field solvers and cheaper and faster computers has completely eliminated the need for closed-form models. The scattering matrix of any multilayer planar transmission can be characterized in no time using affordable full-wave solvers. Figure 3.41 and 3.42 show the three-dimensional (3-D) view of a microstripline step junction and the computed scattering matrix, respectively. The added advantage of a full EM analysis-based approach is that it also characterizes 3-D discontinuities in multilayer planar transmission lines.

One has to keep in mind that the existing closed-form empirical models for microstrip discontinuities are valid not only with poor accuracy but also do not account for the enclosure effects. Use of full-wave solvers offers a complete picture of

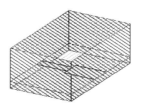

Figure 3.41 Microstripline step discontinuity (substrate thickness = 2000 μm, ε_r = 2.22).

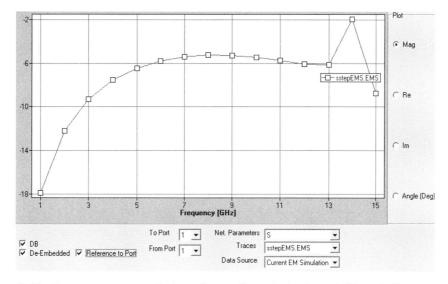

Figure 3.42 Frequency response of microstrip step discontinuity shown in Figure 3.41 as computed by EM3DS [58]. Computation time: 2.0 seconds.

not only a single discontinuity but also the interactions among more than the same or different type of discontinuities.

3.9 Single-Conductor Closed Transmission Lines

3.9.1 Hollow Metallic Waveguides

Hollow metallic tubes, usually with a rectangular or circular cross section, are ideally suited for high-power and low-loss microwave applications. Such transmission lines are called waveguides. Due to the absence of any center conductor, waveguides support only TE and TM modes. Depending on the cross-section geometry of the waveguides, shown in Figure 3.43(a), simultaneous existence of two different modes with the same phase velocity is possible. Such modes are called degenerate modes. In multimoded waveguides more than one mode is allowed to propagate simultaneously. Due to the non-TEM nature of the supported mode, the waveguide exhibits a high-pass-like frequency response. Also due to the nonuniqueness of the voltage and current definitions, the characteristic impedance, Z_0, of a waveguide cannot be defined uniquely. Z_0 may be defined in terms of the voltage-current ratio, the power-current ratio, or the power-voltage ratio, as shown below.

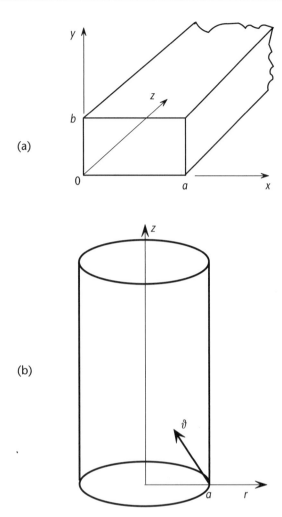

Figure 3.43 (a) Rectangular waveguide, and (b) circular waveguide.

$$\left.\begin{aligned} Z_0(V,I) &= V/I \\ Z_0(P,I) &= 2P/II^* \\ Z_0(P,V) &= VV^*/2P \end{aligned}\right\} \quad (3.279)$$

In the above equations, the definitions of the voltage and current are somewhat arbitrary, leading to different values of the characteristic impedance. Following the above definition, we get for the fundamental TE_{10} mode of a rectangular waveguide, the expressions given in Section 2.1. For a detailed derivation of waveguide field expressions, the reader should see any standard text book on electromagnetics engineering [59]. For ready reference, such expressions for rectangular and circular waveguides are shown in Tables 3.6 and 3.8, respectively [60].

Figure 3.44 shows the sequence in which various modes come into existence as the operating frequency is raised in a rectangular waveguide. For the TE_{mn} mode, the attenuation constant in a rectangular waveguide is given by

3.9 Single-Conductor Closed Transmission Lines

Table 3.6 TE and TM Mode Fields of a Rectangular Waveguide

Field Component	TE_{mn}	TM_{mn}
E_x	$-\Phi_{mn}\dfrac{n\pi}{b}\cos\left(\dfrac{m\pi}{a}x\right)\sin\left(\dfrac{n\pi}{b}y\right)e^{-j\beta_{mn}z}$	$-\Phi_{mn}\dfrac{m\pi}{a}\cos\left(\dfrac{m\pi}{a}x\right)\sin\left(\dfrac{n\pi}{b}y\right)e^{-j\beta_{mn}z}$
E_y	$\Phi_{mn}\dfrac{m\pi}{a}\sin\left(\dfrac{m\pi}{a}x\right)\cos\left(\dfrac{n\pi}{b}y\right)e^{-j\beta_{mn}z}$	$\Phi_{mn}\dfrac{n\pi}{b}\sin\left(\dfrac{m\pi}{a}x\right)\cos\left(\dfrac{n\pi}{b}y\right)e^{-j\beta_{mn}z}$
E_z	0	$\Phi_{mn}\dfrac{k_{c,mn}^2}{k}\sin\left(\dfrac{m\pi}{a}x\right)\sin\left(\dfrac{n\pi}{b}y\right)e^{-j\beta_{mn}z}$
H_x	$-\Phi_{mn}\dfrac{m\pi}{a}\sin\left(\dfrac{m\pi}{a}x\right)\cos\left(\dfrac{n\pi}{b}y\right)e^{-j\beta_{mn}z}$	$\Phi_{mn}\dfrac{n\pi}{b}\sin\left(\dfrac{m\pi}{a}x\right)\cos\left(\dfrac{n\pi}{b}y\right)e^{-j\beta_{mn}z}$
H_y	$-\Phi_{mn}\dfrac{n\pi}{b}\cos\left(\dfrac{m\pi}{a}x\right)\sin\left(\dfrac{n\pi}{b}y\right)e^{-j\beta_{mn}z}$	$-\Phi_{mn}\dfrac{m\pi}{a}\cos\left(\dfrac{m\pi}{a}x\right)\sin\left(\dfrac{n\pi}{b}y\right)e^{-j\beta_{mn}z}$
H_z	$\Phi_{mn}\dfrac{k_{c,mn}^2}{k_0}\cos\left(\dfrac{m\pi}{a}x\right)\cos\left(\dfrac{n\pi}{b}y\right)e^{-j\beta_{mn}z}$	0

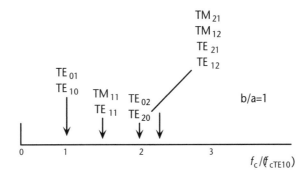
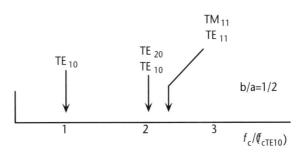

Figure 3.44 Mode sequence in a rectangular waveguide.

$$\alpha_{mn} = \frac{R_s}{b\eta_0\sqrt{1-\left(\dfrac{\lambda_0}{\lambda_c}\right)^2}}\left[\dfrac{\left\{1+\dfrac{b}{a}\right\}\left(\dfrac{\lambda_0}{\lambda_c}\right)^2 +}{\dfrac{b}{a}\left\{\dfrac{\varepsilon_{0n}}{2}-\left(\dfrac{\lambda_0}{\lambda c}\right)^2\right\}\dfrac{m^2ab+n^2a^2}{m^2b^2+n^2a^2}}\right] \text{(Nepers/unit length)} \quad (3.280)$$

where R_s is the surface resistance and $\eta_0 = 120\pi$ Ω is the free space impedance. $\varepsilon_{0n} = 1$ for $n = 0$ and 0 otherwise. For the fundamental TE_{10} mode, the equation for attenuation constant reduces to

$$\alpha_{c(TE_{m0})} = \frac{0.00019\sqrt{f\varepsilon_r}}{b} \frac{\left\{1 + \frac{2b}{a}\left(\frac{\lambda}{\lambda_{c(TE_{m0})}}\right)^2\right\}}{\sqrt{1 - \left(\frac{\lambda}{\lambda_{c(TE_{m0})}}\right)^2}} \text{ (dB/unit length)} \quad (3.281)$$

For a TM mode, the attenuation constant is given by

$$\alpha_{mn} = \frac{R_s}{b\eta_0 \sqrt{1 - \left(\frac{\lambda_0}{\lambda_c}\right)^2}} \left[\frac{m^2 b^3 + n^2 a^3}{m^2 b^2 a + n^2 a^3}\right] \text{ (Nepers/unit length)} \quad (3.282)$$

The eigenvalue problem for the TE and the TM modes of a rectangular waveguide satisfies the following Helmholtz's equations

$$\nabla^2_{xy} \Phi_{mn} + k^2_{mn} \Phi_{mn} = 0 \quad (3.283)$$

with the boundary conditions
$\Phi_{mn} = 0$, for TM modes
and
$\frac{\partial \Phi_{mn}}{\partial n} = 0$, for TE modes
on the waveguide walls. The solutions to (3.282) is given by

$$\Phi^{TE}_{mn} = \Phi_{mn} \cos\left(\frac{m\pi}{a}x\right)\cos\left(\frac{n\pi}{b}y\right) \quad (3.284a)$$

and

$$\Phi^{TM}_{mn} = \Phi_{mn} \sin\left(\frac{m\pi}{a}x\right)\sin\left(\frac{n\pi}{b}y\right) \quad (3.284b)$$

for the TE and the TM modes, respectively. The corresponding electric and magnetic field components are shown in Table 3.6 [60,61].
For a normalized mode the following condition holds:

$$\iint_S |\Phi_{mn}|^2 dxdy = \frac{1}{k^2_{c,mn}} \quad (3.285)$$

3.9 Single-Conductor Closed Transmission Lines

where S is the cross-sectional area of the waveguide and the corresponding mode field is given by

$$\Phi_{mn} = \frac{\sqrt{\varepsilon_{0m}\varepsilon_{0n}}}{\pi}\left[\frac{ab}{(mb)^2 + (na)^2}\right]^{\frac{1}{2}} \quad (3.286)$$

The cutoff wavenumber is given by

$$k_{c,mn} = \sqrt{\left(\frac{m\pi}{a}\right)^2 + \left(\frac{n\pi}{b}\right)^2} \quad (3.287)$$

and the phase or propagation constant is

$$\beta_{mn} = \sqrt{k_0^2 - k_{c,mn}^2} \text{ for } k_0 > k_{c,mn} \quad (3.288)$$

Below the cutoff frequency the propagation constant is an imaginary quantity

$$\alpha_{mn} = j\sqrt{k_{c,mn}^2 - k_0^2} \text{ for } k_0 > k_{c,mn} \quad (3.289)$$

The corresponding electric and magnetic field components of the TE and TM modes of a circular waveguide are derived from (3.282). For a circular waveguide the mode functions are given by

$$\Phi_{mn}^{TE} = \Phi_{mn} J_m\left(\frac{x'_{mn}}{a}r\right)\begin{Bmatrix}\sin(m\phi)\\\cos(m\phi)\end{Bmatrix} \quad (3.290)$$

$$\Phi_{mn}^{TE} = \Phi_{mn} J_m\left(\frac{x_{mn}}{a}r\right)\begin{Bmatrix}\sin(m\phi)\\\cos(m\phi)\end{Bmatrix} \quad (3.291)$$

where

$$\Phi_{mn} = \frac{1}{k_{c,mn}\sqrt{\Re_{c,mn}}} \quad (3.292a)$$

The corresponding electric and magnetic field components are shown in Table 3.8 [60,61] and

$$\Re_{c,mn} = \begin{cases}\pi a^2\left[J_0^{'2}(q_{0n}) + J_0^2(q_{0n})\right]_{m=0}\\\frac{\pi a^2}{2}\left[J_m^{'2}(q_{mn}) + \left(1 - \frac{m}{q_{mn}}\right)J_m^2(q_{mn})\right]_{m\neq n}\end{cases} \quad (3.292b)$$

The pair of indices m and n gives the number of half-cycle field variations in the Φ and r directions, respectively, and x_{mn} and x'_{mn} are the nth zeros of the Bessel function $J_m(x)$ and the first derivative $J'_m(x)$, respectively. Tables of these functions can be found in [27]. Table 3.7 shows the normalized cutoff frequencies of circular waveguide. Figure 3.45 shows the mode sequence of a circular waveguide.

The cutoff wavelengths of circular waveguide modes can be obtained from

$$\lambda_{c,mn}^{TE} = \frac{2\pi a}{\sqrt{\varepsilon_r \mu_r} \, x'_{mn}} = \frac{2\pi}{k_{c,mn}^{TE}} \tag{3.293}$$

$$\lambda_{c,mn}^{TM} = \frac{2\pi a}{\sqrt{\varepsilon_r \mu_r} \, x_{mn}} = \frac{2\pi}{k_{c,mn}^{TM}} \tag{3.294}$$

The dominant mode of a circular waveguide is the TE_{11} mode. The cutoff wavelength of the dominant mode is given by

$$\lambda_{c,11}^{TE} = \frac{2\pi a}{1.841\sqrt{\varepsilon_r \mu_r}} = \frac{2\pi}{k_{c,11}^{TE}} \tag{3.295}$$

Table 3.7 Normalized Cutoff Frequencies of a Circular Waveguide

f_c/f_{c11}	Mode	m	n	x_{mn}/x'_{mn}
1.000	TE	1	1	1.841
1.306	TM	0	1	2.405
1.659	TE	2	1	3.054
2.081	TM	1	1	3.832
2.081	TE	0	1	3.832
2.828	TE	3	1	4.201

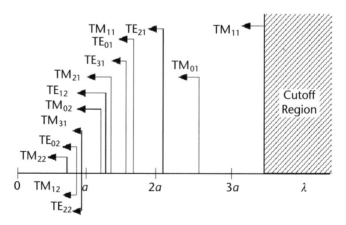

Figure 3.45 Mode sequence of a circular waveguide.

3.9 Single-Conductor Closed Transmission Lines

where $k_{c,mn}^{TE}$ and $k_{c,mn}^{TM}$ are the TE and the TM mode cutoff wavenumbers. The corresponding propagation constants are obtained by combining (3.293) through (3.295) with (3.288) or (3.289). ε_r and μ_r are the dielectric constant and the relative permeability, respectively, of the material completely filling the waveguide. For an air-filled guide, $\varepsilon_r = \mu = 1$.

In Table 3.8 the mode amplitude Φ_{mn} is given by

$$\Phi_{mn} = \Phi_{mn}^{TE} = \sqrt{\frac{\varepsilon_m}{\pi}} \frac{1}{J_m(x'_{mn})\sqrt{x'^2_{mn} - m^2}} \qquad (3.296)$$

$$\Phi_{mn} = \Phi_{mn}^{TM} = \sqrt{\frac{\varepsilon_m}{\pi}} \frac{1}{x_{mn} J'_{mn}(x_{mn})} \qquad (3.297)$$

$$k_{c,mn}^{TE} = \frac{x'_{mn}}{a} \qquad (3.298)$$

$$k_{c,mn}^{TM} = \frac{x_{mn}}{a} \qquad (3.299)$$

Table 3.8 TE and TM Mode Fields of a Circular Waveguide

Field Component	TE_{mn}	TM_{mn}
E_r	$-\Phi_{mn} \frac{m}{r} J_m\left(\frac{x'_{mn}}{a}r\right) \begin{Bmatrix} \cos(m\vartheta) \\ -\sin(m\vartheta) \end{Bmatrix} e^{-\beta_{mn} z}$	$-\Phi_{mn} \frac{x_{mn}}{a} J'_m\left(\frac{x_{mn}}{a}r\right) \begin{Bmatrix} \sin(m\vartheta) \\ \cos(m\vartheta) \end{Bmatrix} e^{-\beta_{mn} z}$
E_ϑ	$\Phi_{mn} \frac{x'_{mn}}{a} J'_m\left(\frac{x'_{mn}}{a}r\right) \begin{Bmatrix} \sin(m\vartheta) \\ \cos(m\vartheta) \end{Bmatrix} e^{-\beta_{mn} z}$	$-\Phi_{mn} \frac{m}{r} J_m\left(\frac{x_{mn}}{a}r\right) \begin{Bmatrix} \cos(m\vartheta) \\ -\sin(m\vartheta) \end{Bmatrix} e^{-\beta_{mn} z}$
E_z	0	$\Phi_{mn} \frac{k_{c,mn}^2}{k} J_m\left(\frac{x_{mn}}{a}r\right) \begin{Bmatrix} \sin(m\vartheta) \\ \cos(m\vartheta) \end{Bmatrix} e^{-\beta_{mn} z}$
H_r	$-\Phi_{mn} \frac{x'_{mn}}{a} J'_m\left(\frac{x'_{mn}}{a}r\right) \begin{Bmatrix} \sin(m\vartheta) \\ \cos(m\vartheta) \end{Bmatrix} e^{-\beta_{mn} z}$	$\Phi_{mn} \frac{m}{r} J_m\left(\frac{x_{mn}}{a}r\right) \begin{Bmatrix} \cos(m\vartheta) \\ -\sin(m\vartheta) \end{Bmatrix} e^{-\beta_{mn} z}$
H_ϑ	$-\Phi_{mn} \frac{m}{r} J_m\left(\frac{x'_{mn}}{a}r\right) \begin{Bmatrix} \cos(m\vartheta) \\ -\sin(m\vartheta) \end{Bmatrix} e^{-\beta_{mn} z}$	$-\Phi_{mn} \frac{x_{mn}}{a} J'_m\left(\frac{x_{mn}}{a}r\right) \begin{Bmatrix} \sin(m\vartheta) \\ \cos(m\vartheta) \end{Bmatrix} e^{-\beta_{mn} z}$
H_z	$\Phi_{mn} \frac{k_{c,mn}^2}{k} J_m\left(\frac{x'_{mn}}{a}r\right) \begin{Bmatrix} \sin(m\vartheta) \\ \cos(m\vartheta) \end{Bmatrix} e^{-\beta_{mn} z}$	0

where the roots x_{mn} and x'_{mn} of the equations $J(x_{mn}) = 0$ and $J'(x'_{mn})$ respectively, are given also in closed form by [62]

$$x_{mn} = \left(m + 2n - \frac{1}{2}\right)\frac{\pi}{2} - \frac{4m^2 - 1}{4\pi\left(m + 2n - \frac{1}{2}\right)} - \frac{(4m^2 - 1)(28m^2 - 3)}{48\pi^2\left(m + 2n - \frac{1}{2}\right)^2} \cdots \quad (3.300)$$

$$x'_{mn} = \left(m + 2n - \frac{3}{2}\right)\frac{\pi}{2} - \frac{4m^2 + 3}{4\pi\left(m + 2n - \frac{3}{2}\right)} - \frac{112m^4 + 328m^2 - 9}{48\pi^2\left(m + 2n - \frac{3}{2}\right)^2} \cdots \quad (3.301)$$

3.9.2 Characteristic Impedance of a Circular Waveguide

As in a rectangular waveguide, the definition of characteristic impedance of a mode in a circular waveguide is not unique. One based on power-current definition for the dominant TE_{11} mode is given by

$$Z_W^{PI}(f) = \frac{354}{\sqrt{1 - \left(\frac{f_{c11}}{f}\right)^2}} \, (\Omega) \quad (3.302)$$

while the wave impedance of a circular waveguide, which is a uniquely defined parameter, is given by

$$Z_W = \frac{\eta_0}{\sqrt{1 - \left(\frac{f_{c11}}{f}\right)^2}} \, (\Omega) \quad (3.303)$$

3.9.3 Attenuation in a Circular Waveguide

The TE mode attenuation constant due to conductor loss in a circular waveguide is given by

$$\alpha_{mn}^{TE} = \frac{R_s}{a\eta_0} \frac{\left[\dfrac{k_{c,mn}^2}{k_0^2} + \dfrac{m^2}{(k_{c,mn}^2 - m^2)}\right]}{\sqrt{1 - \dfrac{k_{c,mn}^2}{k_0^2}}} \, (\text{Nepers/unit length}) \quad (3.304)$$

3.9 Single-Conductor Closed Transmission Lines

and for TM modes

$$\alpha_{mn}^{TM} = \frac{R_s}{a\eta_0} \frac{k_0}{\sqrt{k_0^2 - k_{c,mn}^2}} \text{ (Nepers/unit length)} \quad (3.305)$$

The attenuation constant of the dominant TE_{11} mode of a circular waveguide is given by

$$\alpha_c^{TE_{11}} = \frac{0.00038\sqrt{\varepsilon_r f(GHz)}\left\{\left(\frac{\lambda}{\lambda_{c(TE_{11})}}\right)^2 + 0.42\right\}}{\sqrt{1-\left(\frac{\lambda}{\lambda_{c(TE_{11})}}\right)^2}} \text{ dB/unit length} \quad (3.306)$$

A plot of (3.306) is shown in Figure 3.46 and compared with (3.282) for a rectangular waveguide. It can be seen that the dominant mode conductor loss of a circular waveguide is much lower than that of a rectangular waveguide of comparable dimensions. Figure 3.47 shows the frequency dependence of conductor Q of rectangular and circular waveguides. Analytical equations for the plots are as follows.

Rectangular waveguide:

$$Q_c^{TE_{m0}} = \frac{0.00012 b \sqrt{f(GHz)}}{1 + \frac{2b}{a}\left(\frac{\lambda}{\lambda_{c(TE_{m0})}}\right)^2} \quad (3.307)$$

Circular waveguide:

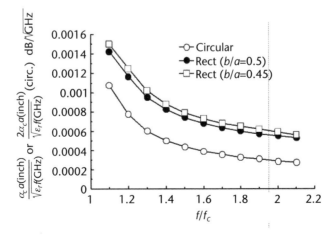

Figure 3.46 Attenuation in rectangular and circular waveguides.

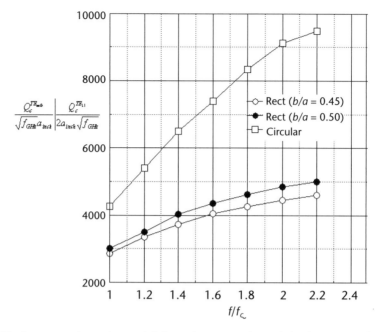

Figure 3.47 Frequency dependence of Q-factor in rectangular and circular waveguides.

Table 3.9 Design Data of Industry Standard Waveguides

Waveguide Size	JAN WG Desig	MIL-W-85 Dash #	Material	Freq Range (GHz)	Freq Cutoff (GHz)	Power (at 1 Atm)		Insertion Loss (dB/100ft)	Dimensions (Inches)	
						CW	Peak		Outside	Wall Thickness
WR284	RG48/U RG75/U	1-039 1-042	Copper Aluminum	2.60 - 3.95	2.08	45 36	7650	.742-.508 1.116-.764	3.000x1.500	0.08
WR229	RG340/U RG341/U	1-045 1-048	Copper Aluminum	3.30 - 4.90	2.577	30 24	5480	.946-.671 1.422-1.009	2.418x1.273	0.064
WR187	RG49/U RG95/U	1-051 1-054	Copper Aluminum	3.95 - 5.85	3.156	18 14.5	3300	1.395-.967 2.097-1.454	1.000x1.000	0.064
WR159	RG343/U RG344/U	1-057 1-060	Copper Aluminum	4.90 - 7.05	3.705	15 12	2790	1.533-1.160 2.334-1.744	1.718x0.923	0.064
WR137	RG50/U RG106/U	1-063 1-066	Copper Aluminum	5.85 - 8.20	4.285	10 8	1980	1.987-1.562 2.955-2.348	1.500x0.750	0.064
WR112	RG51/U RG68/U	1-069 1-072	Copper Aluminum	7.05 - 10.0	5.26	6 4.8	1280	2.776-2.154 4.173-3.238	1.250x0.625	0.064
WR90	RG52/U RG67/U	1-075 1-078	Copper Aluminum	8.2 - 12.4	6.56	3 2.4	760	4.238-2.995 6.506-4.502	1.000x0.500	0.05
WR75	RG346/U RG347/U	1-081 1-084	Copper Aluminum	10.0 - 15.0	7.847	2.8 2.2	620	5.121-3.577 7.698-5.377	0.850x0.475	0.05
WR62	RG91/U RG349/U	1-087 1-091	Copper Aluminum	12.4 - 18.0	9.49	1.8 1.4	460	6.451-4.743 9.700-7.131	0.702x0.391	0.04
WR51	RG352/U RG351/U	1-094 1-098	Copper Aluminum	15.0 - 22.0	11.54	1.2 1	310	8.812-6.384 13.250-9.598	0.590x0.335	0.04
WR42	RG53/U	1-100	Copper	18.0 - 26.5	14.08	0.8	170	13.80-10.13	0.500x0.250	0.04
WR34	RG354/U	1-107	Copper	2.0 - 33.0	17.28	0.6	140	16.86-11.73	0.420x0.250	0.04
WR28	RG271/U	3-007	Copper	26.5 - 40.0	21.1	0.5	100	23.02-15.77	0.360x0.220	0.04

Source: [63].

$$Q_c^{TE_{11}} = \frac{0.000122a\sqrt{f(GHz)}}{\left(\dfrac{\lambda}{\lambda_{c(TE_{11})}}\right)^1 + 0.42} \qquad (3.308)$$

where all dimensions are measured in inches. Equations (3.307) and (3.308) ignore the effects of surface roughness of the waveguide walls[1] [63]. However, such roughness of the walls plays a significant role in waveguide conductor loss and conductor Q. Table 3.9 shows the design data of industry standard rectangular waveguide.

3.9.3.1 Attenuation Caused by Dielectric Loss in a Waveguide

$$\alpha_d = \frac{27.3 \tan\delta}{\lambda}\left(\frac{\lambda_g}{\lambda}\right) \text{(dB/unit length)} \qquad (3.309)$$

where $\tan\delta$ is the loss tangent of the dielectric material completely filling the waveguide, λ is the operating wavelength, and λ_g is the guide wavelength. The unloaded Q-factor of waveguide is given by

$$Q = \left[\frac{1}{Q_c} + \frac{1}{Q_d}\right]^{-1} \qquad (3.310)$$

where Q_c is the conductor Q-factor and Q_d is the dielectric Q-factor. Q_d depends on the dielectric loss only and is given by

$$Q_d = \frac{1}{\tan\delta} \qquad (3.311)$$

and

$$Q_c = \frac{\pi\lambda_g}{\lambda^2\alpha_c} \qquad (3.312)$$

Using $m = 1$ in (3.307), for a completely air-filled copper waveguide, the equation for the dominant TE_{10} mode becomes

$$Q_{c(TE_{10})} = \frac{0.000121b\sqrt{f(GHz)}}{1 + \dfrac{2b}{a}\left(\dfrac{\lambda}{\lambda_{c(TE_{10})}}\right)^2} \qquad (3.313)$$

---

1. Q factor degrades by the factor $[1+(2/\pi)\arctan(1.4(\Delta/\delta_s))]$, where Δ is the surface roughness and δ_s is the skin depth of the material of the wall.

where a and b are in inches.

3.9.4 Maximum Power-Handling Capability of a Waveguide

The maximum peak power-handling capability of a rectangular waveguide supporting the TE_{mn} mode, shown in Figure 3.48, is given by

$$P_{max} = \frac{ab}{2} \frac{\beta_{mn}\left[\left(\frac{m}{a}\right)^2 + \left(\frac{n}{b}\right)^2\right]\left[\left(\frac{b}{n}\right)^2 + \left(\frac{a}{m}\right)^2\right]}{\eta_0 \varepsilon_{0m} \varepsilon_{0n}} |E_{max}|^2 \qquad (3.314)$$

where E_{max} is the breakdown electric field in the material filling the waveguide. For dominant TE_{10} mode in an air-filled waveguide ($E_{max} = 73.66$ kV/inch), (3.314) assumes the form

$$P_{max}(MW) = 3.6 ab \frac{\lambda}{\lambda_g} \qquad (3.315)$$

where a and b are in inches. The maximum average power-handling capability of a particular mode is obtained by multiplying the maximum peak power by the duty cycle of the wave. For example, in an X-band waveguide $a = 0.90''$, $b = 0.40''$, $\lambda = 1.180''$, and $\lambda_g = 1.560''$. Substitution of these values into (3.289) yields $P_{max} = 0.98$ MW. That is the peak pulse power of the fundamental mode (TE_{10}) that can propagate through the waveguide before an electric breakdown occurs. If we assume the pulse is repetitive with a 2% duty cycle, then the maximum average power-handling capability is 19.6 kW. Note that although there is only the vertical component of the electric field in TE_{10} mode in a rectangular waveguide in which $a > b$, P_{max} is

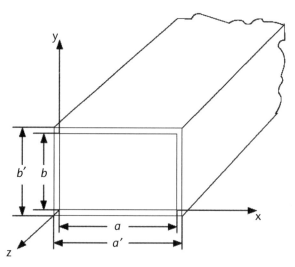

Figure 3.48 Position of waveguide for heat transfer relations.

also a function of the a dimension. However, in high average power microwave systems, average power rating should be considered from a different point of view. For very high power waveguide structures the factor governing the average power rating is the specified rise in waveguide temperature. Let us reconsider (3.282) for the fundamental mode (TE_{10}) mode of a rectangular copper waveguide at 20°C [64]:

$$\alpha_c = \frac{12.68 \times 10^{-5} \frac{\lambda}{\lambda_c} \left[\frac{a}{2b} + \left\{ \frac{\lambda}{\lambda_c} \right\}^2 \right]}{\lambda^{\frac{3}{2}} \sqrt{1 - \left(\frac{\lambda}{\lambda_c} \right)^2}} \text{ dB/ft} \quad (3.316)$$

where

a = the wide inner dimension of the waveguide in meters;
b = the narrow inner dimension of the waveguide in meters;
λ = the operating wavelength in meters;
λ_c = $2a$, the cutoff wavelength in meters.

The average power rating of the waveguide is

$$P_{av} = 1.271 \frac{q}{\alpha_c} \quad (3.317)$$

where q is the rate of heat transfer and is assumed to be composed of thermal convection and thermal radiation, or in other words

$$q = q_c + q_r \text{ Btu/hr} \quad (3.318)$$

$$q_c = (\Delta T)^{\frac{5}{4}} \left[\frac{0.708 A_b}{\sqrt[4]{b'}} + \frac{0.717 A_a}{\sqrt[4]{a'}} \right] \text{ Btu/hr/ft} \quad (3.319)$$

$$A_a = (a' \times 1) \text{ ft}^2 \text{ and } A_b = (b' \times 1) \text{ ft}^2$$

$$q_r = 5.19 \times 10^{-10} A_t \left(T_w^4 - T_a^4 \right) \text{ Btu/hr/ft} \quad (3.320)$$

ΔT = Temperature differential = $T_w - T_a$
T_w = Wall temperature of the waveguide
T_a = Ambient temperature

a' = Outer wide dimension of the waveguide
b' = Outer narrow dimension of the waveguide
$A_t = 2(A_a + A_b)$ = Total outer surface area of 1ft of waveguide in sq ft

An approximate first power equation for heat transfer is

$$q = h_{rc} A_t (T_w - T_a) \text{ Btu/hr/ft} \tag{3.321}$$

where h_{rc} is combined radiation and convection conductance in Btu/hr/(sq.ft)°F.

The above equations describe the average power-handling capability when the waveguide is perfectly matched, or in other words, no voltage standing-wave exists on the line. However, existence of mismatch will give rise to high current points and localized high-temperature zones. Hence one should define a derating factor for the extra temperature rise. There are two possible scenarios to be considered in this context.

Case 1: Equal power delivered to either matched or mismatched loads. This case represents the derating factor for the condition where the same amount of power is transferred to the matched or the mismatched load. This is useful when the transmitter is transmitting power to a matched or a mismatched antenna. In that case the extra rise in temperature is obtained from

$$\Delta T_{S1} \approx q \left\{ \frac{1}{h_{rc} A_t} \frac{S^2 + 1}{2S} + \frac{1}{\sigma_{th} A_c \left(\frac{4\pi}{\lambda} \right)} \frac{S^2 - 1}{2S} \right\} \tag{3.322}$$

where

A_c = cross-sectional area of the waveguide in square feet
σ_{th} = thermal conductivity of waveguide walls
S = VSWR

and

λ = is the wavelength in feet

The corresponding derating factor is

$$DF = \frac{\Delta T}{\Delta T_{S1}} \tag{3.323}$$

Case 2, where not equal power but equal wave amplitudes are delivered to the matched and the mismatched loads. Such a situation is significant when the transmitter faces a balanced duplexer circuit, as shown in Figure 3.49, or a power equalizer circuit where the reflected wave does not affect the loading of the transmitter; for example, in a duplexer circuit the net power delivered to the mismatched load

3.9 Single-Conductor Closed Transmission Lines

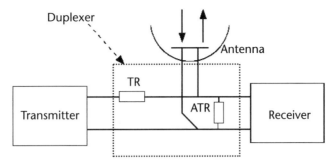

Figure 3.49 Transmitter delivering power to a duplexer.

(TR tube) approaches zero when the duplexer is in the transmit condition. In that case, the extra temperature rise due to mismatch is given by

$$\Delta T_{S2} = q \left\{ \frac{2}{h_{rc}A_c} \frac{S^2+1}{(S+1)^2} + \frac{2}{kA_c \left(\frac{4\pi}{\lambda}\right) + h_{rc}A_c} \frac{S-1}{S+1} \right\} \quad (3.324)$$

The corresponding derating factor is

$$DF = \frac{\Delta T}{\Delta T_{S2}} \quad (3.325)$$

In the past, plotted curves were developed for specific waveguide dimensions, materials, and temperature changes for determination of average power-handling capability of a waveguide. Since the equations involve too many variables it is impossible to plot general curves for any types of waveguides, materials, and temperature change. However, computer programs are available for estimation of the average power-handling capability of a rectangular waveguide for a given set of parameters.

3.9.5 Power-Handling Capability of a Circular Waveguide

The maximum peak power that can be handled by the dominant TE_{11} mode of a circular waveguide is

$$P_{max} = \frac{\pi a^2 \beta_{11}}{1582.3 k_0} |E_{max}|^2 \text{ Watts} \quad (3.326)$$

where E_{max} is the breakdown field of the medium homogeneously filling the waveguide. The maximum sustainable power for the TM_{01} mode is

$$P_{max} = \frac{\pi a^2}{5631} \frac{k}{k_{01}^{TM}} \sqrt{\left(\frac{k}{k_{01}^{TM}}\right)^2 - 1} |E_{max}|^2 \text{ Watts} \quad (3.327)$$

The TE$_{01}$ mode in a circular waveguide is of special interest because its field configuration and the symmetric mode pattern results in an attenuation characteristic that is very useful in communication links. The conductor loss of a circular waveguide's TE$_{10}$ mode decreases with frequency. The maximum power that can be handled by the TE$_{01}$ mode of a circular waveguide is

$$P_{max} = \frac{\pi a^2}{3149} \frac{\beta_{01}}{k} |E_{max}|^2 \text{ Watts} \qquad (3.328)$$

as in rectangular waveguide the maximum average power-handling capability of a circular waveguide is given by (3.317). However, the combined heat transfer coefficient hrc for a circular waveguide has to be obtained using the same procedure as for a rectangular waveguide. Equations (3.315), (3.326), (3,327), and (3.328) are very useful in determining the power-handling capabilities of waveguide filters [15].

3.9.6 Discontinuities in Waveguides and the Circuit Parameters

As mentioned before, no transmission line is useful in microwave design without a discontinuity in it and waveguides are no exception. An accurate result for a step discontinuity in waveguide is very important because it can be shown that any discontinuity in a waveguide is a combination of several basic step discontinuities. For example, an iris in a waveguide is a combination of two back-to-back step discontinuities. The history of waveguide discontinuity dates back to the 1940s and the most significant work in this area was done at the Massachusetts Institute of Technology (MIT) radiation laboratory, during World War II, as a part of radar research. An excellent account of the work is documented in the classic book by Nathan Marcuvitz, Waveguide Handbook, which forms a part of the famous Radiation Laboratory series. Due to lack of digital computers, Marcuvitz [62], Collin [61], Lewin [65], and others developed analytical expressions for waveguide discontinuities using variational calculus, conformal mapping and static field methods. However, the analytical equations are limited in the sense that they do not account for interactions among various waveguide discontinuities that are in close proximity of one another in an actual physical component. The main types of waveguide discontinuities are shown in Figure 3.24. Consider the H-plane step discontinuity in a rectangular waveguide in Figure 3.50.

At the junction plane

$$X = Z_0 \frac{2a}{\lambda_g} \frac{X_{11}\left\{1-\left[1-\left(\frac{2a'}{\lambda}\right)\right]^2 X_0^2\right\}}{1-\left[1-\left(\frac{2a'}{\lambda}\right)^2\right] X_0 X_{22} + \sqrt{1-\left(\frac{2a'}{\lambda}\right)^2}(X_{22}-X_0)} \qquad (3.329)$$

$$\lambda_g = \frac{\lambda}{\sqrt{1-\left(\frac{\lambda}{2a}\right)^2}} \qquad (3.330)$$

3.9 Single-Conductor Closed Transmission Lines

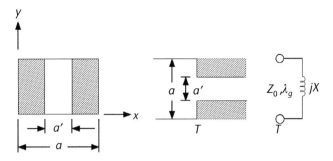

Figure 3.50 Rectangular waveguide step discontinuity and the equivalent circuit.

$$X_{11} = \frac{A}{1 - \frac{1}{4A}\left[(A+1)^2 N_{11} + 2(A+1)CN_{12} + C^2 N_{22}\right]} \tag{3.331}$$

$$X_{22} = \frac{A'}{1 - \frac{1}{4A'}\left[(A'+1)^2 N_{22} + 2(A'+1)CN_{12} + C^2 N_{11}\right]} \tag{3.332}$$

$$X_0 = X_{22} - \frac{X_{12}^2}{X_{11}} \tag{3.333}$$

$$A = \frac{(1+R_1)(1-R_2) + T^2}{(1-R_1)(1-R_2) - T^2} \tag{3.334}$$

$$A' = \frac{(1-R_1)(1+R_2) + T^2}{(1-R_1)(1-R_2) - T^2} \tag{3.335}$$

$$R_1 = -\left[\frac{1-\alpha}{1+\alpha}\right]^\alpha \tag{3.336}$$

$$R_2 = \left[\frac{1-\alpha}{1+\alpha}\right]^{\frac{1}{\alpha}} \tag{3.337}$$

$$C = \frac{2T}{(1-R_1)(1-R_2) - T^2} \tag{3.338}$$

$$T = \frac{4\alpha}{1-\alpha}\left(\frac{1-\alpha}{1+\alpha}\right)^{\left[\frac{1}{2}\left(\alpha+\frac{1}{\alpha}\right)\right]} \tag{3.339}$$

$$\begin{aligned}N_{11} = 2\left(\frac{a}{\lambda}\right)^2 &\left\{\begin{array}{l}1+\dfrac{16R_1}{\pi(1-\alpha^2)}\left[E(\alpha)-\alpha'^2 F(\alpha)\right]\\ \left[E(\alpha')-\alpha^2 F(\alpha')\right]-R_1^2-\alpha^2 T^2\end{array}\right\}\\ &+\frac{12\alpha^2}{(1-\alpha^2)^2}\left[\frac{1-\alpha}{1+\alpha}\right]^{4\alpha}\left[Q-\frac{1}{2}\left(\frac{2a}{3\lambda}\right)^2\right]\\ &+\frac{48\alpha^2}{(1-\alpha^2)^2}\left[\frac{1-\alpha}{1+\alpha}\right]^{\alpha+\frac{3}{\alpha}}\left[Q'-\frac{1}{2}\left(\frac{2a'}{3\lambda}\right)^2\right]\end{aligned} \tag{3.340}$$

$$N_{22} = 2\left(\frac{a}{\lambda}\right)^2 \left\{\begin{array}{l}\alpha^2+\dfrac{16\alpha^2 R_2 E(\alpha)}{\pi(1-\alpha)}\left[F(\alpha')-E(\alpha')-\alpha^2 R_2^2-T^2\right]\\ +\dfrac{48\alpha^2}{(1-\alpha^2)^2}\left[\dfrac{1-\alpha}{1+\alpha}\right]^{3\alpha+\frac{1}{\alpha}}\left[Q'-\dfrac{1}{2}\left(\dfrac{2a'}{3\lambda}\right)^2\right]\end{array}\right\} \tag{3.341}$$

$$\begin{aligned}N_{12} = 2\left(\frac{a}{\lambda}\right)^2 &\left\{\alpha'^2-\frac{4E(\alpha)}{\pi}\left[E(\alpha')-\alpha^2 F(\alpha')\right]-R_1+\alpha R_2\right\}T\\ &+\frac{24\alpha^3}{(1-\alpha^2)^2}\left[\frac{1-\alpha}{1+\alpha}\right]^{\frac{1}{2}\left(7\alpha+\frac{1}{\alpha}\right)}\left[Q-\frac{1}{2}\left(\frac{2a}{3\lambda}\right)^2\right]\\ &+\frac{24\alpha}{(1-\alpha^2)^2}\left[\frac{1-\alpha}{1+\alpha}\right]^{\frac{1}{2}\left(\alpha+\frac{7}{\alpha}\right)}\left[Q'-\frac{1}{2}\left(\frac{2a'}{3\lambda}\right)^2\right]\end{aligned} \tag{3.342}$$

$$\alpha = \frac{a'}{a} = 1-\beta, \qquad \alpha' = \sqrt{1-\alpha^2}$$

$$Q = 1-\sqrt{1-\left(\frac{2a}{3\lambda}\right)^2} \qquad Q' = 1-\sqrt{1-\left(\frac{2a'}{3\lambda}\right)^2}$$

3.9.7 Waveguide Asymmetric H-Plane Step

A waveguide H-plane asymmetric step and its equivalent network are shown in Figure 3.51. The reactance X is given by (3.329). However, the parameters R_1, R_2, T, and N_{11}, N_{22}, and N_{12} for the asymmetric step are as follows:

3.9 Single-Conductor Closed Transmission Lines

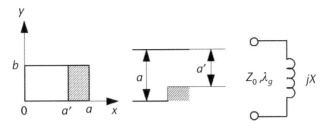

Figure 3.51 Rectangular waveguide asymmetric step discontinuity and its equivalent circuit.

$$R_1 = -\left(\frac{1+3\alpha^2}{1-\alpha^2}\right)\left(\frac{1-\alpha}{1+\alpha}\right)^{2\alpha} \tag{3.343}$$

$$R_2 = \left(\frac{3+\alpha^2}{1-\alpha^2}\right)\left(\frac{1-\alpha}{1+\alpha}\right)^{\frac{2}{\alpha}} \tag{3.344}$$

$$T = \frac{16\alpha^2}{(1-\alpha^2)^2}\left(\frac{1-\alpha}{1+\alpha}\right)^{\alpha+\frac{1}{\alpha}} \tag{3.345}$$

$$N_{11} = 2\left(\frac{a}{\lambda}\right)^2\left[1 + \frac{8\alpha(1-\alpha^2)}{1+3\alpha^2}R_1\ln\frac{1-\alpha}{1+\alpha} + \frac{16\alpha^2}{1+3\alpha^2}R_1 - R_1^2 - \alpha T^2\right]$$

$$+ 2\left[\frac{32}{3}\frac{\alpha^4}{(1-\alpha^2)^2}\left(\frac{1-\alpha}{1+\alpha}\right)^{3\alpha}\right]^2\left[Q - \frac{1}{2}\left(\frac{a}{\lambda}\right)^2\right] \tag{3.346}$$

$$+ 2\left[\frac{32\alpha^2}{(1-\alpha^2)^2}\left(\frac{1-\alpha}{1+\alpha}\right)^{\alpha+\frac{2}{\alpha}}\right]^2\left[Q' - \frac{1}{2}\left(\frac{a'}{\lambda}\right)^2\right]$$

$$N_{22} = 2\left(\frac{a}{\lambda}\right)^2\left[\alpha^2 - \frac{8\alpha(1-\alpha^2)}{3+\alpha^2}R_2\ln\frac{1-\alpha}{1+\alpha} + \frac{16\alpha^2}{1+3\alpha^2}R_2 - \alpha^2 R_2^2 - T^2\right]$$

$$+ 2\left[\frac{32\alpha^2}{(1-\alpha^2)^2}\left(\frac{1-\alpha}{1+\alpha}\right)^{2\alpha+\frac{1}{\alpha}}\right]^2\left[Q - \frac{1}{2}\left(\frac{a}{\lambda}\right)^2\right] \tag{3.347}$$

$$+ 2\left[\frac{32}{3(1-\alpha^2)^2}\left(\frac{1-\alpha}{1+\alpha}\right)^{3\alpha}\right]^2\left[Q' - \frac{1}{2}\left(\frac{a'}{\lambda}\right)^2\right]$$

$$N_{12} = 2\left(\frac{a}{\lambda}\right)^2 \left[\frac{(1-\alpha^2)^2}{2\alpha} \ln\left(\frac{1-\alpha}{1+\alpha}\right) - R_1 - \alpha^2 R_2\right]$$

$$T + \frac{2}{3}\left(\frac{1-\alpha}{1+\alpha}\right)^{5\alpha+\frac{1}{\alpha}} \left[\frac{32\alpha^3}{(1-\alpha^2)^2}\right]^2 \cdot \quad (3.348)$$

$$\left[Q - \frac{1}{2}\left(\frac{a}{\lambda}\right)^2\right] + \frac{2}{3}\left(\frac{1-\alpha}{1+\alpha}\right)^{\alpha+\frac{5}{\alpha}} \left[\frac{32\alpha}{(1-\alpha^2)^2}\right]^2 \left[Q' - \frac{1}{2}\left(\frac{a'}{\lambda}\right)^2\right]$$

The above equations and the equivalent circuit is valid within 1% over the range $\lambda > 2a'$ and $1 < \lambda_c/\lambda < 3$. The second condition guarantees the validity over the full fundamental mode band of the larger guide. However, the first one implies that the equations are valid for apertures only. Outside the range the inaccuracy may shoot up to 10%. The above equations were derived under the assumption that the smaller guide is infinitely long, which means the higher-order modes excited at the discontinuity decay to zero amplitude at a point beyond the step junction. Therefore, we may conclude that the closed-form equations for waveguide discontinuities are of limited validity. Hence, although closed-form equations exist for virtually all types of waveguide discontinuities, we will not present those equations in this treatise.

3.9.8 Mode Matching Method and Waveguide Discontinuities

Today the best way to analyze a waveguide discontinuity is by using the mode-matching method. This powerful and simple-to-implement CAD approach defeats any other approach in terms of computation speed, accuracy, and flexibility. In addition, when the mode-matching method is combined with the boundary element method (BEM), any arbitrarily shaped waveguide discontinuity can be analyzed. The basic method is as follows. Consider the junction between two rectangular waveguides of different sizes, as shown in Figure 3.52.

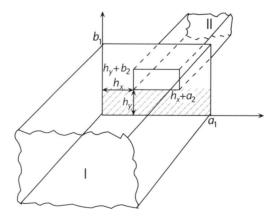

Figure 3.52 EH-plane step junction between two rectangular waveguides.

3.9 Single-Conductor Closed Transmission Lines

Let us assume that there exists a pure fundamental TE_{10} mode in guide one and it propagates in the z-direction and eventually strikes the discontinuity between waveguides 1 and 2. This generates all higher-order modes in both waveguides. The generated waves form a bunch of reflected waves in guide 1 and transmitted waves in guide 2. The mode-matching method generates the overall multimode scattering matrix of the system. However, in most cases we are interested in the fundamental mode-scattering matrix only and fortunately all higher-order modes in both guides decay to insignificant values within a short distance from the junction. The method was first reported by Alvin Wexler [66] in 1967. We describe the method without going into the derivations of the pertinent equations [67,68]. The method is as follows.

The amplitude of the mnth mode's electric field in waveguide $i(i = 1,2)$ due to a unit amplitude pqth mode in waveguide $j(j = 1,2)$ is by definition the (mn, pq) element of the scattering matrix S_{ij}. Moreover, for the smaller waveguide (guide 2):

$$[S_{22}] = [T_2]^{-1}([Y_{02}]+[Y_2])^{-1}([Y_{02}]-[Y_2])[T_2] \tag{3.349}$$

where $[T_i]$ $(i = 1,2)$ is a diagonal matrix which defines the equivalent voltage column vector V_i of guide i in terms of the E-field mode amplitude vector a_i of the same guide.

$$V_i = T_i a_i \tag{3.350}$$

$[Y_{02}]$ is the characteristic admittance matrix of guide 2 and $[Y_2]$ is the input admittance matrix of the junction as seen from guide 2. The principle of conservation of complex power shows that

$$[Y_2] = 2\left[[T_2]^{-1}\right]^T [H]^T [P_1]^T [H][T_2]^{-1} \tag{3.351}$$

where $[P_1]$ is the diagonal matrix whose mnth element represents the complex power carried by the unit amplitude mnth mode of guide 1. The elements of the coupling matrix $[H]$ are given by

$$H_{mn} = Y_{01mn}^* \frac{\iint_{S_a} \vec{e}_{1mn} \cdot \vec{e}_{2pq}}{p_{1mn}} da \tag{3.352}$$

where $\vec{e}_{imn}$ $(i = 1,2)$ is the transverse component of the mnth mode electric field in guide i at $z = 0$, and p_{imn} $(i = 1,2)$ is the complex power carried by the same mode. The integration is carried out over the cross-section S_a of the smaller waveguide. The remaining submatrices of the multimode scattering matrix can be obtained from

$$[S_{22}] = [H]([U]+[S_{22}]) \tag{3.353}$$

$$[S_{21}] = [P_2^*]^{-1}[S_{12}^T][P_1] \tag{3.354}$$

$$[S_{11}] = [H][S_{22}] - [U] \qquad (3.355)$$

Let the modes in each waveguide be divided into TE and TM modes and sequentially ordered. The matrix $[H]$ can then be subdivided into four submatrices

$$[H] = \begin{bmatrix} [H^{TE,TE}] & [H^{TE,TM}] \\ [H^{TM,TE}] & [H^{TM,TM}] \end{bmatrix} \qquad (3.356)$$

where the superscript *TE-TE* stands for coupling between the TE modes in guide 1 and TE modes in guide 2. The elements of various matrices are given by

$$H_{mn,pq}^{TE-TE} = k\left[\frac{nq\pi^2}{b_1 b_2} I_1 I_4 + \frac{mp\pi^2}{a_1 a_2} I_2 I_3\right] \qquad (3.357)$$

$$H_{mn,pq}^{TE-TM} = k\left[\frac{np\pi^2}{b_1 a_2} I_1 I_4 - \frac{mq\pi^2}{a_1 b_2} I_2 I_3\right] \qquad (3.358)$$

$$H_{mn,pq}^{TM-TE} = k\left[\frac{mq\pi^2}{a_1 b_2} I_1 I_4 - \frac{np\pi^2}{b_1 a_2} I_2 I_3\right] \qquad (3.359)$$

$$H_{mn,pq}^{TM-TM} = k\left[\frac{mp\pi^2}{a_1 a_2} I_1 I_4 + \frac{nq\pi^2}{b_1 b_2} I_2 I_3\right] \qquad (3.360)$$

where the integrals $I_i (i = 1,2,3,4)$ are defined as

$$I_1 = \int_{b_x}^{b_x + a_2} \cos\frac{m\pi x}{a_1} \cos\frac{p\pi((x - b_x))}{a_2} dx \qquad (3.361)$$

$$I_2 = \int_{b_y}^{b_y + b_2} \cos\frac{n\pi y}{b_1} \cos\frac{q\pi((y - b_y))}{b_2} dy \qquad (3.362)$$

$$I_3 = \int_{b_x}^{b_x + a_2} \sin\frac{m\pi x}{a_1} \sin\frac{p\pi((x - b_x))}{a_2} dx \qquad (3.363)$$

3.9 Single-Conductor Closed Transmission Lines

$$I_4 = \int_{h_y}^{h_y+b_2} \sin\frac{n\pi y}{b_1} \sin\frac{q\pi\left((y-h_y)\right)}{b_2} dy \tag{3.364}$$

The above integrals can be written as

$$I_1 = \begin{cases} \dfrac{a_2}{2}\cos(u_1 h_x)\Big|_{u_1=u_2} \\ \dfrac{u_1}{u_1^2-u_2^2}\left\{(-1)^p \sin\left[u_1\left(h_x+a_2\right)\right]-\sin(u_1 h_x)\right\} \end{cases} \tag{3.365}$$

$$I_2 = \begin{cases} \dfrac{b_2}{2}\cos(v_1 h_y)\Big|_{v_1=v_2} \\ \dfrac{v_1}{v_1^2-v_2^2}\left\{(-1)^q \sin\left[v_1\left(h_y+b_2\right)\right]-\sin(v_1 h_y)\right\} \end{cases} \tag{3.366}$$

$$I_3 = \begin{cases} \dfrac{a_2}{2}\cos(u_1 h_x)\Big|_{u_1=u_2} \\ \dfrac{u_2}{u_1^2-u_2^2}\left\{(-1)^p \sin\left[u_1\left(h_x+a_2\right)\right]-\sin(u_1 h_x)\right\} \end{cases} \tag{3.367}$$

$$I_4 = \begin{cases} \dfrac{b_2}{2}\cos(v_1 h_y)\Big|_{v_1=v_2} \\ \dfrac{v_2}{v_1^2-v_2^2}\left\{(-1)^q \sin\left[v_1\left(h_y+b_2\right)\right]-\sin(v_1 h_y)\right\} \end{cases} \tag{3.368}$$

where $u_1 = \dfrac{m\pi}{a_1}$, $u_2 = \dfrac{p\pi}{a_2}$, $v_1 = \dfrac{n\pi}{b_1}$, $v_2 = \dfrac{q\pi}{b_2}$

There are some special cases that need to be considered. Those are when m or n = 0 in guide 1 and p or q = 0 in guide 2. For an entirely H-plane discontinuity, such a case does not exist because waveguides do not support TE_{m0} mode with m = 0. Therefore, under the above special conditions (3.365) to (3.368) assume the forms

$$I_1 = \begin{cases} 0\big|_{p=0} \\ \dfrac{a_1}{m\pi}\left\{\sin\left[u_1\left(h_x+a_2\right)\right]-\sin(u_1 h_x)\right\}\big|_{m=0} \end{cases} \tag{3.369}$$

$$I_2 = \begin{cases} 0\big|_{n=0} \\ \dfrac{b_1}{n\pi}\left\{\sin\left[v_1\left(h_y+a_2\right)\right]-\sin(v_1 h_y)\right\}\big|_{q=0} \end{cases} \tag{3.370}$$

Each of the four left-hand side matrices in (3.349) and (3.353) through (3.355) has the form

$$\left[S_{ij(i=1,2;j=1,2)}\right] = \begin{bmatrix} S_{ij}(1,1) & S_{ij}(1,2) & \cdots & \cdots & S_{ij}(1,N) \\ S_{ij}(2,1) & \vdots & \vdots & \vdots & \vdots \\ \vdots & S_{ij}(2,2) & \vdots & \vdots & \vdots \\ \vdots & \vdots & \ddots & \vdots & \vdots \\ S_{ij}(2,N) & \cdots & \cdots & \cdots & S_{ij}(N,N) \end{bmatrix} \quad (3.371)$$

where the subscripts i and j denote the port numbers and the first and the second numerals within parenthesis on the right-hand side denote the mode numbers in guide 1 and guide 2, respectively. For example, if $i = 1$ and $j = 2$ then the term $S_{12}(2,3)$ is the ratio of the amplitude of the third mode at port 1 due to the unit incident complex amplitude of mode 3 in port 2. Therefore, if the first mode or mode 1 in either guide is selected as the fundamental TE_{10} mode, then the fundamental mode reflection and transmission coefficients are $S_{11}(1,1)$ and $S_{21}(1,1)$, respectively. For a detailed account of the mode-matching technique, the reader should to refer to [69].

The elements of the coupling matrix $[H]$ in a mode-matching analysis can be obtained as closed-form analytical expressions only if the cross-sectional shapes of the two waveguides forming the junction are regular (rectangular, circular, etc.) However, in many situations the designer may have to use junctions of nonrectangular shaped waveguides. For any such shape, the expressions for cutoff wavelengths and the corresponding mode field do not have any closed-form expressions. In any such case one has to use more generalized techniques and often numerical approaches such as the two-dimensional finite element method or boundary element method for computation cutoff wavelengths and impedances of the fundamental and higher-order modes. Recently a semianalytical approach based on the use of two-dimensional superquadratic functions has gained considerable importance.

Consider the junction between two waveguides, the cross-sectional shapes of which can be described by generalized superquadratic functions, as shown in Figure 3.53.

The first step in a generalized mode-matching analysis is to obtain the TE and TM mode cutoff frequencies and the corresponding mode fields of the two waveguides. That is accomplished in the following way [69,70].

The mode field is expressed as a series polynomial

$$u(x,y) = \sum_{i=1}^{m} C_i \phi_i(x,y) \quad (3.372)$$

where the basis functions ϕ_i are defined for TM mode as

$$\phi_i(x,y) = \varphi_i(x,y) f_i(x,y) \quad (3.373)$$

and for TE modes as

$$\phi_i(x,y) = f_i(x,y) \quad (3.374)$$

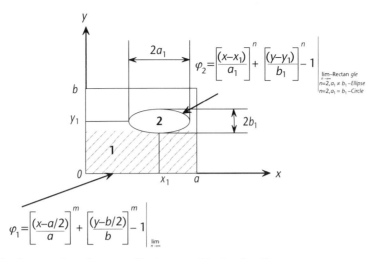

Figure 3.53 Cross section of a generalized waveguide step junction.

The function ϕ_i are chosen such that the Dirichlet boundary condition ($E_z = 0$) is automatically satisfied on the boundary of the waveguide cross section. The Neuman boundary condition $\left(\dfrac{\partial H_z}{\partial n} = 0\right)$ is automatically satisfied by the function in (3.374). Once the basis functions have been defined, the waveguide cutoff frequencies are obtained from the following matrix eigenvalue equation:

$$[K][C]^T = k_c^2 [M][C] \qquad (3.375)$$

The elements of the matrices $[K]$ and $[M]$ are defined as

$$k_{ij} = \iint_S \left(\frac{\partial \phi_i}{\partial x} \frac{\partial \phi_j}{\partial x} + \frac{\partial \phi_i}{\partial y} \frac{\partial \phi_j}{\partial y} \right) dx dy \qquad (3.376)$$

$$m_{ij} = \iint_S \phi_i \phi_j \, dx dy \qquad (3.377)$$

respectively, where S is the cross section of the waveguide.

For t even

$$f_i(x,y) = x^r y^{\frac{t}{2}} \qquad (3.378a)$$

and for t odd

$$f_i(x,y) = x^{\frac{t-1}{2}} y^r \qquad (3.378b)$$

where

$$r = \text{Int}\left(\sqrt{i-1}\right),\ t = (i-1) - r^2$$

Once the eigenfunctions for the various modes are completely known, the component matrices of the coupling matrix $[H]$ are computed from

$$H_{ij}^{TM,TM} = \frac{k_{c2,j}^2}{k_{c2,j}^2 - k_{c1,i}^2} \oint_{C_1} \phi_j^2 \frac{\partial \phi_i^1}{\partial n} dl \qquad (3.379)$$

$$H_{ij}^{TE,TM} = \frac{k_{c1,i}^2}{k_{c1,i}^2 - k_{c2,j}^2} \oint_{C_1} \phi_i^1 \frac{\partial \phi_j^2}{\partial n} dl \qquad (3.380)$$

$$H_{ij}^{TM,TE} = 0 \qquad (3.381)$$

$$H_{ij}^{TE,TE} = \oint_{C_1} \phi_i^1 \frac{\partial \phi_j^2}{\partial n} dl \qquad (3.382)$$

3.9.8.1 Analysis of an Iris

An iris in a waveguide is a back-to-back combination of two-step discontinuities, as shown in Figure 3.54. For an iris of finite thickness there is a finite length of uniform waveguide between the two back-to-back steps. The cross-sectional dimensions of the interposed waveguide are those of the smaller waveguide. The overall scattering matrix consists of three separate scattering matrices. $[S^A]$, $[S^B]$, and $[S^C]$ are the scattering matrices of the junction between guide I and II, guide II as a uniform transmission line, and the junction between guide II and III, respectively. The scattering matrices of the junction between I and II and that of guide II are combined using

$$\left[S_{11}^{AB}\right] = \left[S_{11}^A\right] + \left[S_{12}^A\right]\left[[U] - \left[S_{11}^B\right]\left[S_{22}^A\right]\right]^{-1}\left[S_{11}^B\right]\left[S_{21}^A\right] \qquad (3.383)$$

$$\left[S_{12}^{AB}\right] = \left[S_{12}^A\right]\left[[U] - \left[S_{11}^B\right]\left[S_{22}^A\right]\right]^{-1}\left[S_{12}^B\right] \qquad (3.384)$$

$$\left[S_{21}^{AB}\right] = \left[S_{21}^B\right]\left[[U] - \left[S_{22}^A\right]\left[S_{11}^B\right]\right]^{-1}\left[S_{21}^A\right] \qquad (3.385)$$

$$\left[S_{22}^{AB}\right] = \left[S_{22}^B\right] + \left[S_{21}^B\right]\left[[U] - \left[S_{22}^A\right]\left[S_{11}^B\right]\right]^{-1}\left[S_{22}^A\right]\left[S_{12}^B\right] \qquad (3.386)$$

The overall scattering matrix $[SABC]$ is obtained by using the above equations one more time and replacing the superscripts A with AB and B with C in the above equations. The junction scattering matrices $[S^A]$ and $[S^C]$ are obtained from the

3.9 Single-Conductor Closed Transmission Lines

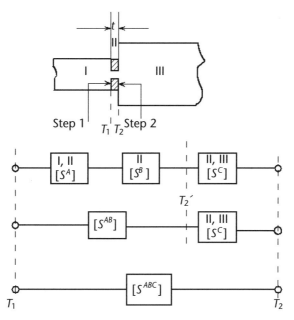

Figure 3.54 An iris discontinuity in a waveguide and the scattering matrix combination steps.

mode-matching analysis of the respective junctions. The scattering matrix of the uniform section of guide II of length t is given by

$$[S^B] = \begin{bmatrix} [L^{TE}] & [\overline{0}] \\ [0] & [L^{TM}] \end{bmatrix}, \quad (3.387)$$

where $[L^{TE}]$ and $[L^{TM}]$ are diagonal matrices of order M^{II} and N^{II}, respectively, assuming the M^{II} TE modes and N^{II} TM modes are considered in guide II.

$$[L^{TE}] = \begin{bmatrix} e^{-j\beta_1^{TE}t} & 0 & \cdots & 0 \\ 0 & e^{-j\beta_2^{TE}t} & \cdots & 0 \\ \vdots & \vdots & \ddots & \vdots \\ 0 & 0 & \cdots & e^{-j\beta_{M^{II}}^{TE}t} \end{bmatrix} \quad (3.388)$$

$$[L^{TM}] = \begin{bmatrix} e^{-j\beta_1^{TM}t} & 0 & \cdots & 0 \\ 0 & e^{-j\beta_2^{TM}t} & \cdots & 0 \\ \vdots & \vdots & \ddots & \vdots \\ 0 & 0 & \cdots & e^{-j\beta_{N^{II}}^{TM}t} \end{bmatrix} \quad (3.389)$$

β_i^{TE} and β_k^{TM} are the propagations constants of the ith TE and the kth TM modes, respectively, in guide II. Usually, β_1 is of the dominant mode, which means β_1^{TE} is the propagation constant of TE10 mode and β_1^{TM} is that of the TM_{11} mode in a rectangular waveguide. $[\bar{0}]$ and $[0]$ are rectangular null matrices of order $(N^{II} \times M^{II})$ and $(M^{II} \times N^{II})$, respectively.

In many situations guides I and II happen to be coaxial and of identical cross-sectional dimensions. In those cases the scattering matrices of guide I and III are related as

$$\begin{aligned} \left[S_{11}^{A} \right] &= \left[S_{22}^{C} \right] \\ \left[S_{12}^{A} \right] &= \left[S_{21}^{C} \right] \\ \left[S_{21}^{A} \right] &= \left[S_{12}^{C} \right] \\ \left[S_{11}^{C} \right] &= \left[S_{22}^{A} \right] \end{aligned} \qquad (3.390)$$

Combining (3.356) to (3.359) with (3.363) gives

$$[S_{22}] = [S_{11}] = \left[S_{22}^{A} \right] + \left[S_{21}^{A} \right][L]\left[S_{11}^{A} \right]\left[[U] - [L]\left[S_{11}^{A} \right][L]\left[S_{11}^{A} \right] \right]^{-1}[L]\left[S_{12}^{A} \right] \qquad (3.391)$$

$$[S_{21}] = [S_{12}] = [S_{21}]\left[[U] - [L]\left[S_{11}^{A} \right][L]\left[S_{11}^{A} \right] \right]^{-1}[L]\left[S_{12}^{A} \right] \qquad (3.392)$$

3.9.8.2 Offset Step Junction

An offset step junction is shown in Figure 3.55. In such a step junction each waveguide is only partially contained in the other. However, the junction can be analyzed using the mode-matching technique by interposing an intermediate section of waveguide that fully contains both waveguides but has the length tending to zero in the limit, as shown in Figure 3.56. Due to such an assumption, the single-junction problem essentially becomes a two-junction problem involving three cascaded networks. Therefore, once we have computed the scattering matrices of the junctions as $[S^A]$ and $[S^C]$ and that of the intermediate section as $[S^B]$ given by (3.360), then we can use the cascading scheme in Figure 3.54 to obtain the overall scattering matrix of the junction.

3.9.9 Waveguide Discontinuity Analysis in General

The mode-matching method described above is well suited for any type of discontinuity in a straight section of waveguide. However, there are a great multitude of applications where a large number of other types of waveguide discontinuities are used. Waveguide discontinuities can be divided into three categories, H-plane, E-plane, and EH-plane. The most efficient and general method for analysis of E- and H-plane discontinuities in a rectangular waveguide is a combination of the mode-matching method and the finite element method [71]. We will present a brief description of the method for H- and E-plane discontinuities. For a detailed description of the method the reader is referred to the book by Pelosi et al. [72].

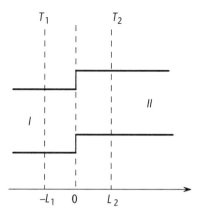

Figure 3.55 Offset step junction.

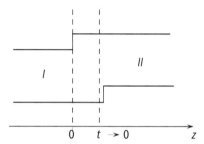

Figure 3.56 Offset step junction with an intermediate section.

3.9.10 Finite Element Modal Expansion Method

3.9.10.1 H-Plane Waveguide Discontinuities

Let us consider the general H-plane junction [72] (no variation in dimension in y-direction) of a rectangular waveguide, as shown in Figure 3.57. The junction is assumed to consist of a perfectly conducting wall whose contour is denoted by Γ_0, while the region enclosed by Γ_0 is denoted by Ω. There are N arbitrarily defined port reference planes Γ_k ($k = 1, N$). Each port is matched to an infinite waveguide whose cross-sectional dimensions are the same as those of the port. Now, if one more of the ports is fed with a TE_{10} mode only the TE_{m0} modes will be excited by the discontinuity and the electric field in the structure will have the E_y component only. Once the scattering has taken place, the general expression for the electric field at the kth port due to the incident field on the jth port is given by

$$E_{ywg}^{(k)}\left(x^{(k)}\right) = \delta k e_1^{(j)}\left(x^{(j)}\right) e^{j\beta_1^{(j)} z^{(j)}} + \sum_{m=1}^{\infty} B_m^{(k)} e_m^{(k)}\left(x^{(k)}\right) e^{-j\beta_m^{(k)} z^{(k)}} \qquad (3.393)$$

$$H_{ywg}^{(k)} = \frac{1}{j\omega\mu_0}\frac{\partial E_{ywg}^{(k)}}{\partial z^{(k)}} \qquad (3.394)$$

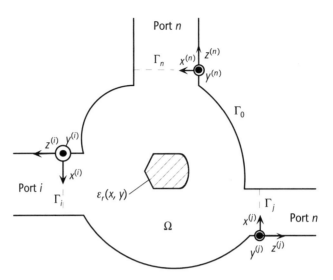

Figure 3.57 General rectangular waveguide H-plane multiport [72].

where $e_m^{(k)}$ is the orthonormal modal function of the TE_{m0} at the kth waveguide port:

$$e_m^{(k)}\left(x^{(k)}\right) = \frac{2}{\sqrt{a^{(k)}b}} \sqrt{\frac{k_0 \eta_0}{\beta_m^{(k)}}} \sin\left(\frac{m\pi}{a^{(k)}} x^{(k)}\right) \quad (3.395)$$

where k_0 and η_0 are the free space propagation constant and the wave impedance, respectively, and $\beta_m^{(k)}$ is the propagation constant of the waveguide attached to the kth port. If the width of that waveguide is $a^{(k)}$ then

$$\beta_m^{(k)} = \sqrt{k_0^2 - \left(\frac{m\pi}{a^{(k)}}\right)^2} \quad (3.396)$$

Since domain Ω has an arbitrary shape, the field inside it cannot be described by an analytic expression involving known functions. Hence, the field is obtained by numerically solving the scalar Helmholtz equation

$$\nabla_t \left(\frac{1}{\mu_r} \nabla_t E_y\right) + k_0^2 \varepsilon_r E_y = 0 \quad (3.397)$$

subject to the boundary conditions

$$E_y = 0 \text{ on } \Gamma_0 \quad (3.398)$$

and

3.9 Single-Conductor Closed Transmission Lines

$$\left.\begin{cases} E_{y|\Gamma_k} = E_{ywg}^{(k)} \\ \dfrac{\partial E_y}{\partial n}\bigg|_{\Gamma_k} = \dfrac{\partial E_{ywg}^{(k)}}{\partial z^{(k)}} \end{cases}\right\} \; k = 1, 2, \ldots N \qquad (3.399)$$

Equation (3.397) can be solved using the finite element method [73], the boundary element method [74], or any other method that can solve the two-dimensional Helmholtz equation. Today, a number of such solvers exist that are available either commercially [75] or on the Internet [76]. Quick_Fem [75] is one such a commercially available solver that is specifically tailored for waveguide discontinuity problems.

3.9.10.2 E-Plane Waveguide Discontinuities

In an E-plane discontinuity [72] none of the port waveguides face any discontinuity in the x-direction, as shown in Figure 3.58. As a result, if the ports are excited by the fundamental TE_{10} mode, only TE_{0n} type higher-order modes are excited at the discontinuities. Also the only field in the x-direction is the h_x field. The field problem in Ω is formulated in terms of h_x, which satisfies the two-dimensional Helmholtz equation.

$$\nabla_t^2 h_x + k_t^2 h_x = 0 \qquad (3.400)$$

where

$$k_t^2 = k_0^2 - \left(\dfrac{\pi}{a}\right)^2 \qquad (3.401)$$

We express the field in each of the port waveguides in terms of $LSE_{1n}^x (n = 0,1\ldots)$ modes only. This set of modes has sn H_x field with a $\sin(\pi x/a)$ type variation.

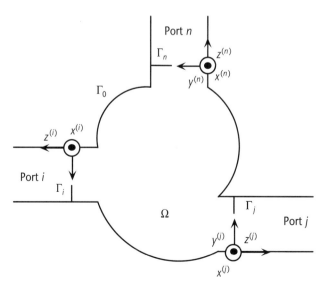

Figure 3.58 General rectangular waveguide E-plane multiport [72].

Therefore, the x-component of the magnetic field in the kth waveguide $h_{xwg}^{(k)}$, when the jth port is excited by the TE_{10} mode, or equivalently the LSE_{10}^x mode, can be expressed as

$$h_{xwg}^{(k)}\left(y^{(k)}\right) = \delta_{kj} h_0^{(j)}\left(y^{(j)}\right) e^{j\beta_{z0}^{(j)} z^{(j)}} - \sum_{n=0}^{\infty} B_n^{(k)} h_n^{(k)}\left(y^{(k)}\right) e^{-j\beta_{zn}^{(k)} z^{(k)}} \qquad (3.402)$$

where $h_n^{(k)}$, the LSE_{1n}^x mode orthonormalized magnetic field in the kth port waveguide, is given by

$$h_n^{(k)}\left(y^{(k)}\right) = \frac{\Lambda_n^{(k)}}{\sqrt{1+\delta_{on}}} \cos\left(\frac{n\pi}{b^{(k)}} y^{(k)}\right) \qquad (3.403)$$

The propagation constant $\beta_{zn}^{(k)}$ and the normalization factor $\Lambda_n^{(k)}$ are given by

$$\beta_{zn}^{(k)} = \sqrt{k_0^2 - \left(\frac{\pi}{a}\right)^2 - \left(\frac{n\pi}{b^{(k)}}\right)^2} \qquad (3.404)$$

$$\Lambda_n^{(k)} = 2\sqrt{\frac{2\eta_0}{ab^{(k)} \beta_{zn}^{(k)} k_t^2}} \qquad (3.405)$$

The associated boundary conditions for the E-plane discontinuities are

$$\frac{\partial h_x}{\partial n} = 0 \text{ on } \Gamma_0 \qquad (3.406)$$

and

$$\left\{\begin{array}{l} h_x|_{\Gamma_k} = h_{xwg}^{(k)} \\ \left.\frac{\partial h_x}{\partial n}\right|_{\Gamma_k} = \frac{\partial h_{xwg}^{(k)}}{\partial z^{(k)}} \end{array}\right\} k = 1, 2, \ldots N \qquad (3.407)$$

Once again, as in the H-plane case, (3.400) can be solved using any two-dimensional numerical Helmholtz equation solver and Quick _FEM [77] is one of those software applications. Having solved (3.400) for h_x, the other field components can be obtained by using Maxwell's equations. However, one has to keep in mind $e_x = 0$. Then the other field components are obtained from

$$h_t = \frac{\frac{\pi}{a}}{k_t^2} \nabla_t h_x \qquad (3.408)$$

$$e_t = j \frac{k_0 \eta_0}{k_t^2} \hat{x} \times h_x \qquad (3.409)$$

The above technique is suitable for the analysis of any kind of either E- or H-plane discontinuity. However, out of that infinite number of waveguide discontinuities, some are more frequently used than the others. Figure 3.59 shows appearances of those discontinuity structures in several waveguide components.

We would like to mention at this point that E-plane counterparts of the H-plane discontinuities shown in Figure 3.59 are also useful. Analyses of all such two-dimensional discontinuities and many other waveguide discontinuities can be efficiently done using the finite element modal matching method.

3.9.11 Typical Three-Dimensional Discontinuity in a Rectangular Waveguide

Figure 3.60 shows an example of a typical three-dimensional waveguide discontinuity. This is a very commonly used typical rectangular waveguide to a double-ridged waveguide step junction. The mode-matching technique with superelliptic functions that represent the various shapes can be used to analyze the junction. Equations (3.349) to (3.356) and (3.372) to (3.382) can be used to calculate the overall multimode scattering matrix [70]. The pertinent ψ-functions needed for the mode-matching analysis are shown in Figure 3.60.

Ridged waveguides are also used in many applications where the knowledge of the fundamental mode cutoff frequency and the characteristic impedance is more than sufficient for an accurate prediction of the circuit behavior. For such cases only the closed-form equation for the fundamental mode parameters are required. Closed-form equations [78] exist for the fundamental mode cutoff wavelength and the characteristic impedance of a ridged waveguide for a symmetrically placed centered ridge. The equations are given by [78,79]

Fundamental mode cutoff wavelength:

$$\lambda_{cr} = 2(a-s) \sqrt{ 1 + \frac{4}{\pi} \left\{ 1 + 0.2 \sqrt{\frac{b}{a-s}} \right\} \left(\frac{b}{a-s} \right) \ln \cos ec \left(\frac{\pi d}{2 b} \right) + \left(2.45 + 0.2 \frac{s}{a} \right) \frac{sb}{d(a-s)} } \qquad (3.410)$$

This formula is accurate to within ±1% in the following range of parameters: $0.01 < d/b \leq 1$, $0 < b/a \leq 1$, $0\ s/a \leq 0.45$.

Characteristic impedance:

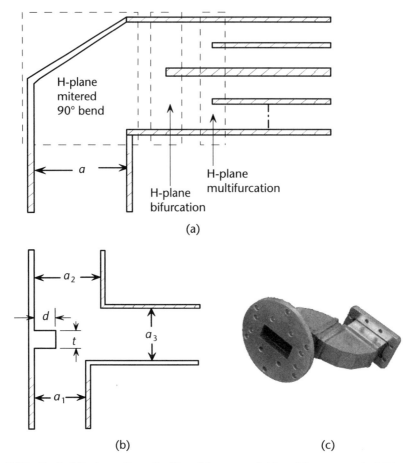

Figure 3.59 Typical H-plane discontinuities: (a) power divider, (b) compensated T-junction for angled bend, and (c) physical discontinuity hardware. (Courtesy of Mr. D. F. Carlberg, M2GLOBAL, San Antonio, TX.)

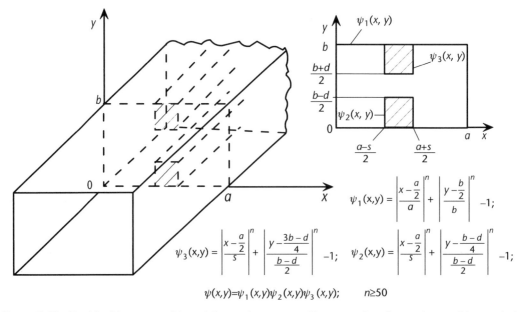

$$\psi_1(x,y) = \left| \frac{x - \frac{a}{2}}{a} \right|^n + \left| \frac{y - \frac{b}{2}}{b} \right|^n - 1;$$

$$\psi_3(x,y) = \left| \frac{x - \frac{a}{2}}{s} \right|^n + \left| \frac{y - \frac{3b-d}{4}}{\frac{b-d}{2}} \right|^n - 1; \quad \psi_2(x,y) = \left| \frac{x - \frac{a}{2}}{s} \right|^n + \left| \frac{y - \frac{b-d}{4}}{\frac{b-d}{2}} \right|^n - 1;$$

$$\psi(x,y) = \psi_1(x,y)\psi_2(x,y)\psi_3(x,y); \quad n \geq 50$$

Figure 3.60 Double ridge waveguide and the pertinent superellipse equations for mode-matching analysis.

3.9 Single-Conductor Closed Transmission Lines

$$Z_0 = \frac{Z_{0\infty}}{\sqrt{1-\left(\frac{\lambda}{\lambda_{cr}}\right)^2}} \quad (3.411)$$

where

$$Z_{0\infty} = \frac{120\pi^2\left(\frac{b}{\lambda_{cr}}\right)}{\left\{\frac{b}{d}\sin\left(\frac{\pi s}{\lambda_{cr}}\right)+\left[\left(\frac{2b}{\lambda_{cr}}\right)\ln\cos ec\left(\frac{\pi d}{2b}\right)+\tan\left(\frac{\pi}{2}\frac{(a-s)}{\lambda_{cr}}\right)\right]\cos\left(\frac{\pi s}{\lambda_{cr}}\right)\right\}\sqrt{1-\left(\frac{\lambda}{\lambda_{cr}}\right)^2}} \quad (3.412)$$

The above equation is based on the voltage-current definition of characteristic impedance, where the voltage is the line integral of the electric field between the ridges at the center of the guide ($x = a/2$), while the current is the total longitudinal surface current on the top wall of the guide that includes the surface of the upper ridge. Equations (3.410) to (3.412) are also applicable to a single-ridge waveguide with the following interpretation. In the expression for the cutoff frequency (3.410), b is twice the height of the single-ridge guide and d is twice the spacing between the ridge and the bottom wall. The same interpretation goes for the characteristic impedance (3.384) and (3.385). The computed impedance should be divided by two to obtain the impedance of the single ridge waveguide.

Since ridge waveguides are used in many passive waveguide components as transmission lines and resonators, the attenuation constant of a ridge waveguide is very important. An approximate expression for ridge waveguide attenuation has been given by Hopfer [79] and δ is the skin depth in meters. The parameter of interest is the bandwidth between the fundamental TE_{10} mode and the next higher-order mode. If the waveguide and the excitation system are symmetric, then the TE_{20} is not excited. As a result, the next higher-order mode becomes the TE_{30} mode. A relative bandwidth of 8 is achievable for $s/a=0.5$ and $d/b=0.1$. Note that the maximum relative fundamental-mode bandwidth achievable in a rectangular waveguide is 2. Table 3.10 shows the industrial specifications of ridged waveguides.

$$\alpha = 8.686\frac{\frac{\pi\lambda_{cr}}{b\lambda^2}+\Lambda}{\sqrt{\left(\frac{\lambda_{cr}}{\lambda}\right)^2-1}} \quad (dB/m) \quad (3.413)$$

where

$$\Lambda = \frac{\left[2\pi\frac{\iota-\beta}{\iota}\delta^2\Lambda_1+\frac{4\pi^2}{k}B^2\frac{\iota-\beta}{\iota}\right]\tan\left[\frac{2\pi\rho}{k}\right]+\pi\Lambda_2}{ka^2\left[\frac{2\iota}{\beta}\Lambda_1\tan^2\left(\frac{2\pi\rho}{k}\right)-2\tan\left(\frac{2\pi\rho}{k}\right)+\rho\Lambda_2\right]} \quad (3.414)$$

Table 3.10 Ridged Waveguide Specifications

Waveguide Size	MIL-W-23351 Dash #	Material	Freq Range (GHz)	Freq Cutoff (GHz)	Power (at 1 Atm) CW	Power (at 1 Atm) Peak	Insertion Loss (dB/ft)	Dimensions (Inches) A	B	C	D	E	F
WRD250		Alum Brass Copper Silver Al	2.60 - 7.80	2.093	24	120	0.025 0.025 0.018 0.019	1.655	0.715	2	1	0.44	0.15
WRD350 D24	4-029 4-303 4-031	Alum Brass Copper	3.50 - 8.20	2.915	18	150	0.0307 0.0303 0.0204	1.48	0.688	1.608	0.816	0.37	0.292
WRD475 D24	4-033 4-034 4-035	Alum Brass Copper	4.75 - 11.00	3.961	8	85	0.0487 0.0481 0.0324	1.09	0.506	1.19	0.606	0.272	0.215
WRD500 D36	2-025 2-026 2-027	Alum Brass Copper	5.00 - 18.00	4.222	4	15	0.146 0.141 0.095	0.752	0.323	0.852	0.423	0.188	0.063
WRD650		Alum Brass Copper	6.50 - 18.00	5.348	4	25	0.106 0.105 0.07	0.720	0.321	0.820	0.421	0.173	0.101
WRD750 D24	4-037 4-038 4-039	Alum Brass Copper	7.50 - 18.00	6.239	4.8	35	0.0964 0.0951 0.0641	0.691	0.321	0.791	0.421	0.173	0.136
WRD110 D24	4-041 4-042 4-043	Alum Brass Copper	11.00 - 26.50	9.363	1.4	15	0.171 0.169 0.144	0.471	0.219	0.551	0.299	0.118	0.093
WRD180 D24	4-045 4-046 4-047	Alum Brass Copper	18.00 - 40.00	14.995	0.8	5	0.358 0.353 0.238	0.288	0.134	0.368	0.214	0.072	0.057

[80] *Source:* EW & Radar Systems Engineering Handbook, Naval Air Systems Command, Washington D. C.

$$\Lambda_1 = \tan\left(\frac{\pi\chi}{k}\right) + \frac{\pi\chi}{k}\sec^2\left(\frac{\pi\chi}{k}\right) \qquad (3.415)$$

$$\Lambda_2 = \left(\frac{4\pi}{k}\right)\sec^2\left(\frac{2\pi\delta}{k}\right) \qquad (3.416)$$

$\iota = b/a$, $\chi = s/a$, $k = \lambda_{cr}/a$, $\beta = d/a$, $\rho = (1 - s/a)/2$

We will show in Chapters 5 and 6 how by properly cascading straight sections of empty waveguides, ridged waveguides, and step junctions very wide stopband low-pass and band-pass filters are realized.

This chapter has described the properties and design aspects of various transmission lines used in RF and microwave filter design. In conclusion, we would like to mention that among many other important parameters of any transmission line, the most important one is the insertion loss or the attenuation factor. Figure 3.61 shows the comparison of attenuations in various transmission lines used RF and microwave filter realizations.

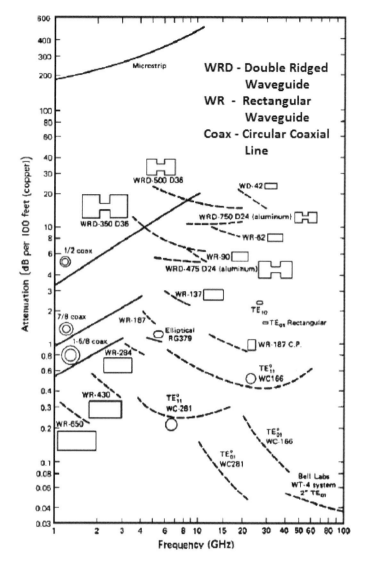

Figure 3.61 Comparison of frequency dependence of attenuation of various transmission lines. (Source: EW & Radar Systems Engineering Handbook, Naval Air Systems Command, Washington DC.)

References

[1] Perlow, S. M., "Analysis of Edge-Coupled Strip and Slab Lines," *IEEE Trans. Microw. Theory Tech.*, Vol. 35, No. 5, May 1987, pp. 522–529.

[2] Marcuvitz, N., *Waveguide Handbook*, IEEE Stevenge, 1986.

[3] Conciauro, G., M. Guglielmi, and R. Sorrentino, *Advanced Modal Analysis*, New York: John Wiley, 2000.

[4] SUPERCOMPACT, Compact Software Inc., Paterson, NJ, 1982.

[5] Sadiku, M. N. O., *Numerical Techniques in Electromagnetics*, Boca Raton, FL: CRC Press, 1982.

[6] Itoh T., and H. Housmand, *Time Domain Methods for Microwave Structures*, New York: IEEE Press, 1998.

[7] Pramanick P., and P. Bhartia, "Hybrids and Couplers" in *Microwave Solid State Circuit Design,* New York: John Wiley and Sons, 1998.

[8] Green, H. E., *Advances in Microwaves,* Vol. 2, New York: Academic Press, 1967, p. 350.

[9] Chang, W., "Accurate Analysis Method for Square Coaxial Lines", *Frequenz,* No. 43, 1989, pp. 271–276.

[10] Allesandri, F. et. al., "Modeling of Square Coaxial Line Structures," *CNET Conf. Digest,* 1990, pp. 278–281.

[11] Lin, W., "Polygonal Coaxial Line with Round Center Conductor," *IEEE Trans. Microw. Theory Tech.,* Vol. MTT-33, No. 6, June 1985.

[12] Barrett, R. M., "Microwave Printed Circuits—The Early Years," *IEEE Trans. Microw. Theory Tech.,* Vol. MTT-32, September 1984, pp. 883–900.

[13] Cohn, S. B., "Characteristic Impedance of Shielded Strip Transmission Lines," *IRE Trans. Microw. Theory Tech.,* Vol. MTT-2, July 1954, pp. 52–55.

[14] Peters, R. W., et al., *Handbook of Triplate Microwave Components,* Nashua, NH: Sanders Associates, 1956.

[15] Matthaei, G., L. Young, and E. M. T. Jones, *Microwave Filters, Impedance Matching Networks and Coupling Structures,* Dedham, MA: Artech House, 1974.

[16] Howe, H. Jr., *Stripline Circuit Design,* Dedham, MA: Artech House, 1974.

[17] Wheeler, H. A., "Transmission Line Properties of a Stripline between Parallel Planes," *IEEE Trans. Microw. Theory Tech.,* Vol. MTT-26, No. 11, November, 1978, pp. 866–876.

[18] Rosloniec, S., *Algorithms for Computer-Aided Design of Linear Microwave Circuits,* Norwood, MA: Artech House, 1990.

[19] TXLINE, MICROWAVE OFFICE, Applied Wave Research, El Segundo, CA, 2000.

[20] EAGLEWARE, Eagleware Corporation, Tucker, GA, 2000.

[21] Hoffman, R. K., *Handbook of Microwave Integrated Circuits,* Norwood, MA: Artech House, 1987.

[22] Perlow, S. M., "Analysis of Edge Coupled Shielded Strip and Slabline Structures," *IEEE Trans. Microw. Theory Tech.,* Vol. MTT-35, May 1987, pp. 522–529.

[23] Matthaei, G. L., L. Young, and E. M. T. Jones, *Microwave Filters Impedance Matching Networks and Coupling Structure,* Dedham, MA: Artech House, 1980.

[24] Oliner, A. A., "Equivalent Circuits for Discontinuities in Balanced Strip Transmission Line," *IRE Trans. Microw. Theory Tech.,* Vol. MTT-3, March 1955, pp. 134–143.

[25] Altschuler H. M., and A. A. Oliner, "Discontinuities in the Center Conductor of Strip Transmission Line," *IRE Trans. Microw. Theory Tech.,* Vol. 8, 1960, pp. 328–339.

[26] Gupta, K. C., R. Garg, and R. Chadha, *Computer Aided Design of Microwave Circuits,* Norwood, MA: Artech House, 1981.

[27] Jhanke, E., and F. Emde, *Tables of Functions,* p. 16, Dover Publications, NY, 1945.

[28] AWR, Applied Wave Research, El Segundo, CA, 2000.

[29] WAVECON, Escondido, CA, 2000.

[30] Ansoft Designer, Ansoft Corporation, San Jose, CA, 2000.

[31] EMPIRE, IMST, GmbH, Germany, 2004.

[32] HFSS, Ansoft Corporation, San Jose, CA, 2000.

[33] Ishii, J., *Data on Transmission Line Filters,* Chapter IV, "Microwave Filters and Circuits, Advances in Microwaves," p. 117, New York: Academic Press, 1970.

[34] Cohn, S. B., "Shielded Coupled-Strip Transmission Line," *Trans. Microw. Theory Tech.,* Vol. 3, October 1955, pp. 29–38.

[35] Delmastro, P., *TRANSLIN: Transmission Line Analysis and Design,* Artech House, Norwood: MA, 1999.

[36] Perlow, S. M., "Analysis of Edge Coupled Shielded Strip and Slabline Structures," *IEEE Trans. Microw. Theory Tech.*, Vol. 35, No. 5, May 1987, pp. 522–529.

[37] Cohn, S. B., "Characteristic Impedances of Broadside Coupled Strip Transmission Lines," *IRE Trans. Microw. Theory Tech.*, Vol. 8, June 1960, pp. 633–637.

[38] Bhartia P., and P. Pramanick, "Computer Aided Design Models for Broadside Coupled Striplines and Suspended Substrate Microstrip Lines," *IEEE Trans. Microw. Theory Tech.*, Vol. 36, No. 11, November 1988, pp. 1476–1481.

[39] Cohn, S. B., "Characteristic Impedances of Broadside Coupled Strip Transmission Lines," *IRE Trans. Microw. Theory Tech.*, Vol. 8, June 1960, pp. 633–637.

[40] Hillberg, W., "From Approximations to Exact Relations for Characteristic Impedances," *IEEE Trans. Microw. Theory Tech.*, Vol. 17, No. 11, November 1969, pp. 259–265.

[41] Shelton, J. P., "Impedances of Offset Parallel Coupled Strip Transmission Lines," *IEEE Trans. Microwave Theory Tech.*, Vol. 14, No. 1, January 1966, pp. 7–15.

[42] Yamamoto, S., T. Azakami, and K. Itakura, "Slit Coupled Strip Transmission Lines," *IEEE Trans. Microwave Theory Tech.*, Vol. 14, No. 1, January 1966, pp. 542–553.

[43] Abramowitz, M., and Stegun, I. A., *Handbook of Mathematical Functions*, New York: Dover Publications, 1964, Section 16.27ff.

[44] Crystal, E., "Coupled Circular Cylindrical Rods between Parallel Ground Planes," *IEEE Trans. Microw. Theory Tech.*, Vol. 12, No. 7, July 1964, pp. 428–438.

[45] Stracca, G. B., G. Maccchiardla, and M. Politi, "Numerical Analysis of Various Configurations of Slablines," *IEEE Trans. Microw. Theory Tech.*, Vol. 34, No. 3, March 1986, pp. 359–363.

[46] Rosloniec, S., "An Improved Algorithm for Computer Aided Design of Coupled Slablines," *IEEE Trans. Microw. Theory Tech.*, Vol. 37, No. 1, January 1989, pp. 258–261.

[47] Rosloniec, S., *Algorithms for Computer Aided Design of Linear Microwave Circuits*, Norwood, MA: Artech House, 1990, pp. 205–206.

[48] March, S. L., "Empirical Formulas for Impedance and Effective Dielectric Constant of Covered Microstrip for Use in the Computer Aided Design of Microwave Integrated Circuits," *Proc. European Microwave Conference*, Kent, UK, 1981, pp. 671–676.

[49] Wells, G., *Modeling of Shielded Microstrip Transmission Lines*, M.Sc. thesis, Department of Electrical Engineering, University of Saskatchewan, Canada, 1996.

[50] Edwards, T. C., *Foundations for Microstrip Circuit Design*, New York: John Wiley, 1981, pp. 84–86.

[51] Wells, G., and P. Pramanick, "An Accurate Dispersion Expression for Shielded Microstrip Lines," *Int. J. Microwave Mill.*, Vol. 5, No. 4, 1995, pp. 287–295.

[52] Pozar, D., and N. C. Das, PCAAMT: *Personal Computer Aided Analysis of Multilayered Transmission Lines*, computer software, Amherst, MA: Antenna Associates, 1989.

[53] Pramanick, P., and P. Bhartia, "Phase Velocity Dependence of Frequency Dependent Characteristic impedance of Microstrip," *Microwave J.*, June 1989.

[54] Kirsching, M., and R. H. Jansen, "Accurate Wide-Range Design Equations for the Frequency Dependent Characteristic of Coupled Microstrip Lines," *IEEE Trans. Microw. Theory Tech.*, Vol. 32, No. 1, January 1984, pp. 83–90; also Vol. 33, No. 3, March 1985, p. 288.

[55] Pramanick, P., and P. Bhartia, "Computer Aided Design Models for Millimeter-Wave Finlines and Suspended Substrate Microstrip Lines," *IEEE Trans. Microw. Theory Tech.*, Vol. 33, No. 12, 5, January, 1985, pp. 1429–1435.

[56] Bhat, B., and S. Koul, "Unified Approach to Solve a Class of Strip and Microstrip Like Transmission Lines," *IEEE Trans. Microw. Theory Tech.*, Vol. 30, No. 5, May 1982, pp. 679–686.

[57] Pramanick, P., and P. Bhartia, "Computer Aided Design Models for Broadside Coupled Striplines and Suspended Substrate Microstrip Lines," *IEEE Trans. Microw. Theory Tech.*, Vol. 36, No. 11, November 1988, pp. 1476-1481.

[58] Farina, M., EM3DS-Intregral Equation Based EM Solver for Layered Structures, MEM Research, Italy, 2002.

[59] Ramo, S., J. R. Whinnery, and T. Van Duzar, *Fields and Waves in Communication Electronics*, New York: John Wiley & Sons, 1994.

[60] Conciauro, G., M. Guglielmi, and R. Sorrentino, *Advanced Modal Analysis-CAD Techniques for Waveguide Components and Filters*, New York: John Wiley & Sons, 2000.

[61] Collin, R. E., *Field Theory of Guided Waves*, Second Edition, Piscataway, NJ: IEEE Press, 1991.

[62] Marcuvitz, N., *Waveguide Handbook*, New York: Dover Publications, 1965.

[63] EW Warfare and Radar Systems Engineering Handbook, Naval Air Systems Command, Washington, DC.

[64] King, H. E., "Rectangular Waveguide Theoretical CW Average Power Rating," *IEEE Trans. Microw. Theory Tech.*, Vol. 11, No. 7, July 1961, pp. 349–357.

[65] Lewin, L., *Theory of Waveguides*, London: Newness-Butterworth, 1975.

[66] Wexler, A., "A. Solution of Waveguide Discontinuities by Modal Analysis," *IEEE Trans. Microw. Theory Tech.*, Vol. 15, 1967, pp. 508–517.

[67] Masterman P. H., and P. J. B. Clarricoats, "Computer Field Matching Solution of Waveguide Discontinuities," *IEE Proc.*, Vol. 118, 1971, pp. 51–63.

[68] James, G. L., "Analysis and Design of Corrugated Cylindrical Waveguide Mode Converters," *IEEE Trans. Microw. Theory Tech.*, Vol. 29, 1981, pp. 1059–1066.

[69] Uher, J., J. Borenemann, and U. Rosenberg, *Waveguide Components for Antenna Feed Systems*, Norwood, MA: Artech House, 1993.

[70] Lin, S. L., L.- W. Li, T.- S. Yeo, and M.- S. Leong, "A Unified Modal Analysis of Off-Centered Waveguide Junctions with Thick Iris," *IEEE Microw. Wirel. Compon. Lett.*, Vol. 11, No. 9, September 2001, pp. 388–390.

[71] Kiyoshi I., and K. Masanori, "Numerical Analysis of H-Plane Waveguide Junctions by Combination of Finite and Boundary Elements," *IEEE Trans. Microw. Theory Tech.*, Vol. 36, No. 12, December 1988, pp. 1343–1351.

[72] Pelosi, G., R. Coccioli, and S. Sellari, *Quick Finite Elements for Electromagnetic Waves*, Norwood, MA: Artech House, 1998.

[73] Jin, J., *The Finite Element Method for Electromagnetics*, New York: John Wiley Interscience, 1993.

[74] Brebbia, C. A., and J. Dominguez, *Boundary Elements: An Introductory Course*, New York: McGraw Hill, 1992.

[75] PDEase2D, Macsyma Inc., Boston, 1992.

[76] Hect, F., O. Pironneau, and K. Ohtsuka, FreeFEM ++, http://www.freefem.org.

[77] Pelosi, G., R. Coccioli, and S. Sellari, *Quick Finite Elements for Electromagnetic Waves*, Norwood, MA: Artech House, 1998.

[78] Hoefer, W. J. R., and M. N. Burton, "Closed-Form Expressions for the Parameters of Finned and Ridged Waveguides," *IEEE Trans. MTT*, Vol. 30, No. 12. December 1982, pp. 2190–2194.

[79] Hopfer, S., "The Design of Ridged Waveguides," *IRE Trans. MTT*, Vol. 3, October, 1955, pp. 20–29.

CHAPTER 4

Low-Pass Filter Design

A low-pass filter is the most basic building block of a host of microwave passive components. Such components not only include any other type of filters such as high-pass, band-pass, and bandstop filters but also components like power dividers, couplers, hybrids, and matching transformers. Therefore, it is worthwhile to present here a brief description of the theory of lumped element low-pass filters. The most primitive method of low-pass filter design is the image-parameter method [1]. However, the method suffers from a serious drawback of unpredictability of skirt selectivity beyond the edges of the passband. Modern network synthesis methods offer a low-pass design method, which is based on the concept of insertion loss and is very versatile in most respects [2].

4.1 Insertion Loss Method of Filter Design

Consider the terminated two-port network shown in Figure 4.1. Let us also assume that the terminating impedances R_g and R_L are purely real quantities. According to the definition of normalized power-wave amplitudes, we can define

$$a_1 = \frac{V_1 + R_g I_1}{2\sqrt{R_g}} \quad (4.1)$$

$$b_1 = \frac{V_1 - R_g I_1}{2\sqrt{R_g}} \quad (4.2)$$

$$a_2 = \frac{V_2 + R_L I_2}{2\sqrt{R_L}} \quad (4.3)$$

$$b_2 = \frac{V_2 - R_L I_2}{2\sqrt{R_L}} \quad (4.4)$$

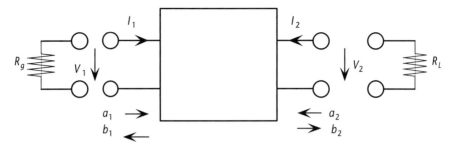

Figure 4.1 Terminated two-port network.

The above normalized power-wave amplitude parameters are related via the following equations (see Section 2.4):

$$b_1 = S_{11}a_1 + S_{12}a_2 \tag{4.5a}$$

$$b_2 = S_{21}a_1 + S_{22}a_2 \tag{4.5b}$$

or in matrix form

$$b] = [S]a] \tag{4.6a}$$

$$[S] = \begin{bmatrix} S_{11} & S_{12} \\ S_{21} & S_{22} \end{bmatrix} \tag{4.6b}$$

If P_1 and P_2 are the power levels at ports 1 and 2 respectively, then

$$P_1 = |a_1|^2 - |b_1|^2 \tag{4.7}$$

$$P_2 = |a_2|^2 - |b_2|^2 \tag{4.8}$$

If the network is lossless, then the conservation of power requires

$$P_2 - P_1 = |a_2|^2 - |b_2|^2 - |a_1|^2 + |b_1|^2 = 0 \tag{4.9}$$

which in turn leads to the following equations for the scattering parameters

$$|S_{11}|^2 + |S_{21}|^2 = |S_{12}|^2 + |S_{22}|^2 = 1 \tag{4.10}$$

In actual practice R_g is the output impedance of the generator connected to port 1, as shown in Figure 4.2.

From the above network we obtain the following relationships:

The output power

4.1 Insertion Loss Method of Filter Design

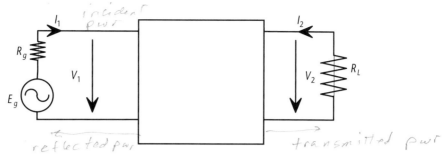

Figure 4.2 Two-port network with a generator.

$$P_L = \frac{|V_2|^2}{R_L} \quad \text{(4.11)}$$

The input impedance

$$Z_{in} = \frac{V_1}{I_1} \quad \text{(4.12)}$$

The maximum available power from the generator under matched conditions is

$$P_m = \frac{|E_g|^2}{4R_g} \quad \text{(4.13)}$$

From the definition of scattering parameters

$$S_{11} = \left.\frac{b_1}{a_1}\right|_{a_2=0} = \frac{\frac{V_1 - R_g I_1}{2\sqrt{R_g}}}{\frac{V_2 + R_L I_2}{2\sqrt{R_L}}} = \frac{Z_{in} - R_g}{Z_{in} + R_g} \quad \text{(4.14)}$$

$$S_{21} = \left.\frac{b_2}{a_1}\right|_{a_2=0} = \frac{2V_2}{E_g}\sqrt{\frac{R_g}{R_L}} \quad \text{(4.15)}$$

$$|S_{21}|^2 = \frac{P_L}{P_m} \quad \text{(4.16)}$$

The right-hand side of the above equation is the ratio of the power absorbed by the load and the maximum available power from the generator. This is also known as the insertion loss of the two-port network.

Let us consider the following example in Figure 4.3:

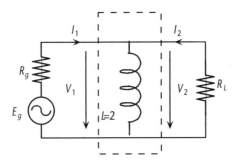

Figure 4.3 Network with one shunt inductor.

$$Z_{in} = \frac{V_1}{I_1} = \frac{sLR_L}{sL + R_L} \quad (4.17)$$

$$V_2 = \frac{Z_{in} E_g}{Z_{in} + R_g} \quad (4.18)$$

Let us assume, without loss of generality, $R_g = R_L = 1\ \Omega$ and $E_g = 1$ volt (rms)

$$S_{11} = \frac{Z_{in} - 1}{Z_{in} + 1} = \frac{-1}{4s + 1} \quad (4.19)$$

Combining (4.15) and (4.19), we obtain

$$S_{21} = \frac{-1}{4s + 1} \quad (4.20)$$

Since the network is symmetrical, the overall scattering matrix is given by

$$[S] = \frac{1}{4s+1}\begin{bmatrix} -1 & 4s \\ 4s & -1 \end{bmatrix} \quad (4.21)$$

4.2 Belevitch Matrix and Transfer Function Synthesis

Belevitch [3] showed that the scattering matrix of a loss-less passive two-port network has the general form

$$[S] = \frac{1}{h(s)}\begin{bmatrix} f(s) & g(s) \\ g(s) & -\xi f^*(s) \end{bmatrix} \quad (4.22)$$

where $f(s)$, $g(s)$, and $h(s)$ are real polynomials of s, * denotes the complex conjugate, and $\xi = \pm 1$. The (+) sign is used when $g(s)$ is an even polynomial and the (−) sign is used otherwise. Also, $h(s)$ is strictly a Hurwitz polynomial and the conservation of power is guaranteed by the following equation:

$$h(s)h^*(s) = f(s)f^*(s) + g(s)g^*(s) \tag{4.23}$$

Comparing (4.6b) and (4.23) gives

$$S_{11}(s) = \frac{f(s)}{h(s)} \tag{4.24}$$

and

$$S_{21}(s) = \frac{g(s)}{h(s)} \tag{4.25}$$

We define the characteristic function $\vartheta(s)$ as

$$\vartheta(s) = \frac{S_{11}(s)}{S_{21}(s)} \tag{4.26}$$

Combining (4.26) and (4.10) gives

$$\left|S_{21}(s)\right|^2 = \frac{1}{1+\left|\vartheta(s)\right|^2} \tag{4.27}$$

The above equation shows how the characteristic function is related to the insertion loss of the two-port network. We are now in a position to reveal the significance of the characteristic function and how it enables us to synthesize the two-port network on the basis of its insertion loss response in the frequency domain. Equation (4.27) in the radian frequency domain assumes the form

$$\left|S_{21}(j\omega)\right|^2 = \frac{1}{1+\left|\vartheta(j\omega)\right|^2} \tag{4.28}$$

which can also be written as

$$\left|S_{21}(j\omega)\right|^2 = \frac{1}{1+F(\omega^2)} \tag{4.29}$$

where the function *F* is known as the approximation function. The form of *F* can be chosen based on the type of low-pass filter response desired.

4.2.1 Butterworth Approximation

The most fundamental and the simplest such function is the Butterworth approximation function. The function has the form [4]

$$F(\omega^2) = \omega^{2N} \tag{4.30}$$

where N identifies the degree of the approximation function. N plays a role in the skirt selectivity or the attenuation roll off of the filter in the stopband and the variation of the insertion loss in the passband. Combining (4.29) and (4.30) gives

$$|S_{21}(j\omega)|^2 = \frac{1}{1+\omega^{2N}} \tag{4.31}$$

Figure 4.4 shows a graphical representation of (4.31). The Butterworth response is also known as the maximally flat response.

The locations of the poles of a Butterworth filter in the complex frequency plane or *s*-plane are obtained by equating the denominator of the equation

$$|S_{21}(j\omega)|^2 = \frac{1}{1+(-js)^{2N}} = \infty \tag{4.32a}$$

or

$$1+(-js)^{2N} = 0 \tag{4.32b}$$

The roots of the above equation are given by

$$s_M = je^{j\theta_M} = -\sin\theta_M + j\cos\theta_M = \sigma_M + j\omega_M; \quad M = 1, 2, \cdots 2N \tag{4.33}$$

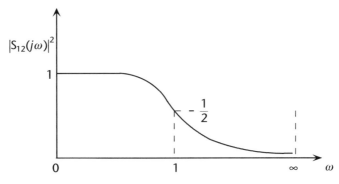

Figure 4.4 Normalized Butterworth response.

with

$$\theta_M = \frac{(2M-1)\pi}{2N} \tag{4.34}$$

Figure 4.5 shows the locations of the roots of a third-order ($N = 3$) Butterworth filter in the complex frequency or s plane.

According to (4.33) and (4.34), the roots are

$$\begin{aligned} s_1 &= -1 \\ s_2 &= -\frac{1}{2} + j\sqrt{3} \\ s_3 &= -\frac{1}{2} - j\sqrt{3} \end{aligned} \tag{4.35a}$$

and

$$\begin{aligned} s_4 &= +1 \\ s_5 &= +\frac{1}{2} + j\sqrt{3} \\ s_6 &= +\frac{1}{2} - j\sqrt{3} \end{aligned} \tag{4.35b}$$

Therefore, (4.32a) can be written as

$$S_{21}S_{21}^* = \frac{1}{(s-s_1)(s-s_2)(s-s_3)} \cdot \frac{1}{(s-s_4)(s-s_5)(s-s_6)} \tag{4.36}$$

In order to satisfy the realizability conditions we consider the poles in the left half of the complex frequency or s plane and obtain

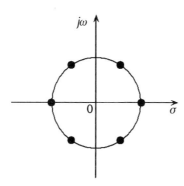

Figure 4.5 Poles of $N = 3$ Butterworth filter.

$$S_{21} = \frac{1}{(s-s_1)(s-s_2)(s-s_3)} = \frac{1}{1+2s+2s^2+s^3} \qquad (4.37)$$

Therefore, from (4.22) and (4.37), we have

$$h(s) = 1 + 2s + 2s^2 + s^3 \qquad (4.38)$$

and

$$g(s) = 1 \qquad (4.39)$$

$$\xi = +1 \qquad (4.40)$$

Combining (4.39), (4.40) and (4.23) gives

$$ff^* = hh^* - gg^* = (1 - s^6) - 1 = -s^6 = (s^3)((-s)^3) \qquad (4.41)$$

or

$$f = s^3 \qquad (4.42)$$

From (4.24), we get

$$S_{11}(s) = \frac{f(s)}{g(s)} = \frac{s^3}{1+2s+2s^2+s^3} \qquad (4.43)$$

At this point we assume (4.31) to be the frequency response of a low-pass filter whose output/input power ratio drops to half of the value at zero frequency of the flat passband at $\omega = 1$ radian/sec. We also assume that the actual filter is a two-port passive network, which is terminated by a 1Ω real impedance at both ends, as shown in Figure 4.2 with $R_g = R_L = 1\Omega$. Then the input impedance of the network is given by

$$Z_1(s) = \frac{1+S_{11}(s)}{1-S_{11}(s)} = \frac{1+2s+2s^2+2s^3}{1+2s+2s^2} \qquad (4.44)$$

Equation (4.44) representing the input impedance of the terminated passive two-port can be realized by using inductors and capacitors and adopting either Cauer-I or Cauer-II synthesis techniques [5]. Before we proceed with the demonstration of the synthesis procedure, the driving point impedance $Z_1(s)$ and the corresponding input reflection coefficient have to fulfill certain conditions known as positive realizability condition, as mentioned in Chapter 1 (Section 1.5). The properties are summarized in Table 4.1.

4.2 Belevitch Matrix and Transfer Function Synthesis

Table 4.1 Positive Reliazibility Conditions for $Z_1(s)$ and $S_{11}(s)$

Item	$Z_1(s)$	$S_{11}(s)$
1	Coefficients of $Z_1(s)$ must be real	Coefficients of $S_{11}(s)$ must be real
2a	Behavior of $Z_1(s)$ as $s \to \infty$ must be one of the forms ks, k or $\dfrac{k}{s}$, k being a real positive constant	Behavior of $S_{11}(s)$ as $s \to \infty$ must be of one of the forms: $\left[-1+\dfrac{2R_1}{ks}\right]$, $\left[\dfrac{R_1-k}{R_1+k}\right]$ or $\left[1-\dfrac{2k}{R_1 s}\right]$ and these apply respectively to the three forms of $Z_1(s)$
b	$Z_1(s)$ can have no poles on the right half s-plane	$|S_{11}(s)| \neq 1$ and $\mathrm{Re}(s) > 0$ and finite R_1
c	Any pole of $Z_1(s)$ s-plane $j\omega$ axis must be simple and have real and positive residue.	Approach of $S_{11}(s)$ to -1 due to s-plane $j\omega$ axis poles of $Z_1(s)$ must be of the form $\underset{s \to s_0}{Lt}|S_{11}(s)| \to -1 + \dfrac{2R_1}{k}(s-s_0)$ where k is the residue of the pole at s_0
3	$\mathrm{Re}[Z_1(s)] \geq 0$ for $s = j\omega$	$|S_{11}(s)| \leq 1$ for $s = j\omega$

Source: From Peter M. Kelly, Ph.D thesis, California Institute of Technology, 1960.

4.2.2 Cauer Synthesis

Equation (4.44) shows that the input impedance of a two-port low-pass filter to be synthesized can be represented in general by a rational function:

$$Z_1(s) = \frac{N(s)}{D(s)} = \frac{m_1(s) + n_1(s)}{m_2(s) + n_2(s)} \tag{4.45}$$

In the above equation $m_1(s)$ and $n_1(s)$ are the even and odd parts of the numerator polynomial and $m_2(s)$ and $n_2(s)$ are the even and odd parts, respectively, of the denominator polynomial. From (4.45), we get

$$\mathrm{Re}\, Z(j\omega) = Ev\, Z(s)\big|_{j\omega} = \frac{m_1(j\omega)m_2(j\omega) - n_1(j\omega)n_2(j\omega)}{m_2^2(j\omega) - n_2^2(j\omega)} \tag{4.46a}$$

$$\mathrm{Im}\, Z(j\omega) = Odd\, Z(s)\big|_{j\omega} = \frac{n_1(j\omega)m_2(j\omega) - m_1(j\omega)n_2(j\omega)}{m_2^2(j\omega) - n_2^2(j\omega)} \tag{4.46b}$$

Since $Z_1(s)$ is the input or driving point impedance of a passive lossless network, it has no resistive part or real part, or $\mathrm{Re}Z(j\omega)$ is zero for all ω. Therefore, equating the right-hand side of (4.46a) to zero gives the following possibilities:

1. $m_1(j\omega) = n_2(j\omega) = 0$

2. $n_1(j\omega) = m_2(j\omega) = 0$
3. $m_1(j\omega)m_2(j\omega) = n_1(j\omega)n_2(j\omega)$

From the above possibilities we draw the following conclusion:
Either

$$Z(s) = \frac{m(s)}{n(s)} \quad \text{or} \quad Z(s) = \frac{n(s)}{m(s)} \quad (4.47)$$

where the degrees of the polynomials $m(s)$ and $n(s)$ differ by unity because $Z(s)$ and $Y(s)$ can have a simple pole or zero at $s = \infty$. Such an impedance function can be expanded in the form of a continued fraction like

$$Z(s) = \frac{D(s)}{N(s)} = \frac{a_n s^n + a_{n-2} s^{n-1} + \cdots}{a_{n-1} s^{n-1} + a_{n-3} s^{n-3} + \cdots} = g_1 s + \cfrac{1}{g_2 s + \cfrac{1}{g_3 s + \cfrac{1}{g_4 s + \cfrac{1}{\ddots}}}} \quad (4.48a)$$

or

$$Y(s) = \frac{D(s)}{N(s)} = \frac{a_n s^n + a_{n-2} s^{n-1} + \cdots}{a_{n-1} s^{n-1} + a_{n-3} s^{n-3} + \cdots} = g_1 s + \cfrac{1}{g_2 s + \cfrac{1}{g_3 s + \cfrac{1}{g_4 s + \cfrac{1}{\ddots}}}} \quad (4.48b)$$

This gives the first Cauer canonical form of networks, as shown in Figure 4.6.

In the above first Cauer form of network (Figure 4.6(a)), successive pole removals at infinity from the impedance and admittance functions was used [6]. In the second Cauer form of network, successive pole removals at zero frequency are used from impedance and admittance functions, as follows:

$$Z(s) = \frac{N(s)}{D(s)} = \frac{a_0 + a_2 s + a_4 s + \cdots}{a_1 s + a_3 s + a_5 s + \cdots} = \cfrac{1}{g_1 s} + \cfrac{1}{\cfrac{1}{g_2 s} + \cfrac{1}{\cfrac{1}{g_3 s} + \cfrac{1}{\cfrac{1}{g_4 s} + \ddots}}} \quad (4.49a)$$

or

4.2 Belevitch Matrix and Transfer Function Synthesis

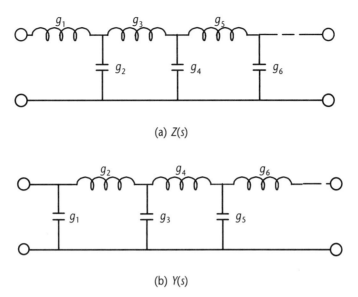

(a) $Z(s)$

(b) $Y(s)$

Figure 4.6 First Cauer form of networks.

$$Y(s) = \frac{N(s)}{D(s)} = \frac{a_0 + a_2 s + a_4 s + \cdots}{a_1 s + a_3 s + a_5 s + \cdots} = \frac{1}{g_1 s} + \cfrac{1}{\cfrac{1}{g_2 s} + \cfrac{1}{\cfrac{1}{g_3 s} + \cfrac{1}{\cfrac{1}{g_4 s} + \cdots}}} \quad (4.49\text{b})$$

The corresponding networks are shown in Figure 4.7. We will illustrate the above procedure by realizing the impedance function in (4.44) corresponding to the third-order Butterworth response. A continued fraction expansion of the right-hand side of (4.44) yields

$$\begin{array}{r} 2s^2 + 2s + 1) \, 2s^3 + 2s^2 + 2s + 1 \, (s \\ \underline{2s^3 + 2s^2 + s} \\ s + 1) \, 2s^2 + 2s + 1 \, (2s \\ \underline{2s^2 + 2s} \\ 1) \, s + 1 \, (s + 1 \\ \underline{s + 1} \end{array}$$

or in the first Cauer form of network:

$$Z_1(s) = s + \cfrac{1}{2s + \cfrac{1}{s+1}} \quad (4.50)$$

The corresponding network is shown in Figure 4.8.

Figure 4.9 shows the computed frequency response of the designed third-order Butterworth low-pass filter computed with those of second- and fourth-order filters.

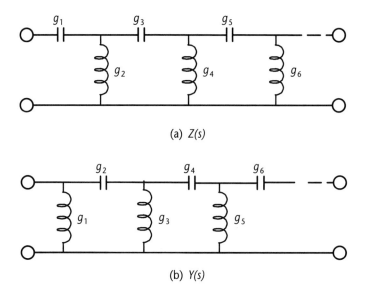

Figure 4.7 Second form of a Cauer network.

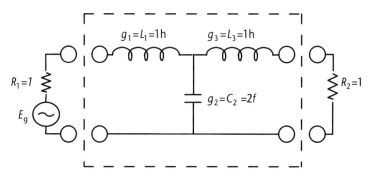

Figure 4.8 First Cauer form of a third-order Butterworth low-pass filter of equal unity terminations and 1 rad/sec cutoff frequency.

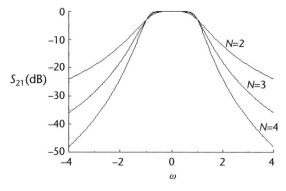

Figure 4.9 Comparison of third-order Butterworth response with second- and fourth-order responses.

Figure 4.9 shows that all three responses have the value of −3.0 dB at the cutoff frequency $\omega = \pm 1$ radian/sec. However, the slopes of the responses become steeper beyond the cutoff frequencies as N increases. Eventually the slope approaches the value of $-6N$ dB per octave or $-10N$ dB per decade.

4.2.3 General Solution for Butterworth Low-Pass Filter Response

When the terminating impedances are unequal and not necessarily equal to unity and the radian cutoff frequency is not equal to unity but ω_0 rads/sec, then the frequency response of an Nth order Butterworth filter can be shown to be of the form [8]

$$|S_{21}(j\omega)|^2 = \frac{K_0}{1+\left(\dfrac{\omega}{\omega_0}\right)^{2N}} \tag{4.51}$$

where

$$K_0 = \frac{4R_g R_L}{\left(R_g + R_L\right)^2} \tag{4.52}$$

The corresponding power reflection coefficient is obtained from (4.51) as

$$|S_{11}(j\omega)|^2 = 1 - |S_{21}(j\omega)|^2 = \frac{1 - K_0 + \left(\dfrac{\omega}{\omega_0}\right)^{2N}}{1+\left(\dfrac{\omega}{\omega_0}\right)^{2N}} \tag{4.53}$$

Using the theorem of analytic continuation and replacing $-j\omega$ by s in the above equation gives [6]

$$|S_{11}(s)||S_{11}(-s)| = \delta^{2N} \frac{1+(-1)^N x^{2N}}{1+(-1)^N y^{2N}} \tag{4.54}$$

where

$$\delta = \sqrt[2N]{1 - K_0} \tag{4.55}$$

and

$$y = \frac{s}{\omega_0} \quad \text{and} \quad x = \frac{y}{\delta\omega_0} \tag{4.56}$$

The zeros and the poles of the right-hand side of (4.54) are the solutions of the following equations:

$$(-1)^N x^{2N} + 1 = 0 \tag{4.57a}$$

$$(-1)^N y^{2N} + 1 = 0 \tag{4.57b}$$

The poles are located on a unit circle in the s-plane as

$$s_k = \omega_0 e^{j\frac{(2k+N-1)}{2N}}, \quad k = 1, 2, \cdots, 2N \tag{4.58}$$

Once the locations of the zeros and poles are known, we consider the poles on the left half of the s-plane so that the denominator of $S_{11}(s)$ is strictly a Hurwitz polynomial [7]. A Hurwitz polynomial $H(s)$ has the following properties:

1. $H(s)$ is real when s is real
2. Roots of $H(s)$ have real parts, which are zero or negative.

Although the numerator of $Z_1(s)$ is not necessarily a Hurwitz polynomial, the coefficients of the polynomial must be real. Therefore, we should consider the real zeros and complex conjugate pairs of zeros only. This gives

$$S_{11}(s) = \pm \frac{\prod_{i=1}^{N}(x - x_i)}{\prod_{i=1}^{N}(y - y_i)} = \pm \delta^N \frac{a_0 + a_1 x + \cdots + a_{N-1} x^{N-1} + x^N}{a_0 + a_1 y + \cdots + a_{N-1} y^{N-1} + x^N} \tag{4.59}$$

where

$$a_N = 1 \tag{4.60a}$$

and

$$a_k = a_{k-1} \left[\frac{\cos\left\{(k-1)\frac{\pi}{2N}\right\}}{\sin\left\{\frac{k\pi}{2N}\right\}} \right]; k = 1, 2, \cdots, N \tag{4.60b}$$

For $R_2 > R_1$, we choose the minus sign in (4.59), then

$$Z_{in}(s) = R_1 \left[\frac{1 + S_{11}(s)}{1 - S_{11}(s)} \right] \tag{4.61}$$

The ladder realization or Cauer form using continued fraction expansion [6] of the right-hand side of (4.61) yields the following set of equations for the low-pass-prototype parameters as [8]

$$g_1 = \frac{2R_1 \sin\left(\frac{\pi}{2N}\right)}{\omega_0 (1-s)} \tag{4.62a}$$

4.2 Belevitch Matrix and Transfer Function Synthesis

$$g_{2m-1}g_{2m} = \frac{4\sin(\theta_{4m-3})\sin(\theta_{4m-1})}{\omega_0^2(1-2s\cos(\theta_{4m-2})+\delta^2)} \quad (4.63b)$$

$$g_{2m+1}g_{2m} = \frac{4\sin(\theta_{4m-1})\sin(\theta_{4m+1})}{\omega_0^2(1-2s\cos(\theta_{4m})+\delta^2)} \quad (4.63c)$$

for $m = 1, 2, \ldots, \left[\dfrac{N}{2}\right]$, where $\theta_m = \dfrac{m\pi}{2N}$ and N is the largest integer less than or equal to $\dfrac{N}{2}$.

The last element is

$$g_N = \frac{2R_2\sin(\theta_1)}{\omega_0(1+\delta)}; \text{ for odd } N \quad (4.63d)$$

$$g_N = \frac{2\sin(\theta_1)}{\omega_0(1+\delta)R_2}; \text{ for even } N \quad (4.63e)$$

Ladder networks corresponding to odd and even N are shown in Figure 4.10. Note that the first element is a series inductor, for an odd N the last element is a series inductor, and for an even N the last element is a shunt capacitor.

For $R_2 > R_1$, we choose the plus sign in (4.59), then [6]

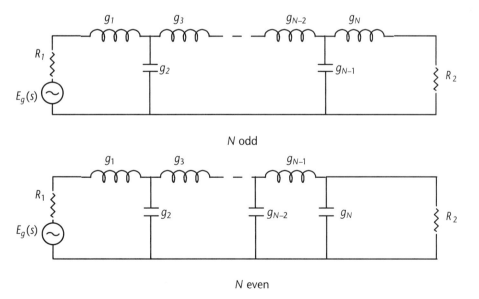

Figure 4.10 Cauer network realization of Butterworth response ($R_2 > R_1$).

$$Z_{in}(s) = \left[\frac{1 - S_{11}(s)}{1 + S_{11}(s)}\right] R_1 \quad (4.64)$$

The element values are given by (4.63) and Figure 4.11 shows the Cauer network realizations.

4.2.3.1 Required Order of a Butterworth Filter

The number of elements needed in a Butterworth filter is determined by the following three parameters:

1. $\gamma = \frac{\omega_s}{\omega'}$, the stopband to passband ratio (S = stopband frequency; ω' = cutoff frequency. $\omega' = 1$ for normalized case)
2. IL = stopband attenuation (dB) at ω_S
3. RL = passband return loss (dB)

The required order is given by [9]

$$N \geq \frac{IL + RL}{20 \log \gamma} \quad (4.65)$$

For example, if $IL = -60$ dB, $RL = -26$ dB, and $\gamma = 3$, then $N \geq 9.01$. Therefore, one should choose $N = 10$.

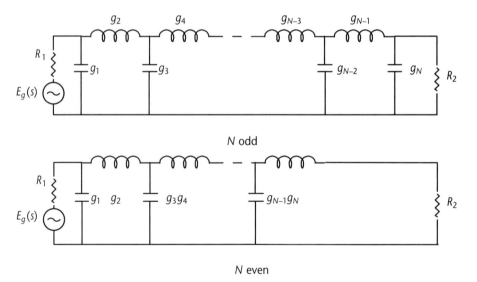

Figure 4.11 Cauer network realization for $R_2 > R_1$.

4.2.3.2 Time Delay of a Butterworth Low-Pass Filter

An important parameter of any filter is its frequency-dependent time delay. It can be shown that the time delay in seconds for the passage of a single frequency signal through a Butterworth low-pass filter of order N is given by [10]

$$T_d = \frac{\sum_{r=0}^{N-1} \frac{\omega^{2N}}{\sin\left\{(2r+1)\frac{\pi}{2N}\right\}}}{1+\omega^{2N}} \text{(sec)} \qquad (4.66)$$

For $N=2$

$$T_d = \frac{\sqrt{2}+\sqrt{2}\omega^2}{1+\omega^4} \text{(sec)} \qquad (4.67)$$

Figure 4.12 shows the time delay response plot of the frequency dependence of a second-order Butterworth filter.

4.2.4 Chebyshev Approximation

In a Butterworth low-pass filter approximation, the frequency response characteristic is closest to that of an ideal low-pass filter at $\omega = 0$. When we approach the cutoff at frequency ω_0, the approximation becomes poorer. Unlike in the Butterworth filter, if one allows some ripple in the passband, the approximation becomes better than the Butterworth case. Finally, it is best when the ripple has equal amplitudes in the passband. This is also known as equiripple passband approximation. Such equiripple approximation is realized using Chebyshev's minimax theorem [11] and the corresponding analytical function. Owing to its equiripple property, the filter passband is equally good from $\omega = 0$ to $\omega = \omega_0$. The Chebyshev polynomial of order N is defined by

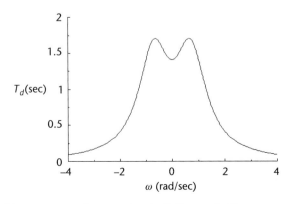

Figure 4.12 Time delay response of a second-order Butterworth filter.

$$T_N(\omega) = \cos\left[N\cos^{-1}(\omega)\right] \quad for |\omega| < 1 \tag{4.68a}$$

and

$$T_N(\omega) = \cosh\left[N\cosh^{-1}(\omega)\right] \quad for |\omega| > 1 \tag{4.68b}$$

$$C_0(\omega) = 1 \tag{4.69a}$$

$$C_1(\omega) = \omega \tag{4.69b}$$

Higher-order Chebyshev polynomials are obtained from the following recursive formula:

$$C_k(\omega) = 2\omega C_{k-1}(\omega) - C_{k-1}(\omega) \tag{4.70}$$

Using (4.70) and (4.69) gives

$$C_2(\omega) = 2\omega C_1(\omega) - C_0(\omega) = 2\omega^2 - 1 \tag{4.71}$$

The behavior of $C_3(\omega)$ and $C_4(\omega)$ are shown in Figure 4.13.

The important properties of Chebyshev polynomials are

1. The magnitude of $C_N(\omega)$ remains within ±1 within the interval $|\omega|<1$
2. For $|\omega| > 1$, $C_N(\omega)$ increases rapidly with increasing values of $|\omega|$.
3. The zeros of $C_N(\omega)$ are situated within the interval $|\omega| < 1$.

Considering the above characteristics of a Chebyshev polynomial, the transfer function of a Chebyshev low-pass filter is defined as

$$|S_{12}(\omega)|^2 = \frac{1}{1+\varepsilon^2 C_N^2(\omega)} \tag{4.72}$$

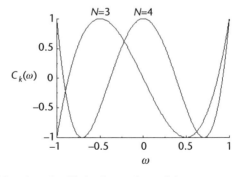

Figure 4.13 Third- and fourth-order Chebeshev polynomials.

Figure 4.14 shows the frequency response of a fourth-order Chebyshev low-pass filter. In accordance with the properties of a Chebyshev polynomial, the response oscillates between the maximum 1 and $1/(1+\varepsilon^2)$. Beyond the cutoff frequency, which in the present case is 1, the response approaches zero very rapidly. The distance between maximum and minimum in the pass band is the

$$\text{Passband ripple} = 1 - \frac{1}{1+\varepsilon^2} \tag{4.73}$$

At cutoff, for any value of N

$$|S_{12}(\omega)| = \frac{1}{\sqrt{1+\varepsilon^2}} \tag{4.74}$$

Sufficiently above cutoff $|\omega| \gg 1$

$$|S_{12}(\omega)| \cong \frac{1}{\varepsilon C_N(\omega)} \tag{4.75}$$

The attenuation in decibels

$$\alpha(dB) \cong 20\log(\varepsilon) + 20\log\left[C_N(\omega)\right] \tag{4.76}$$

For large ω, $C_N(\omega)$ can be approximated by $2^{N-1}\omega^N$. Consequently, the attenuation becomes

$$\alpha(dB) = 20\log(\varepsilon) + 20N\log(\omega) + 6(N-1) \tag{4.77}$$

Therefore, beyond the passband, the Chebyshev response falls off at the rate of 20N dB/decade after an initial fall of $6(N-1)+20\log(\varepsilon)$ dB. Usually ε has a very low value (less than unity). As a result, the term $20\log(\varepsilon)$ is negative. Therefore, the filter

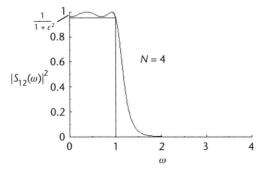

Figure 4.14 Fourth-order Chebyshev approximation of a low-pass filter.

designer should use a sufficiently large value of N in order to compensate for the decrease in loss due to ε.

4.2.4.1 Required Order of a Chebyshev Filter

Unlike the Butterworth response, the Chebyshev response depends on two factors, ε and N. The required order of a Chebyshev filter is given by [9]

$$N \geq \frac{IL + RL + 6}{20\log(\gamma + \sqrt{\gamma^2 - 1})} \tag{4.78}$$

The various parameters in (4.78) are described in Figure 4.15.

The procedure can be illustrated by the following example. Let $IL = 40$ dB, $RL = 20$ dB, and $\gamma = 3.5$, as shown in Figure 4.15. Then (4.78) gives $N = 3.9476$. Therefore, one should choose $N = 4$. The frequency response of a general Chebyshev filter can be written as [8].

$$\left|S_{21}\left(\frac{\omega}{\omega_0}\right)\right|^2 = \frac{K_0}{1 + \varepsilon^2 C_N^2\left(\frac{\omega}{\omega_0}\right)} \tag{4.79}$$

where

$$K_0 = \left|S_{21}(0)\right|^2 = \frac{4R_1R_2}{(R_1 + R_2)^2} \tag{4.80a}$$

for N odd, and

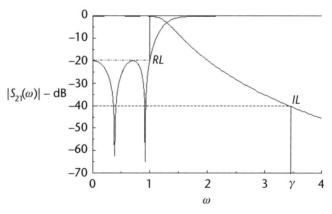

Figure 4.15 Pertinent parameters of a Chebyshev low-pass filter.

4.2 Belevitch Matrix and Transfer Function Synthesis

$$K_0 = (1+\varepsilon^2)|S_{21}(0)|^2 = (1+\varepsilon^2)\frac{4R_1R_2}{(R_1+R_2)^2} \qquad (4.80\text{b})$$

for N even.
Now,

$$\left|S_{11}\left(\frac{\omega}{\omega_0}\right)\right|^2 = 1 - \left|S_{21}\left(\frac{\omega}{\omega_0}\right)\right|^2 = \frac{1 - K_0 + \varepsilon^2 C_N^2\left(\frac{\omega}{\omega_0}\right)}{1 + \varepsilon^2 C_N^2\left(\frac{\omega}{\omega_0}\right)} \qquad (4.81)$$

Replacing ω/ω_0 by ϖ in the above equations gives

$$\left|S_{11}(\varpi)\right|^2 = 1 - \left|S_{21}(\varpi)\right|^2 = \frac{1 - K_0 + \varepsilon^2 C_N^2(\varpi)}{1 + \varepsilon^2 C_N^2(\varpi)} \qquad (4.82)$$

$$S_{11}(j\varpi)S_{11}(-j\varpi) = \frac{1 - K_0 + \varepsilon^2 C_N^2(\varpi)}{1 + \varepsilon^2 C_N^2(\varpi)} \qquad (4.83)$$

Replacing ϖ by $-js$ gives

$$S_{11}(s)S_{11}(-s) = (1 - K_0)\frac{1 + \tilde{\varepsilon}^2 C_N^2(-js)}{1 + \varepsilon^2 C_N^2(-js)} \qquad (4.84)$$

where

$$\tilde{\varepsilon} = \frac{\varepsilon}{\sqrt{1 - K_0}} \qquad (4.85)$$

The poles of the right-hand side of (4.84) are given by

$$s_r = -\sinh(\tilde{\eta})\sin\left[\frac{2r-1}{2N}\pi\right] + j\cosh(\tilde{\eta})\cos\left[\frac{2r-1}{2N}\pi\right] \qquad (4.86\text{a})$$

where

$$\tilde{\eta} = \frac{1}{N}\sinh^{-1}\left(\frac{1}{\varepsilon}\right) \qquad (4.86\text{b})$$

while the zeros $\tilde{s}_r$ are obtained by using the same equations but $\tilde{\eta}$ is replaced by

$$\tilde{\tilde{\eta}} = \frac{1}{N}\sinh^{-1}\left(\frac{1}{\tilde{\varepsilon}}\right) \tag{4.86c}$$

From the above equations we conclude that the poles of a Chebyshev transfer function lie on an ellipse whose center is at the origin with the semimajor and semiminor axes $\cosh(\tilde{\eta})$ and $\sinh(\tilde{\eta})$, respectively. Similarly, the zeros are located on an ellipse whose center is at the origin and whose semimajor and semiminor axes are $\cosh(\tilde{\tilde{\eta}})$ and $\sinh(\tilde{\tilde{\eta}})$, respectively. Figure 4.16 shows the locus of the poles of a Chebyshev filter. The corresponding locus of zeros is obtained by replacing $\tilde{\eta}$ by $\tilde{\tilde{\eta}}$ in Figure 4.16.

Once the locations of the poles and the zeros in the complex frequency plane are known, one can form the reflection coefficient function $S_{11}(s)$ using the same line of arguments as was done in case of the Butterworth filter. That is, we consider the poles on the left-hand side of the s-plane and complex conjugate poles on the $j\omega$ axis. Also, the zeros are chosen either in the left-half s-plane or the right-half s-plane so that all the coefficients of the numerator of the reflection coefficient function are real. This gives us

$$S_{11}(s) = \pm\frac{\prod_{r=1}^{N}(s-\tilde{s}_r)}{\prod_{r}^{N}(s-s_r)} \tag{4.87}$$

For $R_2 > R_1$, we choose the minus sign in (4.59), then

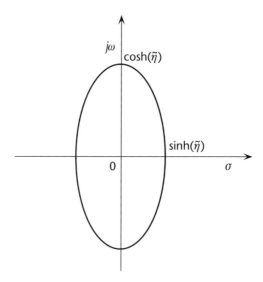

Figure 4.16 Locus of poles of a Chebyshev low-pass filter.

4.2 Belevitch Matrix and Transfer Function Synthesis

$$Z_{in}(s) = R_1 \left[\frac{1 + S_{11}(s)}{1 - S_{11}(s)} \right] \quad (4.88)$$

The ladder realization or Cauer form of realization using continued fraction expansion of the right-hand side of (4.88) yields the following set of equations for the low-pass-prototype parameters [8].

For $R_2 > R_1$, we choose the minus sign in front of the right-hand side of (4.87). Upon expanding, the polynomial and subsequent continued fraction expansion gives

$$g_1 = \frac{2R_1 \sin \theta_1}{\omega_0 \left[\sinh(\eta) - \sinh(\tilde{\eta}) \right]} \quad (4.89a)$$

where

$$\theta_r = \frac{r\pi}{2N} \quad (4.89b)$$

$$g_{2r-1} g_{2r} = \frac{4 \sin(\theta_{4r-3}) \sin(\theta_{4r-1})}{\omega_0^2 \Theta_{2r}} \quad (4.89c)$$

$$g_{2r+1} g_{2r} = \frac{4 \sin(\theta_{4r-1}) \sin(\theta_{4r+1})}{\omega_0^2 \Theta_{2r}} \quad (4.89d)$$

for $r = 1, 2, \ldots, \left[N/2 \right]$, and

$$\Theta_{2r} = \sinh^2(\eta) + \sinh^2(\tilde{\eta}) + \sin^2(\theta_{2r}) - 2 \sinh(\eta) \sinh(\tilde{\eta}) \cos(\theta_{2r}) \quad (4.89e)$$

and for odd N

$$g_N = \frac{2R_2 \sin \theta_1}{\omega_0 \left[\sinh(\eta) + \sinh(\tilde{\eta}) \right]} \quad (4.89f)$$

for even N

$$g_N = \frac{2 \sin \theta_1}{\omega_0 \left[\sinh(\eta) + \sinh(\tilde{\eta}) \right] R_2} \quad (4.89g)$$

The network topologies corresponding to $R_2 > R_1$ and $R_1 > R_2$ are shown in Figures 4.10 and 4.11, respectively.

4.2.4.2 Time Delay of a Chebyshev Filter

It can be shown that the time delay in seconds for the passage of a single frequency signal through a Chebyshev low-pass filter of order N is given by [10]

$$T_d = \frac{\varepsilon^2 \sum_{r=0}^{N-1} U_{2r}(\varpi) \sinh(2N-2r-1)\eta}{\sin(2r+1)\theta_1} \Big/ \left(1+\varepsilon^2 T_N^2(\varpi)\right) \quad (4.90a)$$

where

$$U_r(\varpi) = \frac{\sin(r+1)\phi}{\sin\phi} \quad (4.90b)$$

and

$$\phi = \cos^{-1}(\varpi) \quad (4.90c)$$

Figure 4.17 shows a plot of the frequency dependence of the group delay for a third-order Chebyshev low-pass filter.

4.2.5 Elliptic Function Approximation

Butterworth and Chebyshev approximations are so-called all pole approximations because the transmission zeros of such functions are at infinite frequency. Unlike such filters, Cauer parameter filters or Elliptic function filters have their transmission

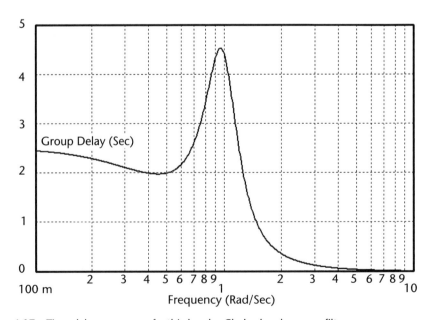

Figure 4.17 Time delay response of a third-order Chebyshev low-pass filter.

zeros at finite frequencies. Consequently, unlike the Chebyshev approximation, it has a ripple not only in the passband but also in the stopband. The advantage of an elliptic filter is that it offers a much steeper skirt selectivity, immediately beyond the cutoff frequency, than a Chebyshev filter of the same order. However, such an advantage is achieved at the cost of a more complex filter configuration than that of a Chebyshev filter. For example, a Butterworth or Chebyshev filter can be realized using nonresonant elements and a simple Cauer form of ladder network using simple inductors and capacitors, as shown in the preceding sections. But, an elliptic filter needs either shunt or series resonant elements, series or shunt elements, or cross-coupled elements between nonadjacent nodes in the ladder. In what follows, we will first discuss the basic theory of an elliptic low-pass filter approximation and then the lumped element synthesis technique.

4.2.5.1 Theory of Elliptic Filter Approximation

The transfer function of an elliptic low-pass filter is given by (see Figure 4.2)

$$|S_{21}(\omega)|^2 = \frac{K_0}{1+\varepsilon^2 \Psi_N^2 \left(\frac{\omega}{\omega_0}\right)} \tag{4.91}$$

Without loss of generality, we may assume K_0 and ω_0 to be unity. Hence the normalized transfer function of an elliptic low-pass filter can be written as

$$|S_{21}(\omega)|^2 = \frac{1}{1+\varepsilon^2 \Psi_N^2(\omega)} \tag{4.92a}$$

The properties of function $\Psi_N(\omega)$ and the synthesis of an elliptic function low-pass filter have been studied by various researchers using different techniques [9,10,12,13]. However, the best ones are by Baher [14] and Zhu and Chen [8]. The main properties of $\Psi_N(\omega)$ are

1. $\Psi_N(\omega)$ has N zeros within $|\omega| < 1$
2. $\Psi_N(\omega)$ oscillates between ± 1 in the interval $|\omega| < 1$
3. At cutoff $\Psi_N(\omega) = 1$
4. $\Psi_N(\omega)$ oscillates between $\pm 1/k_1$ within $|\omega| > 1/k$

where

$$k = \frac{\omega_0}{\omega_s} \tag{4.92b}$$

and the discrimination parameter

$$k_1 = \sqrt{\frac{10^{\alpha_{max}/10} - 1}{10^{\alpha_{min}/10} - 1}} \qquad (4.92c)$$

Figure 4.18 shows the graphical representation of (4.91).

The above properties of $\Psi_N(\omega)$ forces it to satisfy the following first-order ordinary differential equation [8,14]:

$$\frac{d\Psi_N(\omega)}{\sqrt{[1-\Psi_N^2(\omega)][1/k_1^2 - \Psi_N^2(\omega)]}} = \frac{C d\omega}{\sqrt{[1-\omega^2][1/k^2 - \omega^2]}} \qquad (4.93a)$$

with

$$\left.\frac{d\Psi_N(\omega)}{d\omega}\right|_{\Psi_N(\omega)=1, |\omega| \neq 1} = 0 \qquad (4.93b)$$

and

$$\left.\frac{d\Psi_N(\omega)}{d\omega}\right|_{\Psi_N(\omega)=k_1, |\omega| \neq \omega_s} \qquad (4.93c)$$

where C is a constant. The forms of the function $\Psi_N(\omega)$ that satisfy the above conditions are given by, for N-odd,

$$\Psi_N(\omega) = h_0 \frac{\omega(\omega^2 - \omega_1^2)\cdots(\omega^2 - \omega_r^2)\cdots(\omega^2 - \omega_q^2)}{(1 - k^2\omega_1^2\omega^2)\cdots(1 - k^2\omega_r^2\omega^2)\cdots(1 - k^2\omega_q^2\omega^2)} \qquad (4.94)$$

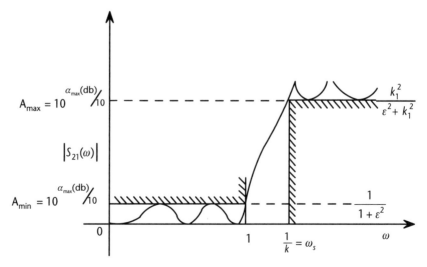

Figure 4.18 Frequency response of a fifth-order elliptic filter; $\omega_0 = 1$, $\omega_s = 1/k$.

4.2 Belevitch Matrix and Transfer Function Synthesis

where various $\omega'_r s$ are defined by

$$\omega_r = \omega_0 sn\left\{\frac{2rK(k)}{N}, k\right\}, \quad r = 1, 2, \cdots \frac{1}{2}(N-1) \tag{4.95}$$

and for N-even

$$\Psi_N(\omega) = h_e \frac{(\omega^2 - \omega_1^2)\cdots(\omega^2 - \omega_r^2)\cdots(\omega^2 - \omega_q^2)}{(1 - k^2\omega_1^2\omega^2)\cdots(1 - k^2\omega_r^2\omega^2)\cdots(1 - k^2\omega_q^2\omega^2)} \tag{4.96}$$

where ω_r are given by

$$\omega_r = \omega_0 sn\left\{\frac{(2r-1)K(k)}{N}, k\right\} \tag{4.97}$$

h_o and h_e are chosen such a way that the condition $\Psi_N(1) = 1$ is satisfied, which gives

$$h_o = \frac{\prod_{r=1}^{\frac{1}{2}(N-1)}(1 - k^2\omega_r^2)}{\prod_{r=1}^{\frac{1}{2}(N-1)}(1 - \omega_r^2)} \tag{4.98}$$

and

$$h_e = \frac{\prod_{r=1}^{\frac{1}{2}(N-1)}(1 - k^2\omega_r^2)}{\prod_{r=1}^{\frac{1}{2}(N-1)}(1 - \omega_r^2)} \tag{4.99}$$

Equations (4.95) and (4.97) involve the so-called elliptic integral function $K(k)$ and the Jacobian sine elliptic function $sn(u, k)$. The functions are defined as [15]

$$K(k) = \int_0^{\frac{\pi}{2}} \frac{dx}{\sqrt{1 - k^2 \sin^2 x}} \tag{4.100}$$

and

$$sn(u, k) = \frac{2\pi\sqrt{v}}{kK(k)} \sum_{r=1}^{\infty} \frac{v^{r-1}}{1 - v^{2r-1}} \sin\left\{\frac{(2r-1)u}{K(k)}\frac{\pi}{2}\right\} \tag{4.101}$$

where

$$v = e^{-\frac{\pi K(k')}{2K(k)}} \quad (4.102)$$

with

$$k' = \sqrt{1-k^2} \quad (4.103)$$

For $0 < k \leq 0.173$

$$\frac{K(k')}{K(k)} = \frac{4}{\pi}\ln\left(\frac{\sqrt{2}}{k}\right) \quad (4.104a)$$

and, for $0.173 < k \leq 1$

$$\frac{K(k')}{K(k)} = \frac{\pi}{\ln(2) + 2\tan^{-1}\left(\sqrt{k}\right)} \quad (4.104b)$$

4.2.5.2 Order of an Elliptic Filter

The required order of an elliptic filter is given by [8]

$$N \geq \frac{\frac{K(k)}{K(k')}}{\frac{K(k_1)}{K(k_1')}} \quad (4.105)$$

where the parameters k and k_1 have been defined by (4.92b) and (4.92c), respectively. Consider the following example. We are required to design an elliptic low-pass filter with the following specifications:

1. $\frac{\omega_s}{\omega_0} = 2$
2. Passband ripple $\varepsilon = -0.01$ dB
3. Stopband isolation at $\omega = \omega_s$ is -50 dB

Using (4.92a) and (4.105) gives the filter order $N = 5$. Using (4.95), we obtain

$$\omega_1 = sn\left(\frac{2K}{N}, k\right) = 0.6152 \text{ rad/sec} \quad (4.106a)$$

4.2 Belevitch Matrix and Transfer Function Synthesis

$$\omega_2 = sn\left(\frac{4K}{N}, k\right) = 0.9573 \text{ rad/sec} \tag{4.106b}$$

and

$$\varepsilon = \sqrt{10^{\frac{0.01}{10}} - 1} = 0.048 \tag{4.107}$$

Therefore, using the above parameters in (4.91) gives the normalized transducer power gain

$$|S_{21}(\omega)|^2 = \frac{1}{1 + \varepsilon^2 \left(\dfrac{k^5}{\aleph}\right) \dfrac{\omega^2 \left(\omega^2 - \omega_1^2\right)^2 \left(\omega^2 - \omega_2^2\right)^2}{\left(1 - k^2\omega_1^2\omega^2\right)^2 \left(1 - k^2\omega_2^2\omega^2\right)^2}} \tag{4.108}$$

where

$$\aleph = k^5 \left[\frac{\left(1 - \omega_1^2\right)^2 \left(1 - \omega_2^2\right)^2}{\left(1 - k^2\omega_1^2\right)^2 \left(1 - k^2\omega_2^2\right)^2}\right]^2 \tag{4.109}$$

Using $\omega = -js$ in (4.108) and (109) and using the relationship

$$S_{11}(s)S_{11}(-s) = 1 - S_{21}(s)S_{21}(-s) \tag{4.110}$$

gives [8]

$$S_{11}(s)S_{11}(-s) = \frac{-\varepsilon^2 \dfrac{k^2}{\aleph} s^2 \left(s^2 + \omega_1^2\right)^2 \left(s^2 + \omega_2^2\right)^2}{\left(1 + k^2\omega_1^2 s^2\right)^2 \left(1 + k^2\omega_2^2 s^2\right)^2 - \varepsilon^2 \dfrac{k^2}{\aleph} s^2 \left(s^2 + \omega_1^2\right)^2 \left(s^2 + \omega_2^2\right)^2} \tag{4.111}$$

From the above equation, the poles and zeros of $S_{11}(s)$ which lead to a realizable input impedance of the filter are (obtained by using MATLAB) ([16])

$$\begin{cases} p_1 = -0.921 \\ p_2 = -0.6402 \pm j0.8527 \\ p_3 = -0.1960 \pm j1.2002 \end{cases} \tag{4.112a}$$

and

$$\begin{cases} z_1 = 0.00 \\ z_2 = \pm j0.6152 \\ z_3 = \mp j0.9610 \end{cases} \quad (4.112b)$$

respectively. Using the above results, the input reflection coefficient is obtained as

$$S_{11}(s) = \frac{d_4 s^4 + d_2 s^2 + d_0}{e_5 s^5 + e_4 s^4 + e_3 s^3 + e_2 s^2 + e_1 s + e_0} \quad (4.113)$$

where d_i ($i = 4, 2, 0$) and e_i ($i = 5,4,3.2,1,0$) numerical constants.

The corresponding input impedance is obtained by using

$$Z_{in}(s) = \frac{1 + S_{11}(s)}{1 - S_{11}(s)} \quad (4.114)$$

Once $Z_{in}(s)$ is known in terms of s, the corresponding network can be synthesized using midshunt or midseries form. Darlington's [17] synthesis procedure based on continued fraction expansion and Lin and Tocad's [18] algorithm can be used to obtain the element values (see Appendix 4A for Darlington's method). Figure 4.19 shows the synthesized filter in two different forms. The filters are synthesized using the above procedure and a Fortran code [19]. The corresponding frequency responses are shown in Figures 4.20.

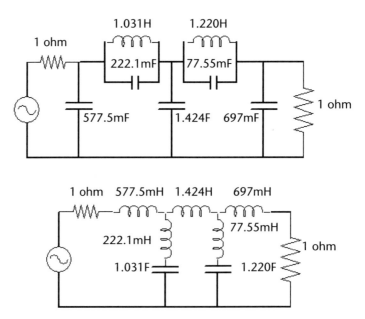

Figure 4.19 Synthesized fifth-order elliptic low-pass filter.

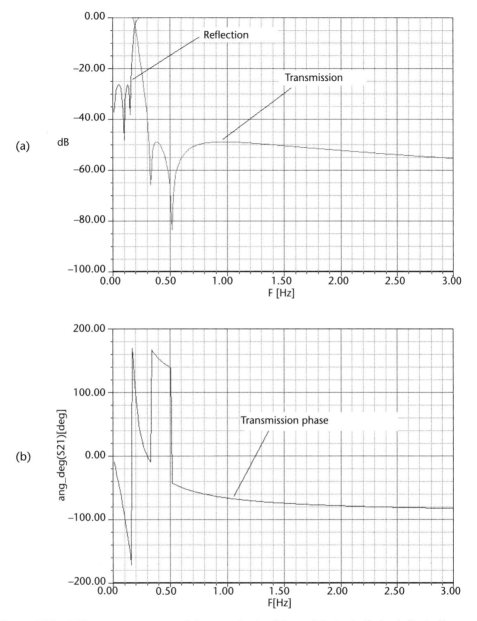

Figure 4.20 (a) Frequency response of the magnitude of S_{11} and S_{21} in decibels of elliptic filter and (b) frequency response of the phase of S_{12} in decibels of elliptic filter analyzed by Ansoft Designer (Ansoft Corporation, Bethelehem, PA, 2005).

4.2.6 Generalized Chebyshev Low-Pass Filters

In a generalized Chebyshev low-pass filter [20], the designer has more degrees of freedom than what is available in a Chebyshev or elliptic function filter. The designer has separate control over the passband ripple and the location of the attenuation poles in the stopband and skirt selectivity. These flexibilities allow a designer to realize many special properties like flat group delay in the passband and more compact cross-coupled filters. Let us consider (4.92a) in order to understand the theory behind the operation of a generalized Chebyshev low-pass filter.

$$|S_{21}(\omega)|^2 = \frac{1}{1+\varepsilon^2 \Psi_N^2(\omega)} \quad (4.115)$$

Also, as before since the conservation of power holds,

$$|S_{11}(\omega)|^2 + |S_{21}(\omega)|^2 = 1 \quad (4.116)$$

where the function $\Psi_N(\omega)$ is defined as

$$\Psi_N(\omega) = \cosh\left\{\sum_{n=1}^{N} \cosh^{-1}\left(\frac{\omega - \frac{1}{\omega_n}}{1 - \frac{\omega}{\omega_n}}\right)\right\} \quad (4.117)$$

where $\omega_n = -js_n$ is the position of the nth attenuation pole in the stopband in the complex s-plane. It can be very easily verified that (4.117) satisfies all the conditions, besides others, for a Chebyshev filter and it degenerates into an actual Chebyshev filter when all the prescribed attenuation poles approach infinity. Determination of the transducer power gain as a rational function of ω is as follows.

In (4.119), we write

$$x_n = \frac{\omega - \frac{1}{\omega_n}}{1 - \frac{\omega}{\omega_n}} \quad (4.118)$$

and use the identity

$$\cosh^{-1}(x_n) = \ln\left\{x_n + \sqrt{x_n^2 - 1}\right\} = \ln(a_n + b_n) \quad (4.119)$$

where

$$a_n = x_n \quad (4.120)$$

$$b_n = \sqrt{x_n^2 - 1} \quad (4.121)$$

Using the above equations gives

$$\Psi_N(\omega) = \frac{\prod_{n=1}^{N}\left\{\left(\omega - \frac{1}{\omega_n}\right) + \sqrt{\left(1 - \frac{1}{\omega_n^2}\right)}\omega'\right\} + \prod_{n=1}^{N}\left\{\left(\omega - \frac{1}{\omega_n}\right) - \sqrt{\left(1 - \frac{1}{\omega_n^2}\right)}\omega'\right\}}{2\prod_{n=1}^{N}\left(1 - \frac{\omega}{\omega_n}\right)} \quad (4.122)$$

4.2 Belevitch Matrix and Transfer Function Synthesis

$$\omega' = \sqrt{\omega^2 - 1} \tag{4.123}$$

We can write

$$\Psi_N(\omega) = \frac{E_N(\omega) + E'_N}{2H_N(\omega)} \tag{4.124}$$

where

$$E_N(\omega) = \prod_{n=1}^{N} \left\{ \left(\omega - \frac{1}{\omega_n}\right) + \sqrt{\left(1 - \frac{1}{\omega_n^2}\right)} \omega' \right\} \tag{4.125a}$$

$$E'_N(\omega) = \prod_{n=1}^{N} \left\{ \left(\omega - \frac{1}{\omega_n}\right) - \sqrt{\left(1 - \frac{1}{\omega_n^2}\right)} \omega' \right\} \tag{4.125b}$$

and

$$H_N(\omega) = \prod_{n=1}^{N} \left(1 - \frac{\omega}{\omega_N}\right) \tag{4.125c}$$

Using the above functions, $\Psi_N(\omega)$ can be determined using a recursive procedure as follows:

$$E_k(\omega) = E_{k-1} \left\{ \left(\omega - \frac{1}{\omega_k}\right) + \omega'\sqrt{1 - \frac{1}{\omega_k^2}} \right\} \tag{4.126a}$$

$$E'_k(\omega) = E'_{k-1} \left\{ \left(\omega - \frac{1}{\omega_k}\right) - \omega'\sqrt{1 - \frac{1}{\omega_k^2}} \right\} \tag{4.126b}$$

with

$$E_1(\omega) = \left(\omega - \frac{1}{\omega_1}\right) + \omega'\sqrt{1 - \frac{1}{\omega_1^2}} \tag{4.127a}$$

$$E'_1(\omega) = \left(\omega - \frac{1}{\omega_1}\right) - \omega'\sqrt{1 - \frac{1}{\omega_1^2}} \tag{4.127b}$$

Following the above recursive procedure, it can be shown that the numerator of $\Psi_N(\omega)$ is equal to [21]

$$Num[\Psi_N(\omega)] = E_N(\omega) + E'_N(\omega) = \frac{1}{2}\left[\{U_N(\omega) + V_N(\omega)\} + \{U'_N(\omega) + V'_N(\omega)\}\right] \quad (4.128)$$

Also

$$U_N(\omega) = U'_N(\omega) \quad (4.129a)$$

$$V_N(\omega) = -V'_N(\omega) \quad (4.129b)$$

Using the above relationships we obtain

$$|S_{21}(\omega)|^2 = \frac{1}{1+\varepsilon^2\Psi_N^2(\omega)} = \frac{H_N^2(\omega)}{H_N^2(\omega) + \varepsilon^2 V_N^2(\omega)} \quad (4.130)$$

$$|S_{11}(\omega)|^2 = 1 - |S_{21}(\omega)|^2 = \frac{\varepsilon^2 V_N^2(\omega)}{H_N^2(\omega) + \varepsilon^2 V_N^2(\omega)} \quad (4.131)$$

Using the above procedure, $S_{11}(s)$ can be expressed in the form given by the right-hand side of (4.113). For example [21], if ω_1 and ω_2 are 1.3217 rad/sec and 1.8082 rad/sec, respectively, and the desired passband return loss is −22 dB ($\varepsilon = 0.00635$, using $\varepsilon = \sqrt{10^{RL/10} - 1}$), then

$$\Psi_N(\omega) = \frac{6.0637\omega^4 - 4.6032\omega^3 - 4.7717\omega^2 + 3.2936\omega + 0.1264}{0.4183\omega^2 - 1.3096\omega + 1} \quad (4.132)$$

Once $\Psi_N(\omega)$ is obtained, the rest of the procedure is the same as that followed for Butterworth, Chebyshev, and Elliptic filter syntheses. Figure 4.21 shows the computed frequency response of the filter.

Since the generalized low-pass filter is basically an extracted pole filter where it has transmission zeros at real frequencies like in an elliptic filter, it has the same topology as the elliptic filter shown in Figure 4.19. The appendix summarizes the network synthesis steps involved in realizing a low-pass prototype from the transfer function in the frequency domain.

4.3 Concept of Impedance Inverter

Realization of ladder networks like those shown in Figure 4.10 is in most cases impractical in microwave circuits. An impedance inverter is a versatile element that overcomes this problem. Ideally, it is a lossless reciprocal, frequency-independent two-port passive network having the ABCD matrix

4.3 Concept of Impedance Inverter

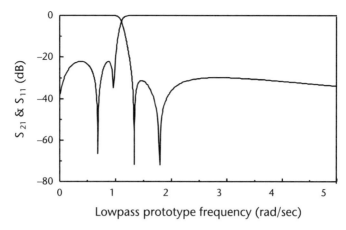

Figure 4.21 Frequency response of generalized Chebyshev low-pass filter ($\omega_1 = 1.3217$, $\omega_2 = 1.8082$ rad/sec, and $RL = -22$ dB).

$$\begin{bmatrix} A & B \\ C & D \end{bmatrix} = \begin{bmatrix} 0 & jK \\ \dfrac{j}{K} & 0 \end{bmatrix} \quad (4.133)$$

Figure 4.22(a) shows a schematic representation of an impedance inverter. Let us terminate the impedance inverter by a load Z_L, as shown in Figure 4.22(b). The input impedance of the terminated inverter as a two-port network is given by

$$Z_{in}(s) = \frac{A + \dfrac{B}{Z_L}}{C + \dfrac{D}{Z_L}} = \frac{K^2}{Z_L} \quad (4.134)$$

Equation (4.134) shows that the input impedance of an impedance inverter is the reciprocal of the terminating impedance multiplied by the square of K. Let us consider the third-order Butterworth low-pass filter synthesis example in Section 4.2.1. The desired driving point impedance to be synthesized is

$$Z_1(s) = \frac{1 + S_{11}(s)}{1 - S_{11}(s)} = \frac{1 + 2s + 2s^2 + 2s^3}{1 + 2s + 2s^2} \quad (4.135)$$

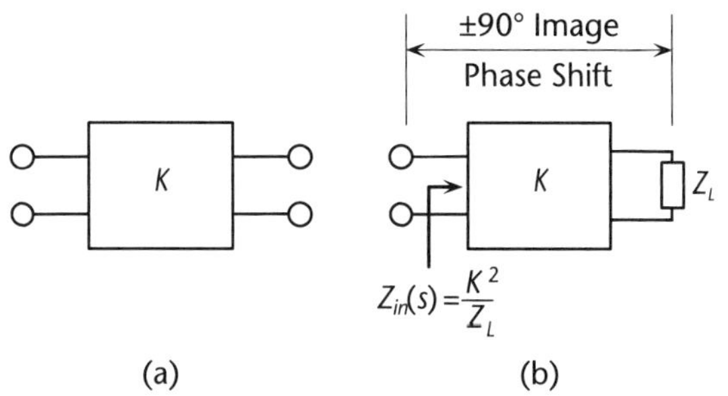

Figure 4.22 (a) Impedance inverter and (b) impedance inverter terminated by a load Z_L.

Extracting a 1h series inductor gives

$$Z_2(s) = Z_1(s) - s = \frac{1+2s+2s^2+2s^3}{1+2s+2s^2} - s = \frac{1+s}{1+2s+2s^2} \quad (4.136)$$

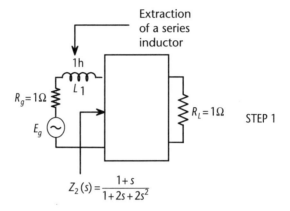

STEP 1

Extraction of an impedance inverter $K = 1$ gives

$$Z_3(s) = \frac{1}{Z_2(s)} = \frac{1+2s+2s^2}{1+s} \quad (4.137)$$

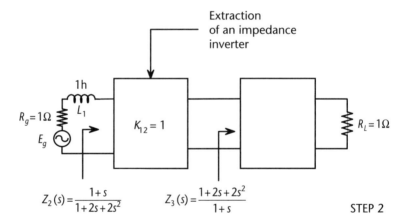

STEP 2

Extraction of a 2h series inductor gives

$$Z_4(s) = Z_3(s) - 2s = \frac{1+2s+2s^2}{1+s} - 2s = \frac{1}{1+s} \quad (4.138)$$

4.3 Concept of Impedance Inverter

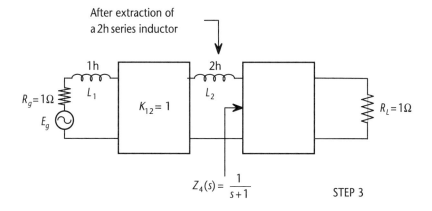

STEP 3

Extraction of a second impedance inverter of $K_{23} = 1$ gives

$$Z_5(s) = \frac{1}{Z_4(s)} = s+1 \qquad (4.139)$$

$Z_1(s) = \frac{1+2s+2s^2+2s^3}{1+2s+2s^2}$ $Z_3(s) = \frac{1+2s+2s^2}{1+s}$ $Z_4(s) = \frac{1}{s+1}$ $Z_5(s) = 1+s$

$Z_2(s) = \frac{1+s}{1+2s+2s^2}$

STEP 4

The final network after extraction
of 3 inductor and 2 impedance inverters

General formulas for an Nth-order inverter-coupled Butterworth filter are [9] (see Figure 4.2 and (4.51))

1. Matched case ($R_L = R_g = 1$, $K_0 = 1$)

$$g_r = 2\sin\left\{\frac{2r-1}{2N}\right\} \qquad (4.140)$$

$$\begin{array}{c} K_{r,r+1} = 1 \\ (r = 1, 2, \cdots, N) \end{array} \qquad (4.141)$$

2. Minimum phase $S_{11}(s)$

$$L_r \text{ or } C_r = \frac{2\sin\left\{(2r-1)\dfrac{\pi}{2N}\right\}}{1-\alpha}, (r = 1, 2, \cdots, N) \quad (4.142)$$

$$K_{r,r+1} = J_{r,r=1} = \frac{\sqrt{1 - 2\alpha\cos\left(\dfrac{r\pi}{N}\right) + \alpha^2}}{1-\alpha}, \quad (r = 1, \cdots, N-1) \quad (4.143)$$

$$R_L = \frac{1+\alpha}{1-\alpha} \quad (4.144)$$

$$\alpha = \sqrt[2N]{1-K_0} \quad (4.145)$$

3. Singly terminated case or zero source impedance ($R_g = 0$, $R_L = 1$)

$$L_r \text{ or } C_r = \sin\left\{\frac{(2r-1)\pi}{2N}\right\}, (r = 1, 2, \ldots, N) \quad (4.146)$$

$$K_{r,r+1} \text{ or } J_{r,r=1} = \sin\left(\frac{r\pi}{2N}\right), (r = 1, 2, \ldots, (N-1)) \quad (4.147)$$

One can start synthesizing a ladder network from the input admittance $Y_1(s)$ instead of the input impedance $Z_1(s)$ as above. In that case the series inductances become shunt capacitances, as shown in Figure 4.23. The values of the capacitances are obtained from (4.140), by replacing g by C and the values of the J-inverters are obtained from (4.141) by replacing K by J.

The corresponding equations for the Chebyshev filter are [9]

1. Matched case ($K_0 = 1$, $R_L = R_g = 1$)

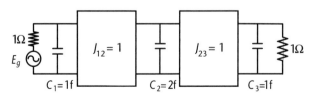

Figure 4.23 Admittance inverter-coupled Butterworth low-pass filter.

4.3 Concept of Impedance Inverter

$$K_{r,r+1} \text{ or } J_{r,r+1} = \frac{\sqrt{\eta^2 + \sin^2\left\{\dfrac{r\pi}{N}\right\}}}{\eta}, \quad (r = 1, 2, \cdots, N-1) \tag{4.148}$$

$$L_r \text{ or } C_r = \frac{2}{\eta} \sin\left\{\frac{(2r-1)\pi}{2N}\right\}, \quad (r = 1, 2, \cdots, N) \tag{4.149}$$

$$\eta = \sinh\left\{\frac{1}{N}\sinh^{-1}\left(\frac{1}{\varepsilon}\right)\right\} \tag{4.150}$$

2. Minimum phase $S_{11}(s)$

$$L_r \text{ or } C_r = \frac{2\sin\left\{(2r-1)\dfrac{\pi}{2N}\right\}}{\eta - \xi}, \quad (r = 1, 2, \cdots, N) \tag{4.151}$$

$$K_{r,r+1} \text{ or } J_{r,r+1} = \frac{\sqrt{\xi^2 + \eta^2 - 2\eta\xi\cos\left(\dfrac{r\pi}{N}\right) + \sin^2\left(\dfrac{r\pi}{N}\right)}}{\eta - \xi}, \quad (r = 1, 2, \cdots, N-1) \tag{4.152}$$

$$R_L = \frac{\eta + \xi}{\eta - \xi} \tag{4.153}$$

$$\xi = \sinh\left\{\frac{1}{N}\sinh^{-1}\left(\frac{\sqrt{1 - K_0}}{\varepsilon}\right)\right\} \tag{4.154}$$

3. Singly terminated case ($R_L = 1\Omega$, $S_{12} = Z_{12}$)

$$L_r \text{ or } C_r = \frac{2\sin\left\{\dfrac{(2r-1)\pi}{2N}\right\}}{\eta}, \quad (r = 1, 2, \cdots, N) \tag{4.155}$$

$$K_{r,r+1} \text{ or } J_{r,r+1} = \sin\left(\frac{r\pi}{2N}\right)\frac{\sqrt{\eta^2 + \cos^2\left(\dfrac{r\pi}{N}\right)}}{\eta}, \quad (r = 1, 2, \cdots, N-1) \tag{4.156}$$

The impedance inverter coupled elliptic low-pass filter configuration is shown in Figure 4.24 [9].

The impedance inverter coupled prototype parameters are [9]

1. Minimum phase $S_{11}(s)$
2.

$$C_r = \frac{(\xi-\eta)ds\left\{(2r-1)\dfrac{K(k)}{N}\right\}dn\left\{(2r-1)\dfrac{K(k)}{N}\right\}}{2\xi\eta(1-k^2)} \quad (4.157)$$

$$B_r = C_r cd\frac{(2r-1)K(k)}{N}, \quad (r=1,2,\cdots,N) \quad (4.158)$$

$$X_r = \frac{-\left\{sn\left[2(r-1)\dfrac{K(k)}{N}\right]+sn\left(\dfrac{2rK(k)}{N}\right)\right\}\left\{k^2\xi\eta cd\left[(2r-1)\dfrac{K(k)}{N}\right]cd\left(\dfrac{K(k)}{N}\right)-1\right\}}{\xi-\eta} \quad (4.159a)$$

$$K_{r,r+1} = \frac{\sqrt{\begin{array}{l}\left[\xi^2\left[1+\eta^2k^2sn^2\left(\dfrac{2rK(k)}{N}\right)\right]-2\eta\xi cn\left(\dfrac{2rK(k)}{N}\right)\right]\\ dn\left(\dfrac{2rK(k)}{N}\right)+\eta^2+sn^2\left(\dfrac{2rK(k)}{N}\right)\end{array}}}{\eta-\xi} \quad (4.159b)$$

$(r=1,2,\cdots,N-1)$

$$R_L = \frac{\xi+\eta}{\xi-\eta} \quad (4.160)$$

$$sn\left\{\frac{NK(k_1^2)}{K(k^2)}U_0\big|k_1^2\right\} = \frac{j}{\varepsilon} \quad (4.161)$$

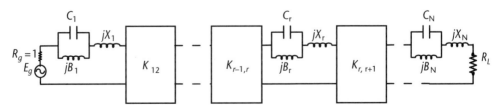

Figure 4.24 Inverter-coupled prototype for an elliptic filter.

4.3 Concept of Impedance Inverter

$$\eta = -jsn(U_0|k_1^2) \tag{4.162}$$

$$sn\left\{\frac{NK(k_1^2)}{K(k^2)}U_0\bigg|k_1^2\right\} = \frac{j}{\varepsilon\sqrt{1-K_0^2}} \tag{4.163}$$

$$\xi = -jsn(U_0|k_1^2) \tag{4.164}$$

3. Matched case ($K_0 = 1$, $\xi = \infty$, $R_L = 1$)

$$C_r = \frac{ds\left[(2r-1)\frac{K(k)}{N}\right]dn\left[(2r-1)\frac{K(k)}{N}\right]}{2\eta(1-k^2)}, \quad r = (1,2,\ldots,N) \tag{4.165}$$

$$B_r = C_r cd\left[\frac{(2r-1)K(k)}{N}\right] \tag{4.166}$$

$$X_r = -\eta k^2\left[sn\left(\frac{2(r-1)K(k)}{N}\right) + sn\left(\frac{2rK(k)}{N}\right)\right]cd\left(\frac{K(k)}{N}\right)cd\left(\frac{(2r-1)K(k)}{N}\right) \tag{4.167}$$

$$K_{r,r+1} = \sqrt{1+\eta^2 k^2 sn^2\left(\frac{2rK(k)}{N}\right)}, \quad r = (1,2,\ldots,N-1) \tag{4.168}$$

4. Singly terminated case ($RL = 1$, $S_{12} \to Z_{12}$)

$$C_r = \frac{ds\left[(2r-1)\frac{K(k)}{N}\right]dn\left[(2r-1)\frac{K(k)}{N}\right]}{\eta(1-k^2)} \tag{4.169}$$

$$B_r = C_r cd\left[\frac{(2r-1)K(k)}{N}\right] \tag{4.170}$$

$$X_r = \frac{-\left\{sn\left[2(r-1)\frac{K(k)}{N}\right] + sn\left(2r\frac{K(k)}{N}\right)\right\}\left\{k^2\eta^2 cd\left(\frac{K(k)}{N}\right)cd\left[(2r-1)\frac{K(k)}{N}\right] - 1\right\}}{2\eta} \quad (4.171)$$

$(r = 1, 2, \cdots, N)$

$$K_{r,r+1} = \frac{sn\left(\frac{2rK(k)}{N}\right)\sqrt{\left[1 + \eta^2 k^2 cd^2\left(\frac{rK(k)}{N}\right)\right]\left[1 + \eta^2 dc^2\left(\frac{rK(k)}{N}\right)\right]}}{2\eta} \quad (4.172)$$

$(r = 1, 2, \cdots, N-1)$

The elliptic functions appearing in the above equations are defined as [15] and $sn(u,k)$ is defined in (4.100) and (4.101).

$$cn(u,k) = \sqrt{1 - sn^2(u,k)}$$
$$dn(u,k) = \sqrt{1 - k^2 sn^2(u,k)}$$
$$cd(u,k) = \frac{cn(u,k)}{dn(u,k)}$$
$$dc(u,k) = \frac{1}{cd(u,k)}$$
$$ds(u,k) = \frac{dn(u,k)}{sn(u,k)}$$

4.3.1 Physical Realization of Impedance and Admittance Inverters

The impedance and admittance inverters described in the preceding sections were assumed to be frequency-independent and ideal, which provides a ±90° phase shift and an impedance or admittance inversion with respect to the value K or J of the inverter according to (4.134). Let us consider the ABCD matrix of a section of transmission line of electrical length θ and characteristic impedance Z_0

$$\begin{bmatrix} A & B \\ C & D \end{bmatrix} = \begin{bmatrix} \cos(\theta) & jZ_0 \sin[\theta] \\ \frac{\sin(\theta)}{jZ_0} & \cos(\theta) \end{bmatrix} \quad (4.173)$$

Now, if $\theta = 90°$, then

$$\begin{bmatrix} A & B \\ C & D \end{bmatrix} = \begin{bmatrix} 0 & jZ_0 \\ \frac{1}{jZ_0} & 0 \end{bmatrix} \quad (4.174)$$

which is the ABCD matrix of an impedance inverter of value $K = Z_0$. Similarly it can be shown that the same transmission line is also an admittance inverter of value $J = Y_0 = 1/Z_0$. However, one has to keep in mind that the condition $\theta = 90°$ is fulfilled precisely at a single frequency only. Such an ideal inverter does not exist in reality. However, for a couple of percentage of frequency variation around the frequency at which the transmission line is exactly 90° long, the deviation from an ideal inverter behavior can be accepted for all practical purposes. A filter designed on the basis of ideal inverter behavior would have a frequency response, which is very close to theoretical response for that case. For larger bandwidths or larger frequency variations around the nominal frequency, departure of a 90° line from the ideal one can be accommodated into the design by splitting the line into an ideal one and one with two additional short lengths of transmission line, one on either side, which account for the excess or deficit in phase shift from the ideal phase shift of 90° [22]. Such an inverter is shown in Figure 4.25. In most cases, the compensating elements are absorbed in adjoining networks. For example, if the K-inverter is a coupling element in a band-pass filter, the length of the compensating elements shortens the adjoining transmission line resonator elements.

Like transmission line inverters, lumped element inverters exist that consist of pure reactive elements like inductors and capacitors. Such inverters are shown in Figure 4.26 [22]. The negative elements are absorbed in adjoining networks as shown in Figure 4.27 [22]. In general, a transmission line discontinuity can be used as an inverter in which the compensating elements appear as extra phase shifts. For example, let us consider the H-plane iris discontinuity in a rectangular waveguide, shown in Figure 4.28. The iris has the fundamental mode scattering matrix

$$[S] = \begin{bmatrix} S_{11} & S_{12} \\ S_{21} & S_{22} \end{bmatrix} \tag{4.175}$$

The scattering matrix can be used to derive the equivalent T-network consisting of two series reactive elements and a shunt element together with the phase shifts of $\phi/2$.

The values of the K-inverter and the phase shift are given by

$$\frac{K}{Z_w} = \left| \tan\left(\frac{\phi}{2}\right) + \tan^{-1}\left(\frac{x_s}{Z_w}\right) \right| \tag{4.176}$$

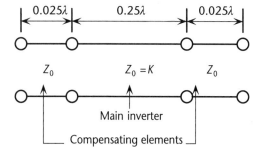

Figure 4.25 Compensated transmission line impedance inverter.

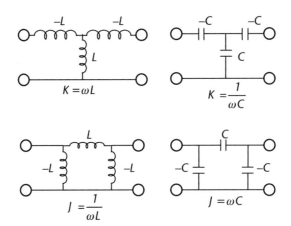

Figure 4.26 Lumped element inverters.

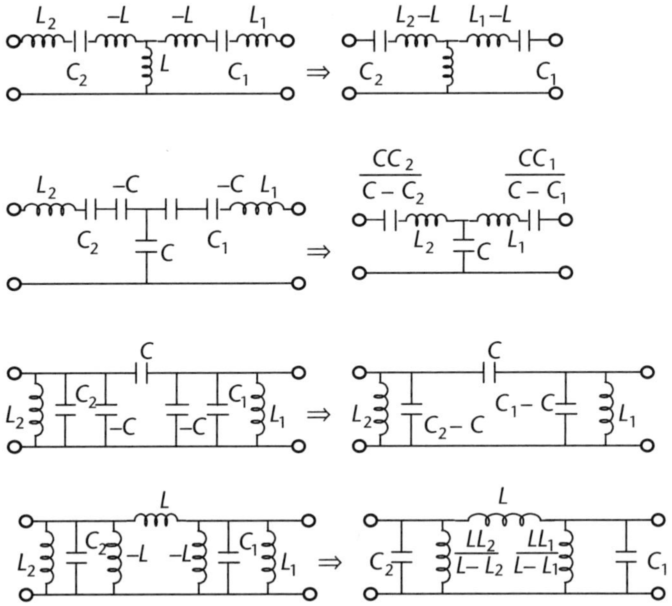

Figure 4.27 Absorption of negative elements of an inverter in adjoining networks.

$$\phi = -\tan^{-1}\left(\frac{2x_p}{Z_w} + \frac{x_s}{Z_w}\right) - \tan^{-1}\left(\frac{x_s}{Z_w}\right) \qquad (4.177)$$

Note that the extra phase shift $\phi/2$ is equivalent to the negative elements in the lumped circuit impedance inverters in Figure 4.26. Therefore, if the phase shift is positive then it has to be subtracted from the adjacent networks. We will illustrate the procedure in Chapters 5 and 6 while discussing waveguide low-pass and band-pass filters.

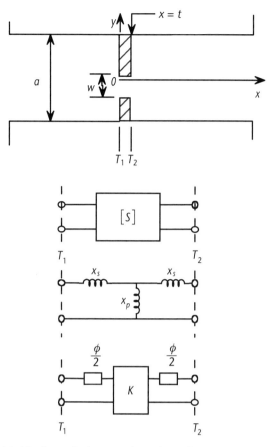

Figure 4.28 Waveguide iris discontinuity as an impedance inverter.

4.4 Low-Pass Prototype Using Cross-Coupled Networks

One of the most significant low-pass prototype configurations that dominates the modern-day electrical filter design is the cross-coupled low-pass prototype. In principle, the cross-coupled configuration can be based on any of the four basic frequency response types such as Butterworth, Chebyshev, elliptic, and the generalized Chebyshev response. We will demonstrate the method based on the generalized Chebyshev response.

Consider the generalized Chebyshev response given by (4.92a), which is

$$|S_{21}(\omega')|^2 = \frac{1}{1+\varepsilon^2 \Psi_N^2(\omega')} \quad (4.178)$$

where ε is related to the passband ripple via (4.73) and the passband return loss R_L via

$$\varepsilon = \frac{1}{\sqrt{10^{\frac{RL}{10}}-1}} \quad (4.179)$$

$$\left.\begin{array}{l}\Psi_N(\omega') = \cosh\left(\sum_{n=1}^{N}\cosh^{-1}(x_n)\right) \\ x_n = \dfrac{\omega' - 1/\omega'_n}{1 - \omega'/\omega'_n}\end{array}\right\} \quad (4.180)$$

where, $s_n = j\omega'_n$ is the location of the nth transmission zero in the complex s-plane. For all values of N, $|\Psi_N(\omega' = \pm 1)| = 1$. Also, the rational function $\Psi_N(\omega')$ can be expressed as [21]

$$\Psi_N(\omega') = \frac{P_N(\omega')}{\prod_{n=1}^{N}\left(1 - \dfrac{\omega'}{\omega'_n}\right)} = \frac{P_N(\omega')}{D_N(\omega')} \quad (4.181)$$

where

$$P_{N+1}(\omega') = -P_N(\omega')\left(1 - \frac{\omega'}{\omega'_N}\right)^2 \sqrt{\frac{1 - \dfrac{1}{\omega'^2_{N+1}}}{1 - \dfrac{1}{\omega'^2_N}}} + P_N(\omega')\left[\omega' - \frac{1}{\omega'_{N+1}} + \left(\omega' - \frac{1}{\omega'_N}\right)\sqrt{\frac{1 - \dfrac{1}{\omega'^2_{N+1}}}{1 - \dfrac{1}{\omega'^2_N}}}\right] \quad (4.182)$$

with

$$P_0(\omega') = 1, \quad P_1(\omega') = \omega' - \frac{1}{\omega'_1} \quad (4.183)$$

Obviously the above procedure is an alternative to Cameron's [20], which is described in Section 4.2.6. Now, let us consider the coupled line form of low-pass prototype circuit shown in Figure 4.29.

The coupling values M_{ij} ($i = 1,2,..., N$ and $j = 1,2,..., N$) completely define the network. The loop analysis of the network yields the following equation

$$\begin{bmatrix} e_s \\ 0 \\ 0 \\ 0 \\ \bullet \\ \bullet \\ \bullet \\ 0 \end{bmatrix} = \begin{bmatrix} S + R_1 & -j\omega M_{12} & -j\omega M_{13} & \bullet & \bullet & \bullet & \bullet & -j\omega M_{1N} \\ -j\omega M_{12} & S & -j\omega M_{23} & \bullet & \bullet & \bullet & \bullet & \bullet \\ -j\omega M_{13} & -j\omega M_{23} & S & \bullet & \bullet & \bullet & \bullet & \bullet \\ \bullet & \bullet & \bullet & \bullet & \bullet & \bullet & \bullet & \bullet \\ \bullet & \bullet & \bullet & \bullet & \bullet & \bullet & \bullet & \bullet \\ \bullet & \bullet & \bullet & \bullet & \bullet & \bullet & \bullet & \bullet \\ \bullet & \bullet & \bullet & \bullet & \bullet & \bullet & \bullet & \bullet \\ -j\omega M_{1N} & \bullet & \bullet & \bullet & \bullet & \bullet & -j\omega M_{N-1,N} & S + R_N \end{bmatrix} \begin{bmatrix} i_1 \\ i_2 \\ i_3 \\ \bullet \\ \bullet \\ \bullet \\ \bullet \\ i_N \end{bmatrix} \quad (4.184)$$

or,

4.4 Low-Pass Prototype Using Cross-Coupled Networks

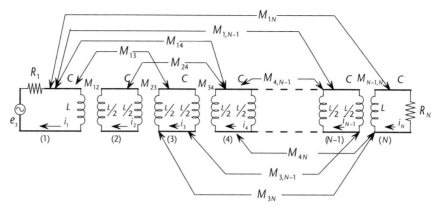

Figure 4.29 Generalized coupled-resonator network.

$$[e] = [A][I] \tag{4.185}$$

For $L = 1$ Henry, $S = s = j\omega$ and the two-port scattering parameters are given by [21]

$$S_{21} = -j2\sqrt{R_1 R_2}\left\{[A]^{-1}\right\}_{N1} \tag{4.186}$$

$$S_{11} = 1 + j2R_1\left\{[A]^{-1}\right\}_{11} \tag{4.187}$$

The normalized mutual inductances or coupling coefficients are obtained by following an optimization procedure by Amari [23] as follows.

We define a cost function

$$K = \sum_{i=1}^{N}\left|S_{11}(\omega_{zi})\right|^2 + \sum_{i=1}^{N}\left|S_{21}(\omega_{pi})\right|^2 + \left(\left|S_{11}(\omega = -1)\right| - \frac{\varepsilon}{\sqrt{1+\varepsilon^2}}\right) + \left(\left|S_{11}(\omega = 1)\right| - \frac{\varepsilon}{\sqrt{1+\varepsilon^2}}\right) \tag{4.188}$$

and a topology matrix $[P]$ where

$$P_{ij} = 1 \quad if \quad M_{ij} \neq 0 \tag{4.189}$$

and

$$P_{ij} = 0 \quad if \quad M_{ij} = 0 \tag{4.190}$$

The topology of the network is determined beforehand and should be enforced at each step of the optimization. For a derivative based optimization, the pertinent derivatives are obtained as

$$\frac{\partial S_{11}}{\partial M_{pq}} = -j4R_1 P_{pq} \left\{[A]^{-1}\right\}_{1p} \left\{[A]^{-1}\right\}_{q1} \qquad (4.191)$$

$$\frac{\partial S_{21}}{\partial M_{pq}} = j2\sqrt{R_1 R_2} P_{pq} \left(\left\{[A]^{-1}\right\}_{Np} \left\{[A]^{-1}\right\}_{q1} + \left\{[A]^{-1}\right\}_{Nq} \left\{[A]^{-1}\right\}_{p1} \right) \qquad (4.192)$$

$$\frac{\partial S_{11}}{\partial M_{pp}} = -j2R_1 P_{pp} \left\{[A]^{-1}\right\}_{p1} \left\{[A]^{-1}\right\}_{p1} \qquad (4.193)$$

$$\frac{\partial S_{11}}{\partial M_{pp}} = j2\sqrt{R_1 R_2} P_{pp} \left\{[A]^{-1}\right\}_{Np} \left\{[A]^{-1}\right\}_{p1} \qquad (4.194)$$

$$\frac{\partial S_{11}}{\partial R_1} = j2 \left\{[A]^{-1}\right\}_{11} + 2R_1 \left(\left\{[A]^{-1}\right\}_{11} \left\{[A]^{-1}\right\}_{11} + \frac{R_2}{R_1} \left\{[A]^{-1}\right\}_{N1} \left\{[A]^{-1}\right\}_{N1} \right) \qquad (4.195)$$

$$\frac{\partial S_{21}}{\partial R_1} = -j2\sqrt{\frac{R_2}{R_1}} \left\{[A]^{-1}\right\}_{N1} + 2R_1 \sqrt{\frac{R_2}{R_1}} \left(\left\{[A]^{-1}\right\}_{N1} \left\{[A]^{-1}\right\}_{11} + \frac{R_2}{R_1} \left\{[A]^{-1}\right\}_{NN} \left\{[A]^{-1}\right\}_{N1} \right) \qquad (4.196)$$

Using the above cost function (4.188) together with the derivatives defined by (4.191) through (4.196), a FORTRAN code can be written for the determination of the coupling parameters M_{ij} in (4.184). IMSL [24] library subroutine NCONG is a suitable optimization subroutine that can be used in the code. Following the definition of group delay in a two-port device [23]

$$\tau_g = -\text{Im}\left[\frac{\partial S_{21}}{\partial \omega}\right] \qquad (4.197)$$

we get from the above equations

$$\tau_g = \text{Im}\left[\frac{\sum_{k=1}^{N} \left\{[A]^{-1}\right\}_{Nk} \left\{[A]^{-1}\right\}_{k1}}{\left\{[A]^{-1}\right\}_{N1}}\right] \qquad (4.198)$$

The above procedure has the following advantages over other conventional methods [25, 26].

1. Enforcement of the predetermined topology matrix excludes the need for similarity transforms.
2. Filters of arbitrary even and odd orders can be synthesized.
3. One can synthesize specific coupling elements of a given sign or within a magnitude range if intended implementation calls for such a constraint.
4. The resulting solution, if one is obtained, is not affected by the problem of round-off errors that plague extraction methods.
5. If an exact solution is not found, an approximate, which may be acceptable, is always given. This happens when the desired prototype response is not within the chosen topology.

4.5 Design of Low-Pass Prototypes Using Optimization and the Method of Least Squares

In the preceding sections we have shown various analytical methods for designing a low-pass prototype that offers a prescribed frequency response, Butterworth, Chebyshev, elliptic, or generalized Chebyshev. Such frequency responses have their respective circuit configurations. The method based on computer optimization and the method of least squares (MLS), on the other hand, is very general and versatile.

The basic method of least squares for adaptive filters is described in [27]. The versatility of the method includes the following features [28]:

1. The filter design is combined by the impedance matching of the input and output impedances;
2. The input and the output impedances can be complex;
3. It can design all kinds of filters including the four fundamental ones (i.e., low-pass, high-pass, band-pass, and bandstop);
4. Specification of any frequency band;
5. Specification of any physically realizable frequency response.
6. Dual-band and multiband frequency response;
7. Any selected cascaded circuit configuration.

Referring to Figure 4.30, the power loss ratio of the lossless two-port network is given by

$$P_{LR} = \frac{1}{|S_{21}|^2} \qquad (4.199)$$

$$S_{21} = \frac{2\sqrt{\operatorname{Re}(Z_S)\operatorname{Re}(Z_L)}}{AZ_L + B + CZ_SZ_L + DZ_S} \qquad (4.200)$$

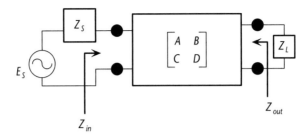

Figure 4.30 Two-port network.

Next, an error function is constructed according to the frequency response in the passband, transition band, and stopband, as shown in Figure 4.31. The filter's frequency response curve is required to pass through the white area and avoid the dashed area. The error function is formed as follows:

$$Error = W_P \times ER_P + W_T \times ER_T + W_S \times ER_S \tag{4.201}$$

where

W_P, W_T, and W_S are weighting functions

and

$$ER_P = \sum_{k \in passband}^{K_P} \frac{1}{2}\left[1 - sign\left(\left|S_{21}^k\right|_{dB} - L_{Ar}\right)\right]\left(\left|S_{21}^k\right|_{dB} - L_{Ar}\right)^2 \tag{4.202}$$

$$ER_S = \sum_{k \in stopband}^{K_P} \frac{1}{2}\left[1 - sign\left(\left|S_{21}^k\right|_{dB} - L_{As}\right)\right]\left(\left|S_{21}^k\right|_{dB} - L_{As}\right)^2 \tag{4.203}$$

$$ER_T = \sum_{k \in transitionband}^{K_P} \frac{1}{2}\left[1 - sign\left(\left|S_{21}^k\right|_{dB} - L_{At}\right)\right]\left(\left|S_{21}^k\right|_{dB} - L_{At}\right)^2 \tag{4.204}$$

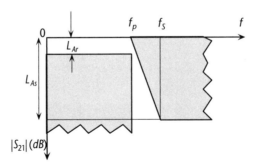

Figure 4.31 Target frequency response.

4.5 Design of Low-Pass Prototypes Using Optimization and the Method of Least Squares

$$L_{At}^k = \frac{L_{As}}{\Delta f} \Delta f_k$$
$$\Delta f_k = f_k - f_p$$
(4.205)

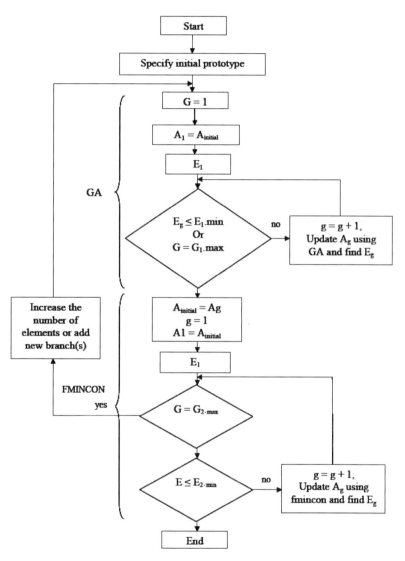

Figure 4.33 Flowchart of MLS algorithm for optimization of filter elements values. (Reproduced courtesy Electromagnetic Academy, United States.)

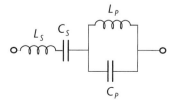

Figure 4.32 The basic unit.

$\Delta f = f_S - f_P$ is the width of the transition band
and

$$\text{sign}(x) = -1 \text{ for } x < 0$$
$$= +1 \text{ for } x \geq 0$$

The process begins with a cascaded connection of several shunt and series units of the form shown in Figure 4.32 [29]. Usually the element next to the source or load is chosen to be a series or a shunt element. Initial values are selected as in (4.206)

$$L = \frac{0.4\{\text{Re}(Z_s) + \text{Re}(Z_L)\}}{\omega'} \quad \text{and} \quad C = \frac{1}{\{\text{Re}(Z_s) + \text{Re}(Z_L)\}\omega'} \quad (4.206)$$

where ω' is cutoff frequency of the low-pass and high-pass filters. It is the center frequency when the method is applied to band-pass and bandstop filters [29].

Next, the computer program determines the optimum values of the circuit elements under the specified constraints. If an inductor becomes too small to be realized, it is short-circuited. Similarly, a capacitor is removed and shorted out if it becomes too large. In the exact same manner, each high-value inductor and low-value capacitor is open-circuited.

Figure 4.33 shows the flowchart of the MLS algorithm for optimization of filter element values [29]. FMINCON is the conjugate gradient subroutine in MATLAB toolbox.

Table 4.2 shows and compares the low-pass prototype element values (see Figure 4.34) computed by MLS optimization and conventional network synthesis method for Chebyshev and elliptic filters. Figure 4.35 shows the corresponding frequency responses. More examples are shown in Figures 4.36 and 4.37. Table 4.3 shows the element values.

Table 4.2 Low Pass Prototype Comparison

		L_{As}	g_1	g_2	g_3	g_4	g_5	g_6	g_7	*Error*
Chebyshev	Literature [34]	41.9 dB	1.1812	1.4228	2.0967	1.5734	2.0967	1.4228	1.812	–
	Designed by MLS		1.115	1.4429	2.055	1.5749	2.14	1.4046	1.245	3.59E-21
			1.21	1.3955	2.11	1.5797	2.07	1.4575	1.14	0
			1.165	1.4541	2.07	1.5720	2.125	1.3945	1.195	0
Elliptic	Literature [34]		1.0481	0.1244	1.2416	1.6843	0.354	1.0031	0.8692	–
	Designed by MLS		1.0385	0.1237	1.245	1.6903	0.3558	1.0000	0.8721	2.00E-16
			1.0816	0.1281	1.22	1.6952	0.3478	1.015	0.8535	5.28E-06
			1.0669	0.1273	1.235	1.6907	03494	1.01	0.8680	2.26E-06

$\text{Re}(Z_S) = \text{Re}(Z_L) = 1$, $\omega_P = 1$ rad/sec, $\omega_S = 1.6129$ rad/sec, $L_{Ar} = 0.1$ dB.

Source: [28].

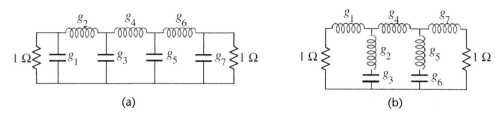

Figure 4.34 Low-pass prototype filter for (a) Chebyshev response, (b) elliptic response. (Reproduced courtesy Electromagnetic Academy, United States.)

Table 4.3a Element Values in Figure 4.36a

L_{Ar}	L_{As}	L_1 (nH)	L_2 (nH)	C_1 (pF)	L_3 (nH)	L_4 (nH)	C_2 (pF)	L_5 (nH)	C_3 (pF)
0.1 dB	30 dB	1.568	0.8294	1.23	2.8615	2.267	0.8701	2.7519	1.12

$f_P = 3.00$ GHz, $f_S = 3.5$ GHz

Source: [28].

Table 4.3b Element Values in Figure 4.36b

L_{Ar}	L_{As}	L_1 (μH)	C_1 (nF)	L_2 (μH)	C_2 (nF)	L_3 (μH)	C_3 (nF)	C_4 (nF)	L_4 (μH)	C_5 (nF)	C_6 (nF)	L_5 (μH)
0.3 dB	25 dB	3.4925	0.3007	2.9459	0.1205	1.5402	0.3403	0.5620	1.4344	0.9852	0.4621	4.5577

$f_P = 4.00$ MHz, $f_S = 4.20$ MHz

Source: [28].

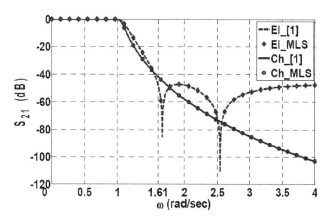

Figure 4.35 Frequency responses of filters shown in Figure 4.34. (Reproduced courtesy Electromagnetic Academy, United States.)

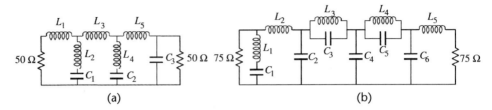

Figure 4.36 Circuit configurations and frequency responses of two extracted pole low-pass filters. (Reproduced courtesy Electromagnetic Academy, United States.)

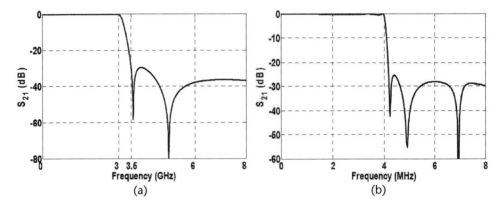

Figure 4.37 Circuit configurations and the frequency responses of two extracted pole low-pass filters designed by the method of least squares. The element values are shown in Tables 4.3a and 4.3b. (Reproduced courtesy Electromagnetic Academy, United States.)

References

[1] Guillemin, E. A., *Communication Networks*, Vol.2, New York: John Wiley & Sons Inc., 1935.

[2] Ragan G. L. (ed.), *Microwave Transmission Circuits*, Section 9.13, New York: McGraw-Hill Book Company, 1948.

[3] Belevitch, V., *Classical Network Theory*, San Francisco: Haolden Day Inc., 1968.

[4] Butterworth, S., "On the Theory of Filter Amplifiers," *Wireless Engineering*, Vol. 7, 1941, pp. 536–541.

[5] Cauer, W., "Synthesis of Linear Communication Networks," New York: McGraw-Hill Co. Inc., 1958.

[6] Van Valkenburg, M. E., *Introduction to Modern Network Synthesis*, New York: John Wiley & Sons, 1960.

[7] Hurwitz, A., "Uber die Bedingungen unter welchen eine Gleichung nur Wurzeln mit negativen reellen Theilen besitzt," *Math Ann.*, Vol. 46, 1895, pp. 273–284.

[8] Zhu, Y. S., and W. K. Chen, *Computer Aided Design of Communication Networks*, Singapore: World Scientific, 2000.

[9] Rhodes, D., *Theory of Electrical Filters*, New York: John Wiley & Sons Inc., 1976.

[10] Wienberg, L., *Network Analysis and Synthesis*, Huntington, NY: Robert E. Kreiger Publishing Co., 1975.

[11] Chebyshev, P. L., "Theorie des mechanismes Connus sous le mom de parallelogrammes," *Oeuvres*, Vol. 18, St. PetersbuRg, 1899.

[12] Herrero, J. L., and G. Willoner, *Synthesis of Filters,* Englewood Cliffs, NJ: Prentice Hall Inc., 1966.

[13] Christian, E., *Introduction to the Design of Transmission Networks,* Raleigh, NC: ITT Telecommunications and North Carolina State University, 1975.

[14] Baher, H., *Synthesis of Electrical Networks,* New York: John Wiley & Sons, 1984.

[15] Jhanke, E., and F. Emde, *Table of Functions,* New York: Dover, 1945.

[16] *MATLAB, Language of Technical Computing,* Natick, MA: Mathworks Inc., 2000.

[17] Darlington, S., "Synthesis of Reactance 4-Poles which Produce Prescribed Insertion Loss Characteristrics," *J. Math. Phys.,* Vol. 18, September 1939, pp. 257–353.

[18] Lin, C. C., and Y. Tokad, "On the Element Values of Mid-Series and Mid-Shunt Lowpass LC Ladder Networks," *IEEE Trans. Circuit Theory,* Vol. CT-15, 1968, pp. 349–353.

[19] Amstutz, P., "Elliptic Approximation and Elliptic Filter Design on Small Computers," *IEEE Trans. Circuits Syst.,* Vol. CAS-25, No. 12, December 1978, pp. 1001–1011. Also correction, p. 776, CAS-30, No. 10, December 1983.

[20] Cameron, R., "Fast Generation of Chebyshev Filter Prototypes with Asymmetrically Prescribed Transmission Zeros," *ESA J,* Vol. 6, No. 1, 1982, pp. 83–95.

[21] Cameron, R., "General Coupling Matrix Synthesis Methods for Chebyshev Filtering Functions," *IEEE Trans. Microw. Theory Techn.,* Vol. 47, No. 4, April, 1999, pp. 433–442.

[22] Collin, R. E., *Foundations for Microwave Engineering,* New York: McGraw-Hill Book Co., 1966.

[23] Amari, S., "Synthesis of Cross-Coupled Resonator Filters Using an Analytical Gradient-Based Optimization Technique," *IEEE Trans. Microw. Theory Techn.,* Vol. 48, No. 9, September 2000, pp. 1559–1564.

[24] IMSL Fortran Library Subroutines, DEC Visual Fortran, 2004.

[25] Cameron, R. J., "General Coupling Matrix Synthesis Methods for Chebyshev Filtering Functions," *IEEE Trans. Microw. Theory Techn.,* Vol. 47, No. 4, April 1999, pp. 433–442.

[26] Orchard, H. J., "Filter Design by Iterated Analysis," *IEEE Trans. Circuits Syst.,* Vol. 32, November 1985, pp. 1089–1096.

[27] Haykin, S., and B. Widrow, *Least-Mean-Square Adaptive Filters,* New York: John Wiley & Sons Inc., 2003.

[28] Oraizi, H., and M. S. Esfahlan, "Optimum Design of Lowpass Filters for Genera; LC Network Configurations by the Method of Least Squares," *PIER Proc. Moscow,* Russia, August 2009, pp. 18–21.

[29] Oraizi, H., and M. S. Esfahlan, "Optimum Design of Lumped Element Filters Incorporating Impedance Matching by the Method of least Squares, *PIER,* Vol. 100, 2010, pp.83–100.

[30] Van Valkenburg, M. E., *Introduction to Modern Network Synthesis,* New York: John Wiley & Sons, Inc., 1960.

[31] Fuzisawa, T., "Realizability Theorem for Mid-Series or Mid-Shunt Low-Pass Ladders without Mutual Induction," *IRE Trans. Circuit Theory,* Vol. 2., No. 4, 1955, pp. 320–325.

[32] Lin, C. C., and Y. Tokad, "On Element Values of Mid-Series and Mid-Shunt Lowpass LC Networks," *IEEE Trans. Circuit Theory,* Vol. CT-15, 1968, pp. 349–353.

[33] Zhu, Y.- S., and W.- K. Chen, Computer-Aided Design of Communication Networks, World Scientific, Singapore, New Jersey, London, Hong Kong, 2000. New York, NY, 1981.

Appendix 4A

The steps involved in Darlington's method for the synthesis of a doubly terminated passive network are as follows [30].

Consider the doubly terminated two-port network shown in Figure A4.1.

Step   Given $|S_{21}|^2$ or equivalent, R_1 and R_2 [30]

1. Test the requirement, $|S_{21}(j\omega)|^2 < 1$ or an equivalent condition such as $\left|\dfrac{V_2(j\omega)}{V_s(j\omega)}\right|^2 \leq \dfrac{R_2}{4R_1}$. If this condition is not satisfied, then it is necessary to scale $|S_{21}|^2$.
2. Obtain

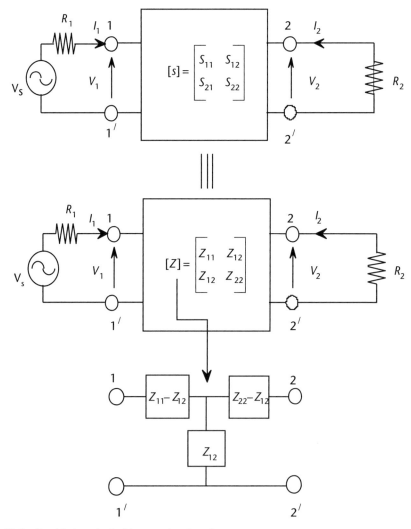

Figure 4A.1 Doubly terminated two-port network.

Appendix 4A

$$|S_{11}(j\omega)|^2 = 1 - |S_{21}(j\omega)|^2 \tag{4A.1}$$

3. Form

$$S_{11}(s)S_{11}(-s) = \frac{C(-s^2)}{B(-s^2)} \tag{4A.2}$$

by letting $j\omega = s$

5. Form $S_{11}(s) = \pm\frac{p_1(s)}{q_1(s)}$ by taking the left-half plane zeros of $B(-s^2)$, and either left-half plane or right-half plane zeros of $C(-s^2)$.

6. Obtain

$$Z_{in}(s) = \frac{m_1 + n_1}{m_2 + n_2} = \frac{1 + S_{11}(s)}{1 - S_{11}(s)} \quad \text{or} \quad \frac{1 - S_{11}(s)}{1 + S_{11}(s)} \tag{4A.3}$$

where m_i, and $n_i (i = 1,2)$ are the even and odd parts of the numerator and denominator polynomials, respectively, of the input impedance function.

7. Examine the roots of $m_1 m_2 - n_1 n_2 = 0$. If all zeros are at the origin or at infinity, then develop $Z_{in}(s)$ as a ladder (see Section 4.2.2). If all zeros are on the imaginary axis, test to see if Fujisawa [31] conditions are satisfied. Then develop the network by Darlington's method shown below.

4A.1.1 Darlington Synthesis [17]

Consider the two-port network in Figure 4A.1. The driving point impedance of the network can be expressed in terms of the Z-parameters as

$$Z_{in}(s) = z_{11}(s)\frac{\left[\frac{1}{y_{22}(s)} + R_2\right]}{z_{22}(s) + R_2} \tag{4A.4}$$

where for a reciprocal network

$$y_{22}(s) = \frac{z_{11}(s)}{z_{11}(s)z_{22}(s) - z_{12}^2(s)} \tag{4A.5}$$

From (4A.3) we distinguish two cases:

$$Z_{in}(s) = \left(\frac{m_1}{n_2}\right)\left\{\frac{\frac{n_1}{m_1}+1}{\frac{m_2}{n_2}+1}\right\} \text{(Case A)} \qquad (4A.6a)$$

$$Z_{in}(s) = \left(\frac{n_1}{m_2}\right)\left\{\frac{\frac{m_1}{n_1}+1}{\frac{n_2}{m_2}+1}\right\} \text{(Case B)} \qquad (4A.6b)$$

If we assume that right-hand side of (4A.3) is a positive real function, then the following function is also positive real:

$$\bar{Z}(s) = \frac{m_1 + n_2}{m_2 + n_1} \qquad (4A.7)$$

Also, the numerators and the denominators of the right-hand sides of (4A.3) and (4A.6) are Hurwitz polynomials [7]. The ratio of the even part to odd part or odd part to even part

$$\left(\frac{m_1}{n_1}\right)^{\pm 1}, \left(\frac{m_2}{n_2}\right)^{\pm 1}, \left(\frac{m_1}{n_2}\right)^{\pm 1}, \left(\frac{m_2}{n_1}\right)^{\pm 1}$$

are all reactance functions. The continued fraction expansions of the functions yield real and positive coefficients. That is the input impedance of a one-port network. Assuming $R_1 = R_2 = 1\Omega$ and comparing (4A.4), (4A.5), and (4A.6) gives the Z-parameters of the network, as shown in Table 4A.1.

In order for $z_{12}(s)$ to be realizable, the quantity under the square root sign must be a full square. Or in other words, its s^2 zeros must be of even multiplicity. In case it is not, it can be remedied by multiplying the ensignant by an auxiliary even polynomial that contains all the first-order factors that occur in the ensignant with odd multiplicity. The multiplying polynomial is assumed to be a Hurwitz polynomial [7]. Let us consider the example

Table 4A.1 z-Parameters for Darlington's Procedure

Case A	Case B
$z_{11}(s) = \dfrac{m_1}{n_2}$	$z_{11}(s) = \dfrac{n_1}{m_2}$
$z_{22}(s) = \dfrac{m_2}{n_2}$	$z_{22}(s) = \dfrac{n_2}{m_2}$
$z_{12}(s) = \dfrac{\sqrt{m_1 m_2 - n_1 n_2}}{n_2}$	$z_{12}(s) = \dfrac{\sqrt{n_1 n_2 - m_1 m_2}}{m_2}$

$$Z(s) = \frac{1+s}{1+4s} \tag{4A.8}$$

in which

$$m_1 m_2 - n_1 n_2 = 1 - 4s^2 = (1+2s)(1-2s)$$

Therefore, we choose

$$m_0^2 - n_0^2 = (1+2s)(1-2s) \tag{4A.9}$$

In the next step, we multiply the numerator and the denominator of the right-hand side of (4A.8) by

$$m_0 + n_0 = 1 + 2s \tag{4A.10}$$

which gives

$$Z(s) = \frac{(1+s)(1+2s)}{(1+4s)(1+2s)} = \frac{1+3s+2s^2}{1+6s+8s^2} \tag{4A.11}$$

Using the equations in Table 4A.1 and (4A.11) gives

$$\left.\begin{array}{l} z_{11}(s) = \dfrac{1+2s^2}{6s} = \dfrac{1}{6s} + \dfrac{s}{3} \\[6pt] z_{22}(s) = \dfrac{1+8s^2}{6s} = \dfrac{1}{6s} + \dfrac{4s}{3} \\[6pt] z_{12}(s) = \dfrac{1-4s^2}{6s} = \dfrac{1}{6s} - \dfrac{2s}{3} \end{array}\right\} \tag{4A.12a}$$

The Darlington realization of the network is shown in Figure 4A.2.
In principle, z_{11}, z_{22}, and z_{12} can be expanded in the forms

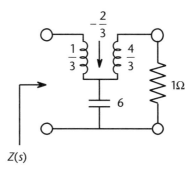

Figure 4A.2 Darlington realized network.

$$z_{11}(s) = \frac{k_{11}^0}{s} + k_{11}^\infty s + \sum_{v=1}^{n} \frac{2k_{11}^v}{s^2 + \omega_v^2}$$

$$z_{22}(s) = \frac{k_{22}^0}{s} + k_{22}^\infty s + \sum_{v=1}^{n} \frac{2k_{22}^v}{s^2 + \omega_v^2} \quad \quad (4A.12b)$$

$$z_{12}(s) = \frac{k_{12}^0}{s} + k_{12}^\infty s + \sum_{v=1}^{n} \frac{2k_{12}^v}{s^2 + \omega_v^2}$$

where, k_{ij}^0, k_{ij}^∞, and k_{ij}^v ($i = 1,2$ and $j = 1,2$) are the residues at the poles at $s = 0$, ∞, $j\omega$, respectively. The realizability condition for $z_{ij}(s)$'s is given by [30]

$$k_{11}^r k_{22}^r - k_{12}^r \geq 0 \quad \quad (4A.13)$$

where $r = 0$, ∞ and v.

When all the zeros of transmission are at the origin and/or at infinity, then the LC ladder network should be found by a continued fraction expansion, as shown in Section 4.2.2 (Butterworth or Chebyshev response). If all the transmission zeros are on the imaginary axis, then the synthesis problem reduces to realizing a $z_{11}(s)$, whose shunt branches resonates at the zeros of $z_{12}(s)$. It can be shown that the zeros of $z_{12}(s)$ are also the transmission zeros of the network [30]. From Table 4A.1 the zeros are also the solution to the equation

$$m_1 m_2 - n_1 n_2 = 0 \quad \quad (4A.14)$$

The realized $z_{11}(s)$ has the network form shown in Figure 4A.3. The network is known as a midseries network. According to Darlington [17] and the form given by Lin and Tokad [32], the values of the components comprising the network are obtained from a partial fraction expansion as follows.

We define

$$F = \frac{z_{11}(s)}{s} \quad \quad (4A.15)$$

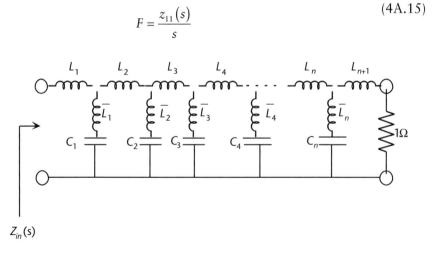

Figure 4A.3 Midseries low-pass configuration.

and

$$\zeta = \frac{1}{\omega^2} \quad (4A.16)$$

We exapnd $F(\zeta)$ as

$$F(\zeta) = \left.\frac{z_{11}(s)}{s}\right|_{\zeta=\frac{1}{-s^2}} = L_1 + \cfrac{1}{\cfrac{C_1}{\zeta_1-\zeta} + \cfrac{1}{L_2 + \cfrac{C_2}{\zeta_2-\zeta} + \cfrac{1}{\ddots \cfrac{C_n}{\zeta_n-\zeta} + \cfrac{1}{L_{n+1}}}}} \quad (4A.17)$$

with

$$\zeta_i = \frac{1}{\omega_i^2} \quad (i=1,2,\cdots,n) \quad (4A.18)$$

$$\left.\begin{array}{l} L_i = F(\zeta_i) \\ C_i = -\dfrac{1}{F'(\zeta_i)} \end{array}\right\} \quad (4A.19)$$

where

$$\overline{\zeta}_i = C_i \overline{L}_i, \quad 0 \le \overline{\zeta}_i \le 1, \text{ for } i=1,2,\cdots,n \quad (4A.20)$$

$$F'(\zeta_i) = \left.\frac{dF(\zeta)}{d\zeta}\right|_{\zeta=\zeta_i} \quad (4A.21)$$

The dual form the network, known as midshunt low-pass prototype, shown in Figure 4A.3 and Figure 4A.4, can also be used to realize the same transfer function.

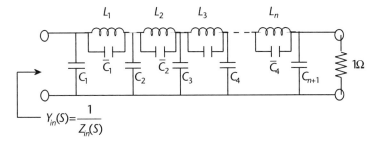

Figure 4A.4 Dual form of midshunt low-pass prototype.

The corresponding continued fraction expansion for the driving point admittance is given by [5, 17]

$$F(\zeta) = \frac{y_{11}(s)}{s}\bigg|_{\zeta=-\frac{1}{s^2}} = C_1 + \cfrac{1}{\cfrac{L_1}{\zeta_1-\zeta} + \cfrac{1}{C_2 + \cfrac{L_2}{\zeta_2-\zeta} + \cfrac{1}{\ddots + \cfrac{L_n}{\zeta_n-\zeta} + \cfrac{1}{C_{n+1}}}}} \qquad (4A.22)$$

where

$$\left.\begin{array}{l} C_i = F(\zeta_i) \\ L_i = -\dfrac{1}{F'(\zeta_i)} \end{array}\right\} \qquad (4A.23)$$

and

$$\zeta_i = L_i \overline{C}_i \quad 0 \leq \zeta_i \leq 1, \text{ for } i = 1, 2, \cdots, n \qquad (4A.24)$$

Also

$$\begin{bmatrix} y_{11}(s) & y_{12}(s) \\ y_{21}(s) & y_{22}(s) \end{bmatrix} = \begin{bmatrix} z_{11}(s) & z_{12}(s) \\ z_{21}(s) & z_{22}(s) \end{bmatrix}^{-1} \qquad (4A.25)$$

If all the transmission zeros are on the imaginary axis, then the synthesis problem reduces to realizing a $y_{11}(s)$, whose series branches resonates at the zeros of $y_{12}(s)$. It can be shown that the zeros of $y_{12}(s)$ are also the transmission zeros of the network [17]. Zhu and Chen [33] present the method and a FORTRAN code.

CHAPTER 5

Theory of Distributed Circuits

The lumped element synthesis techniques described in Chapter 4 are valid for all frequencies. However, physical realization of true lumped elements at microwave frequencies poses a number of obstacles. First, the required largest physical dimension of an element has to be much smaller than the wavelength so that the phase shift over that dimension should be negligible for all practical purposes. Second, the unloaded Q-factor becomes unacceptably small due to the small size of the element. Finally, the power-handling capability of a small element becomes very small due to the size. In order to circumvent the above difficulties, the size of an element is kept large enough and the fact that there occurs appreciable phase shift over any dimension of the element is taken into consideration while designing the element involved.

5.1 Distributed Element Equivalence of Lumped Elements

The discussion of distributed elements begins from the input impedance of a section of a transmission line that has been terminated either by a short circuit or an open circuit. Consider a section of transmission line of length l, characteristic impedance Z_0, and propagation constant β that has been terminated by a load Z_L, as shown in Figure 5.1. The input impedance of the line for $Z_L = 0$ (short circuit) is

$$Z_{in}^{SC} = jZ_0 \tan(\beta l) \tag{5.1}$$

If we write

$$\tan(\beta l) = \Omega \tag{5.2}$$

then (5.1) becomes

$$Z_{in}^{SC} = jZ_0 \Omega \tag{5.3}$$

and has the characteristics of a reactance of an inductor of value Z_0 and frequency Ω. However, like the frequency variable in lumped element network theory, the frequency variable Ω is not a linear function of frequency f. In fact, it is a transcendental function of frequency f. If

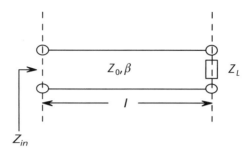

Figure 5.1 Terminated transmission line.

$$\beta = \omega\sqrt{\mu\varepsilon} = 2\pi f\sqrt{\mu\varepsilon} \quad (5.4)$$

where μ and ε are the permeability and the permittivity, respectively, of the medium between the two conductors, then

$$\Omega = \tan\left(2\pi f\sqrt{\mu\varepsilon}l\right) \quad (5.5)$$

Obviously the frequency mapping based on the above equation is not only nonlinear but also periodic. However, when the right-hand side of (5.5) is expanded in a series, it gives

$$\Omega = \left(2\pi l\sqrt{\mu\varepsilon}\right)f + \frac{\left(2\pi l\sqrt{\mu\varepsilon}\right)^3}{3!}f^3 + \cdots \quad (5.6)$$

If the higher-order terms on the right-hand side of the above equation are neglected, then the mapping becomes linear and nonperiodic. But we can do so only if the operating frequency is so small that the term $\left(2\pi l\sqrt{\mu\varepsilon}\right)f$ is sufficiently accurate to approximate the tangent function in (5.5). In other words, the phase shift over the length l of the element should not only be less than, but also be much less than 90°. Under such conditions, (5.3) assumes the form

$$Z_{in}^{SC} = jZ_0 l\sqrt{\mu\varepsilon}(2\pi f) = j\left(Z_0 l\sqrt{\mu\varepsilon}\right)\omega \quad (5.7)$$

which is the reactance of an inductor

$$L = Z_0 l\sqrt{\mu\varepsilon} \quad \text{(henry)} \quad (5.8)$$

Starting our analysis from the input impedance of an open circuit transmission line and using the same line of arguments, it can be shown that the equivalent capacitance of a small section of open-circuited transmission line of characteristic impedance Z_0, propagation constant β, and length l is

$$C = \frac{\sqrt{\mu\varepsilon}}{Z_0}l \quad \text{(farads)} \quad (5.9)$$

5.1 Distributed Element Equivalence of Lumped Elements

Therefore, once again we arrive at the same conclusion that we have to keep the maximum dimension of the element (inductor or capacitor) much smaller than the operating wavelength. One would wonder at this point what price the designer has to pay if a less than quarter-wavelength-long short-circuited transmission line is used to simulate an inductive reactance or a less than quarter-wavelength-long open-circuited transmission line is used to realize a capacitive reactance. Let us consider the lumped element low-pass filter shown in Figure 5.2. The filter is a Chebyshev filter with −0.1-dB ripple and 2-GHz cutoff frequency. The element values can be computed using normalized g-values (lumped element low-pass prototype values described in the previous chapter). The series inductors offer 51.5724 Ω inductive reactance at the cutoff frequency 2 GHz and the shunt capacitor offers 43.58 Ω capacitive reactance. The inductive reactances can be realized by a short-circuited $\lambda/8$-transmission line of characteristic impedance 51.5724 Ω and the capacitive reactance can be realized by an open-circuited $\lambda/8$-transmission line of characteristic impedance 43.58 Ω. The transmission line implementation of the filter is shown in Figure 5.3.

Although the above network is mathematically exact, it will pose two problems when physically implemented. Those are

1. The frequency response will not be that of a true low-pass filter because the stopband will not extend to infinite frequency due to the periodic nature of the reactances offered by the transmission lines. In fact, there will be an infinite number of passbands and stopbands between zero and infinite frequencies, with the first one being the desired passband.

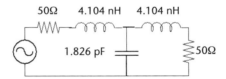

Figure 5.2 Lumped element Chebyshev low-pass filter.

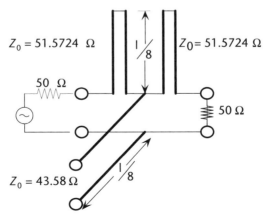

Figure 5.3 Transmission line equivalent of circuit shown in Figure 5.2.

2. In most cases, it may not be possible to fabricate the series short-circuited transmission line sections.

There is no solution to problem 1. This is the price one has to pay in order to use distributed elements in a microwave network. However, the problem becomes a blessing in disguise because a spurious passband of a low-pass filter can be used as the desired passband of a band-pass filter. Similarly a spurious stopband of a low-pass filter may be used as the desired stopband of a bandstop filter. We discuss these issues in later chapters on band-pass and bandstop filter design. In the following sections we will describe the important concepts of Kuroda identity [1] and transmission line elements (TLEs) or unit elements (UEs) [2].

5.2 TLE or UE and Kuroda Identity

By definition, a TLE or UE is a section of transmission line of electrical length θ and characteristic impedance Z_c. The *ABCD* matrix of a unit element is given by

$$[T] = \begin{bmatrix} \cos\theta & jZ_c \sin\theta \\ j\sin\theta/Z_c & \cos\theta \end{bmatrix} \quad (5.10)$$

In terms of the transformed frequency Ω defined by (5.2), the above equation can be written as

$$[T] = \frac{1}{\sqrt{1+\Omega^2}} \begin{bmatrix} 1 & jZ_c\Omega \\ j\Omega/Z_c & 1 \end{bmatrix} \quad (5.11)$$

The uniqueness of the above *ABCD* matrix lies in the fact that the multiplying factor of the matrix $1/\sqrt{1+\Omega^2}$ is an irrational function of the transformed frequency variable Ω. However, when two such UEs are cascaded, the *ABCD* matrix of the resulting network becomes a rational function of . Consequently, the established lumped element network synthesis becomes applicable to distributed element network synthesis when it involves an even number of UEs. Let us consider a UE followed by a series short-circuited transmission line, as shown in Figure 5.4. The *ABCD* matrix of the combined network is given by

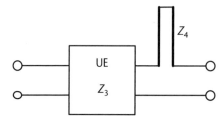

Figure 5.4 UE cascaded to a series short-circuited transmission line.

5.2 TLE or UE and Kuroda Identity

$$[T'] = \begin{bmatrix} 1 & j(Z_3+Z_4)\Omega \\ j\Omega/Z_3 & \left(1 - Z_4\Omega^2/Z_3\right) \end{bmatrix} \quad (5.12a)$$

Let us then consider a combination of a shunt open-circuited transmission line and a UE, as shown in Figure 5.5. The *ABCD* matrix of the overall network is

$$[T''] = \begin{bmatrix} 1 & jZ_2\Omega \\ j\left(\dfrac{1}{Z_1}+\dfrac{1}{Z_2}\right)\Omega & \left(1 - \dfrac{Z_2}{Z_1}\Omega^2\right) \end{bmatrix} \quad (5.12b)$$

In order to establish the equivalence between the networks shown in Figures 5.4 and 5.5 we equate the *ABCD* matrices in (5.12a) and (5.12b) and obtain the following relationships among the element values of the two networks:

$$Z_1 = \dfrac{Z_3^2}{Z_4} + Z_3 \quad (5.13a)$$

$$Z_2 = Z_3 + Z_4 \quad (5.13b)$$

Also solving the above equations for Z_3 and Z_4 gives

$$Z_3 = \dfrac{Z_1 Z_2}{Z_1 + Z_2} \quad (5.14a)$$

$$Z_4 = \dfrac{Z_2^2}{Z_1 + Z_2} \quad (5.14b)$$

The above equivalence is such that the behaviors of the two networks are identical at all frequencies. This is known as Kuroda's identity [1]. Kuroda's identity using

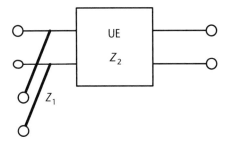

Figure 5.5 Shunt open-circuited transmission line cascaded to a UE.

UEs is extremely useful in converting unrealizable distributed element networks into realizable ones. As an example, let us apply the identity to the low-pass filter shown in Figure 5.3. Figure 5.6(a) shows that we have cascaded a section of transmission line of impedance 50 Ω and length $\lambda/8$ (UEs) at the input and output of the filter, respectively. An application of Kuroda's identity transforms the network into the one shown in Figure 5.6(b), which is physically realizable without any problem. However, the only difference between the networks shown in Figure 5.3 and 5.6 is that the transfer function of the latter has an extra 90° phase shift due to the addition of two $\lambda/8$ transmission lines.

5.3 Effects of Line Length and Impedance on Commensurate Line Filters

In the above sections for distributed networks, we arbitrarily assumed the line lengths to be $\lambda/8$. However, we also had the choice of a shorter or a longer line and correspondingly higher or lower line impedances, respectively. For instance,

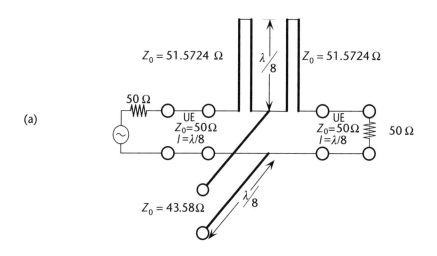

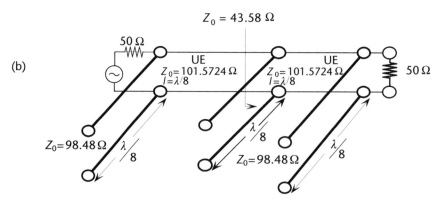

Figure 5.6 (a) Low-pass filter cascaded with UEs at the two ports, and (b) low-pass filter of Figure 5.6(a) after Kuroda's transform.

we could choose all line lengths equal to λ/4 and the characteristic impedance of 69.635 Ω for the first and the last shunt stubs, 30.815 Ω for the center stub, and 71.82 Ω for the two series stubs. In that case, it would be easier to physically realize the transmission lines, especially when planar transmission lines like microstrip or striplines are used. But due to the longer lines, resonance in a line will occur at a much lower frequency spacing and the first spurious passband will be situated at a frequency closer to the desired passband. This will result in a narrower stopband. Therefore, the choice of line length is a trade-off between physical realizability and the stopband width. In a commensurate line filter, all transmission line elements have the same electrical length at any frequency. A type of filter exists in which different TLEs have different electrical lengths at any frequency. Such filters are called noncommensurate line filters and have certain advantages over commensurate line filters. Having discussed a few important points regarding commensurate line filters we will now present the general analytical formulas that are useful in the design of such filters. For the finite commensurate line filter the transfer function can be expressed as a rational polynomial of the form

$$\left|S_{21}\left\{\tanh(\Theta)\right\}\right|^2 = \frac{\sum_{i=1}^{M} a_i \tanh^2(\Theta)}{\sum_{i=1}^{N} b_i \tanh^2(\Theta)} \tag{5.15a}$$

where

$$\Theta = \alpha l + j\beta l = \alpha l + j\omega\sqrt{\varepsilon\mu}\, l = \alpha l + j\theta \tag{5.15b}$$

and $\tanh(\Theta)$ may be regarded as the generalized complex frequency used in distributed line analysis. We may write

$$S = \tanh(\Theta) = \Sigma + j\Omega \tag{5.16}$$

In the case of a lossless line, $\Sigma = 0$. Therefore,

$$S = j\Omega = j\tan\theta = j\tan(\beta l) \tag{5.17}$$

Therefore, (5.15) assumes the form

$$\left|S_{21}\left\{j\tan(\theta)\right\}\right|^2 = \frac{\sum_{i=1}^{M} a_i \left\{\tan^2(\theta)\right\}^i}{\sum_{i=1}^{N} b_i \left\{\tan^2(\theta)\right\}^i} \tag{5.18}$$

Equation (5.18) can be compared with the following equation for the transfer function of a lumped element filter:

$$|S_{21}(\omega)|^2 = \frac{\sum_{i=1}^{M} a_i (\omega^2)^i}{\sum_{i=1}^{N} b_i (\omega^2)^i} \qquad (5.19)$$

However, since the mapping from ω to Ω involves the trigonometric function $\tan(\theta)$, the transfer function $S_{21}\{\tan(\theta)\}$ or $S_{21}(\Omega)$ or will have a periodic behavior in the θ or ω domain. Figure 5.7 compares the frequency response of an ideal lumped element low-pass filter with that of its distributed element counterpart and θ_0 is the electrical length of the lines corresponding to the radian frequency ω_0.

Distributed element networks belong to four categories:

1. Cascaded UE networks;
2. UE and shunt low-pass networks;
3. UE and shunt high-pass networks;
4. UE with inverters.

The first type of circuit does not have any transmission zero at any frequency, while the second type has a transmission zero at $S = 0$, and the third type has a transmission zero at $S = \infty$. It can be shown that a distributed element network must have the transducer gain function of the form [3]

$$|S_{21}(S)|^2 = \frac{K_m S^{2m}(1-S)^n}{P_{m+n+q}(S^2)} \qquad (5.20)$$

The denominator of the right-hand side $P_{m+n+q}(S)$ is a strictly Hurwitz polynomial, m is the number of high-pass elements, n is the number of unit elements, and q is

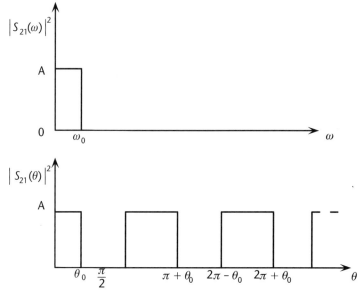

Figure 5.7 Comparison of ideal and distributed low-pass filter responses.

5.3 Effects of Line Length and Impedance on Commensurate Line Filters

the number of low-pass elements. K_m is a constant such that the modulus of the transducer gain is less than or equal to unity.

The above four types of low-pass filters can be realized at microwave frequencies and each has its own merits and demerits. For planar filter realization using microstrip, stripline, suspended stripline, and coaxial line, the cascaded UE filters find the most applications. Such a filter offers wide stopbands if the electrical length of each section is chosen sufficiently small. Using $m = q = 0$ in (5.20) gives

$$|S_{21}(j\Omega)|^2 = |S_{21}(S)|^2 = \frac{K_0(1-S^2)^n}{P_n(S^2)} = \frac{K_0(1+\Omega^2)^n}{\sum_{i=0}^{n} p_i \Omega^{2i}} \qquad (5.21)$$

Using

$$\tan^2(\theta) = \frac{\sin^2(\theta)}{1-\sin^2(\theta)} \qquad (5.22)$$

in (5.21) gives

$$|S_{21}(\Omega)|^2 = \frac{K_0}{\sum_{i=0}^{n} q_i \sin^{2i}(\theta)} \qquad (5.23)$$

For a Butterworth or maximally flat response, the above equation assumes the form

$$|S_{21}(\Omega)|^2 = \frac{K_0}{1+\varepsilon^2 \left\{\frac{\sin^2 \theta}{\sin^2 \theta_c}\right\}^{2n}} \qquad (5.24)$$

where θ_c is the electrical length of the transmission line at the cutoff frequency of the filter. Note that at $\theta = \theta_c$

$$|S_{21}(\Omega)|^2 = \frac{K_0}{1+\varepsilon^2} \qquad (5.25)$$

For a Chebyshev or equal ripple filter

$$|S_{21}(\Omega)|^2 = \frac{K_0}{1+\varepsilon^2 T_n^2 \left\{\frac{\sin^2 \theta}{\sin^2 \theta_c}\right\}} \qquad (5.26)$$

The cascade of unit elements is shown in Figure 5.8. This is an ideal representation of a cascaded transmission line low-pass filter network. The designer has to keep in mind that since it is a cascade of several transmission lines of different characteristic impedances, and adjacent lines have widely different impedance values, significant physical discontinuity occurs between any two consecutive lines. Such discontinuities are inductive and capacitive in nature and of large values that affect the filter performance considerably unless they are compensated for. Design formulas for a cascaded UE low-pass filter or stepped impedance low-pass filter are [4]

Butterworth ($R_g = R_L = 1$)

$$g_r = \frac{2\sin\left[(2r-1)\frac{\pi}{2N}\right]}{\bar{\alpha}}\left\{1 - \frac{\cos\left(\frac{\pi}{N}\right)\bar{\alpha}^2}{4\sin\left[(2r-3)\frac{\pi}{2N}\right]\sin\left[(2r+1)\frac{\pi}{2N}\right]}\right\} \quad (5.27)$$

where

$$\bar{\alpha} = \frac{\alpha}{\sqrt[N]{\varepsilon}} = \frac{\sin\theta_c}{\sqrt[N]{\varepsilon}} \quad (5.28a)$$

$$Y_r = g_r, \text{ for } r \text{ odd} \quad (5.28b)$$

$$Y_r = \frac{1}{g_r}, \text{ for } r \text{ even} \quad (5.28c)$$

Chebyshev

$$g_r = A_r\left[\frac{2\sin\left[(2r-1)\frac{\pi}{2N}\right]}{\alpha} - \frac{\alpha}{4}\left\{\frac{\eta^2 + \sin^2\left(\frac{r\pi}{N}\right)}{\sin\left[(2r+1)\frac{\pi}{2N}\right]} + \frac{\eta^2 + \sin^2\left(\frac{(r-1)\pi}{N}\right)}{\sin\left[(2r-3)\frac{\pi}{2N}\right]}\right\}\right] \quad (5.29a)$$

$$A_r = \frac{\left\{\eta^2 + \sin^2\left(\frac{(r-2)\pi}{N}\right)\right\}\left\{\eta^2 + \sin^2\left(\frac{(r-4)\pi}{N}\right)\right\}\cdots}{\left\{\eta^2 + \sin^2\left(\frac{(r-1)\pi}{N}\right)\right\}\left\{\eta^2 + \sin^2\left(\frac{(r-3)\pi}{N}\right)\right\}\cdots} \quad (5.29b)$$

where the last term $\eta^2 + \sin^2(0)$ is replaced by η. As a result

5.3 Effects of Line Length and Impedance on Commensurate Line Filters

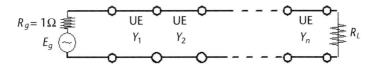

Figure 5.8 UE low-pass prototype.

$$A_2 = \frac{\eta}{\eta^2 + \sin^2\left(\frac{\pi}{N}\right)} \quad (5.29c)$$

and

$$R_L = 1 \text{ for } N \text{ odd}$$

or

$$R_L = \frac{\sqrt{1+\varepsilon^2} - \varepsilon}{\sqrt{1+\varepsilon^2} + \varepsilon} = \tanh^2\left\{\frac{N}{2}\sinh^{-1}\eta\right\} \quad (5.29d)$$

$$Y_r = g_r, \text{ for } r \text{ odd} \quad (5.29e)$$

$$Y_r = \frac{1}{g_r}, \text{ for } r \text{ even} \quad (5.29f)$$

5.3.1 Design Example

Design a stripline stepped impedance low-pass filter with the following specifications:

Passband ripple = 0.043 (−20 dB RL)
Cutoff frequency = 1.00 GHz
Number of sections $N = 5$
Substrate thickness = 0.062"
Section length $\theta_c = 30°$
At cutoff, $\theta = \theta_c = 30°$

Therefore, from (5.27), we get

$$|S_{11}|^2 = 1 - |S_{21}|^2 = 1 - \frac{1}{1+\varepsilon^2} = \frac{\varepsilon^2}{1+\varepsilon^2} \quad (5.30)$$

Also, from the given specifications

$$10\log\left(|S_{11}|^2\right) = -20 \tag{5.31}$$

Solving (5.30) and (5.31) gives

$$\varepsilon \approx 0.1$$

Using (5.29a) through (5.29f), we get

$$Y_1 = Y_5 = 0.0404 \text{ mho} \tag{5.32a}$$

$$Y_2 = Y_4 = 0.0085 \text{ mho} \tag{5.32b}$$

$$Y_3 = 0.0648 \text{ mho} \tag{5.32c}$$

Figure 5.9 shows the computed frequency response of the realized stripline filter using 3-D simulation. Figure 5.10 shows the 3-D view and dimensions of the layout of the filter. The strip widths were obtained by using (3.58) through (3.61). It can be seen that the analyzed filter has a cutoff frequency at 850 MHz instead of 1.00 GHz. This is due to the parasitic inductive reactance caused by the step discontinuities between different adjacent transmission lines.

The error in the cutoff frequency can be corrected by reducing the lengths of the high impedance lines and iteratively checking the response using a full-wave simulator [5]. Figure 5.11 shows the coaxial line version of the same filter. However, an important difference between the two implementations is that in the coaxial filter, the basic design from the low-pass filter has been optimized to have only two levels of impedance. Consequently all high-impedance sections have the same diameter and all low-impedance sections have the same diameter. At the same

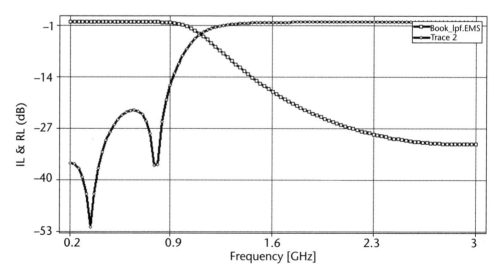

Figure 5.9 Computed frequency response of cascaded UE stripline low-pass filter.

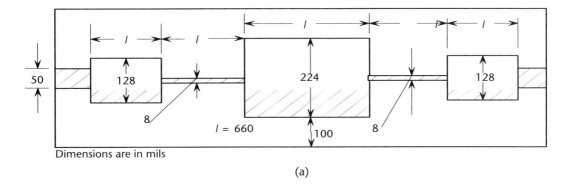

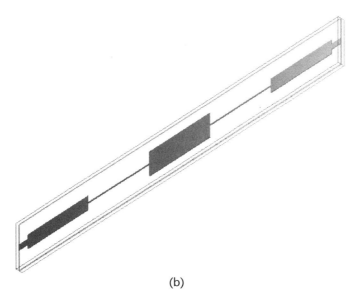

Figure 5.10 (a) Layout of stripline low.pass cascaded UE filter, and (b) 3-D view of a stripline low-pass filter (substrate thickness = 62 mils, $\varepsilon_r = 2.2$).

time, different high-impedance lines have different electrical lengths. Such structures are easier to implement than the ones having more than two impedance levels as obtained directly from the UE prototype. The filter was analyzed and optimized using WAVECON-Procap software [6]. A designer can also design the second type of filter, which is a noncomensurate line filter, directly by using the synthesis option in WAVECON [6].

5.3.2 Low-Pass Filter Prototype with a Mixture of UEs and Impedance Inverters

Cascaded UE low-pass filters are easy to implement using TEM or quasi-TEM transmission lines. For non-TEM transmission lines, a mixture of UEs and impedance inverters is the most efficient way to implement a low-pass filter. Examples of non-TEM transmission lines include waveguides, ridged waveguides, and finlines. Figure 5.12 shows the general configuration of a mixed UE inverter prototype network. The low-pass prototypes for a mixed UE and inverter are [7]

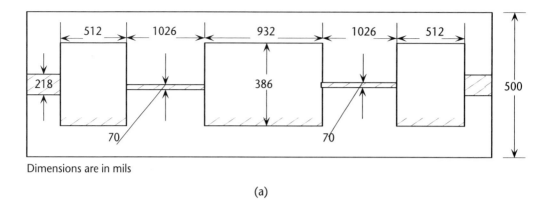

Dimensions are in mils

(a)

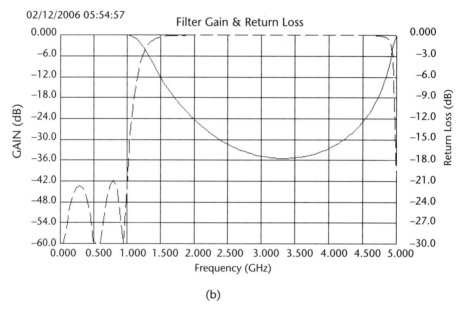

(b)

Figure 5.11 (a) Coaxial line version of a modified cascaded UE low-pass filter, and (b) frequency response of the filter shown in Figure 5.11(a).

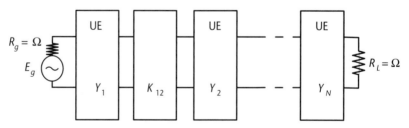

Figure 5.12 Mixed UE inverter distributed prototype network.

Butterworth

$$Y_r = \frac{2\sin\left[(2r-1)\dfrac{\pi}{2N}\right]}{\bar{\alpha}}\left\{1 - \frac{\cos\left(\dfrac{\pi}{N}\right)\bar{\alpha}^2}{4\sin\left[(2r-3)\dfrac{\pi}{2N}\right]\sin\left[(2r+1)\dfrac{\pi}{2N}\right]}\right\}; (r=1,2,\ldots,N) \quad (5.33\text{a})$$

$$K_{r,r+1} = 1 \quad (5.33\text{b})$$

Chebyshev

$$Y_r = \left[\frac{2\sin\left[(2r-1)\dfrac{\pi}{2N}\right]}{\alpha} - \frac{\alpha}{4\eta}\left\{\frac{\eta^2 + \sin^2\left(\dfrac{r\pi}{N}\right)}{\sin\left[(2r+1)\dfrac{\pi}{2N}\right]} + \frac{\eta^2 + \sin^2\left(\dfrac{(r-1)\pi}{N}\right)}{\sin\left[(2r-3)\dfrac{\pi}{2N}\right]}\right\}\right] \quad (5.34\text{a})$$

$$K_{r,r+1} = \frac{\sqrt{\eta^2 + \sin^2\left(\dfrac{r\pi}{N}\right)}}{\eta} \quad (r=1,2,\ldots,N) \quad (5.34\text{b})$$

$$\eta = \sinh\left(\frac{1}{N}\sinh^{-1}\frac{1}{\varepsilon}\right) \quad (5.34\text{c})$$

The most common application of mixed UE inverter distributed prototype networks is a waveguide harmonic-reject low-pass filter. Figure 5.13 shows the height profile of such a filter. The width of each section of the filter remains the same. Each vertical slot acts as a K-inverter.

The filter can also be realized using a ridged waveguide structure, as shown in Figure 5.14. While the full-height waveguide filter is suitable for high-power applications, the ridged-waveguide version offers a compact structure for low-power applications. In the latter version, the slots in the evanescent mode waveguide of reduced height waveguide sections realize the K-inverters.

The *ABCD* matrix of the slot is analyzed by using the mode-matching technique described in Chapter 3. Once the *ABCD* matrix is known, the K-inverter values obtained from (5.33) or (5.34) are normalized as

$$\bar{K}_{r,r+1} = \frac{K_{r,r+1}}{\sqrt{Z_r Z_{r+1}}} \quad (5.35)$$

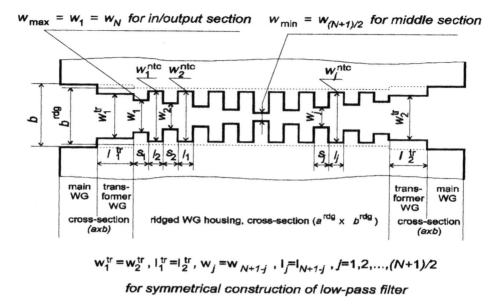

Figure 5.13 Height profile of a mixed UE inverter distributed prototype waveguide low-pass filter.

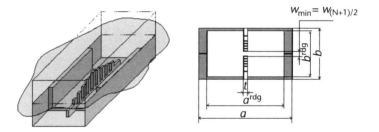

Figure 5.14 Distributed prototype ridged waveguide low-pass filter.

Figure 5.15 shows that the slot discontinuity not only realizes the inverter but also contributes the extra phase lengths ϕ_1 and ϕ_2. Therefore, the implementation of adjoining UEs must take into consideration the phase lengths. For example, if the phase lengths ϕ_1 and ϕ_2 are positive, the UEs on both sides of the slot must be shortened accordingly. The impedance inverter value is calculated from the *ABCD* parameters of the slot using the following equations [8]:

$$\bar{K}_{r,r+1} = \sqrt{Z_r Z_{r+1}} \left\{ \sqrt{L} + \sqrt{L-1} \right\} \qquad (5.36)$$

$$L = 1 + \frac{1}{4}\left\{ (a-d)^2 + (b-c)^2 \right\} \qquad (5.37)$$

$$a = A\sqrt{\frac{Z_{r+1}}{Z_r}} \qquad (5.38a)$$

5.3 Effects of Line Length and Impedance on Commensurate Line Filters

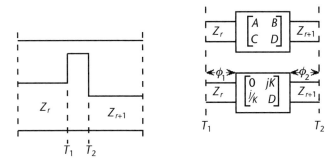

Figure 5.15 (a) Slot and (b) equivalent circuits.

$$b = \frac{B}{\sqrt{Z_r Z_{r+1}}} \tag{5.38b}$$

$$c = C\sqrt{Z_r Z_{r+1}} \tag{5.38c}$$

$$d = D\sqrt{\frac{Z_r}{Z_{r+1}}} \tag{5.38d}$$

$$\tan(2\phi_1) = \frac{2(bd - ac)}{(d^2 - a^2) + (b^2 - c^2)} \tag{5.38e}$$

$$\tan(2\phi_2) = \frac{2(ab - cd)}{(d^2 - a^2) + (b^2 - c^2)} \tag{5.38f}$$

where ϕ_1 and ϕ_2 are in radians. The impedance inverter values can also be expressed directly in terms of complex scattering parameters, as follows:

$$VSWR = \frac{1 + |S_{11}|}{1 - |S_{11}|} \tag{5.39a}$$

where S_{11} is the reflection coefficient seen at port one when the other port is terminated in its characteristic impedance. The reference plane locations are given by

$$\phi_1 = -\frac{\theta_{11}}{2} + \frac{\pi}{2} \tag{5.39c}$$

$$\phi_2 = -\frac{\theta_{22}}{2} + \frac{\pi}{2} \qquad (5.39d)$$

where $S_{11} = |S_{11}|e^{j\theta_{11}}$ and $S_{22} = |S_{22}|e^{j\theta_{22}}$. θ_{22} is the reflection coefficient seen at port two when the other port is terminated in its characteristic impedance. Figure 5.16 shows the analyzed frequency response of a ridged waveguide evanescent mode low-pass filter. The filter was designed and analyzed using WAVEFIL software by Polar Waves Consulting [9]. Such filters can offer stopbands extending above the fourth harmonic of the passband. The only limitation of a ridged waveguide evanescent mode low-pass filter is its power-handling capability. As a result it is useful in low-power applications. For medium-power applications, a corrugated height conventional waveguide low-pass filter, shown in Figure 5.17, is suitable. The design of a tapered corrugated waveguide low-pass filter [10,11] is accomplished as follows.

The tapered corrugated low-pass filter consists of a cascade of generalized impedance inverters connected by short sections of transmission lines, each of electrical length θ_0. The generalized impedance inverter is realized by a capacitive iris. Figure 5.18(a) shows the general profile of a tapered corrugated low-pass filter. Figure 5.18(b) shows the equivalent circuit.

The common performance parameters of the filter are defined by the

- Cutoff frequency f_c

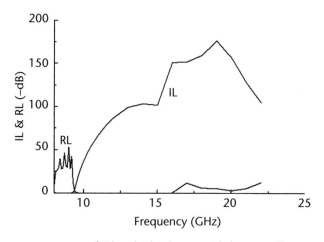

Figure 5.16 Frequency response of X-band ridged waveguide low-pass filter.

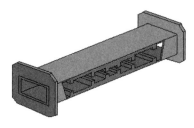

Figure 5.17 Tapered corrugated low-pass filter.

5.3 Effects of Line Length and Impedance on Commensurate Line Filters

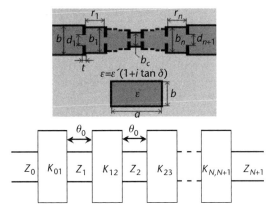

Figure 5.18 Tapered corrugated low-pass filter profile and equivalent network.

- Fractional ripple bandwidth

$$w_g = \frac{4\theta_0}{\pi} \quad (5.40)$$

where θ_0 is the electrical angle corresponding to the cutoff frequency of the low-pass filter.
- Passband return loss, RL(dB)
- Passband ripple, ε
- Isolation bandwidth factor (γ), which is the ratio of the cutoff frequency f_c and the stopband frequency f_s and is given by

$$\gamma = \frac{f_s}{f_c} \quad (5.41)$$

- Stopband isolation L_A at f_s.

In addition to the performance parameters, the cross-sectional dimensions of the waveguide (width and height), taper profile, iris thickness, and the number of modes to be considered form a part of the design specifications. Once the design parameters are obtained, the scaling factor is calculated as

$$\alpha = \sin\left(\frac{\pi w_g}{4}\right) \quad (5.42a)$$

The impedances of the distributed elements are obtained as

$$Z_r = \left[\frac{2\sin\left[(2r-1)\dfrac{\pi}{2N}\right]}{\eta\alpha} - \frac{\alpha}{4\eta}\left\{\frac{\eta^2 + \sin^2\left(\dfrac{r\pi}{N}\right)}{\sin\left[(2r+1)\dfrac{\pi}{2N}\right]} + \frac{\eta^2 + \sin^2\left(\dfrac{(r-1)\pi}{N}\right)}{\sin\left[(2r-3)\dfrac{\pi}{2N}\right]}\right\}\right] \quad (5.42b)$$

for $r = 1, 2, \ldots, N$.

where N is the order of the filter. The K-inverter values are obtained from (5.33b) or (5.34b) and (5.35). However, the inverter values need to be further normalized due to the tapered height profile of the waveguide because each waveguide section has a different characteristic impedance due to a different height or b dimension. Therefore, the normalized k-inverter values are

$$k'_{r,r+1} = \sinh\left[\frac{1}{N}\sinh^{-1}\frac{1}{\varepsilon}\right] \quad (5.42c)$$

In a uniform corrugated waveguide filter, the characteristic impedances of the inductive sections are identical and hence we scale the impedance Z_N to unity and the K-inverters are normalized as

$$K_{r,r+1} = \frac{k'_{r,r+1}}{\sqrt{Z_r Z_{r+1}}}, \; r = 0, 1, \ldots, N \quad (3.42d)$$

$$Z_0 = Z_N = 1 \quad (3.42e)$$

In tapered corrugated waveguide low-pass filters, the characteristic impedances of the inductive sections are not equal, and hence normalization of the K-inverter values with characteristic impedances is done so that the quantity $K/\sqrt{Z_r Z_{r+1}}$ is made invariant when matching a practical network to a prototype. The normalized K-inverter values are

$$K'_{r,r+1} = \frac{K_{r,r+1}}{\sqrt{Z^w_r Z^w_{r+1}}} \quad (3.42f)$$

with

$$r = 1, 2, \ldots, N-1 \quad (5.42g)$$

and

$$Z^w_0 = Z^w_{N+1} = 1$$

Since Z^w_r, the characteristic impedance of the rth waveguide section is given by (2.12a) as

$$Z^w_r = \frac{2b_r}{a}\frac{\eta_0}{\sqrt{1-\left(\frac{\lambda_0}{2a}\right)^2}} \quad (5.42h)$$

Considering the tapering of heights of each waveguide section, (5.35) can be written as

5.3 Effects of Line Length and Impedance on Commensurate Line Filters

$$\bar{\bar{K}}_{r,r+1} = \frac{\bar{K}_{r,r+1}}{\sqrt{\xi_r \xi_{r+1}}} \quad r = 1, 2, \ldots, N \tag{5.42i}$$

with

$$\xi_r = \frac{b_r}{b} \tag{5.42j}$$

A number of tapering profiles are used in the tapered corrugated waveguide low-pass filters. Those are as follows:

1. Square

$$t_i = t_c + \alpha(i-k)^2$$
$$\alpha = \frac{t - t_c}{k^2} \tag{5.43a}$$

2. Cosine

$$t_i = t - (t - t_c)\sin\left\{\frac{i\pi}{(N+1)}\right\} \tag{5.43b}$$

3. Exponential

$$t_i = te^{2\alpha i}$$
$$\alpha = -\frac{\ln\left(\frac{t}{t_c}\right)}{N+1} \tag{5.43c}$$

4. Linear

$$t_i = t - \alpha i$$
$$\alpha = \frac{t - t_c}{N} \tag{5.43d}$$

with

$$t_i = t_{N-i+1} \quad \text{and} \quad k = \frac{N+1}{2}$$
$$i = 1, 2, \ldots, k \tag{5.43e}$$

In the above equations t_c is the height of the central section of the waveguide and t is that of the input and output sections. The scattering matrix of each capacitive iris is analyzed by using a suitable numerical electromagnetic analysis method, preferably the mode-matching method described in Chapter 2. The impedance inverter value is obtained from (5.36) through (5.39). Finally, the section lengths between consecutive irises are obtained from

$$l_r = \frac{\lambda_{g0}}{2\pi}\left[\frac{\pi}{8} - \frac{(\phi_r + \phi_{r+1})}{2}\right] \qquad r = 1, 2, \ldots, N \qquad (5.43f)$$

λ_{g0} is the guide wavelength at the cutoff frequency of the filter and ϕ_r is obtained from (5.38) or (5.39). Table 5.1 shows the specifications and the synthesized dimensions of the filter. Figure 5.19 shows the computed frequency response of a corrugated waveguide low-pass harmonic rejection filter.

Corrugated waveguide harmonic reject filters are quite popular for their moderate power-handling capability and simple configuration. However, the main pitfall lies in the construction of the filter. Any misalignment whatsoever gives rise to higher-order TE_{n0} type modes with significant amplitude in the desired stopband. Consequently, manufacturing tolerance is a significant issue in realization of corrugated waveguide filters. The occurrence of higher-order or mode spikes in the stopband is given by

$$f_n = \sqrt{f_1^2 + (n^2 - 1)f_{c1}^2} \qquad n = 2, 3, \cdots \qquad (5.44)$$

where f_1 is the frequency above f_{01} (cutoff frequency of the low-pass filter), at which the dominant mode attenuation is a few decibels and f_{c1} is the cutoff frequency of the dominant mode of the waveguide. The appearance of undesired higher-order mode problem is compounded when the height (the b dimension) is increased in order to raise the power-handling capability. Figure 5.20 shows a waffle-iron filter and its assembly. In that case not only the TE_{n0} but also the TE_{1n} modes come into play. Higher-order modes in a corrugated waveguide are suppressed by cutting

Table 5.1 Specifications and Dimensions of a Tapered Corrugated Low-Pass Filter*

Number of sections, $N = 9$
Passband ripple level, $Lr = 0.010$ dB
Cutoff frequency, $f_o = 9.000$ GHz
Waveguide width, $a = 22.860$ mm
Waveguide height, $b = 10.160$ mm
Iris thickness, $t = 1.000$ mm

n	d_n (mm)	r_n (mm)	b_n (mm)
0	3.83183	10.04563	10.16000
1	1.29483	7.57539	10.16000
2	0.72952	6.66513	10.16000
3	0.61075	6.42297	10.16000
4	0.58059	6.37095	10.16000
5	0.58059	6.42297	10.16000
6	0.61075	6.66513	10.16000
7	0.72952	7.57539	10.16000
8	1.29483	10.04563	10.16000
9	3.83183	—	—

*Please see Figure 5.18 for definitions of terms

5.3 Effects of Line Length and Impedance on Commensurate Line Filters 229

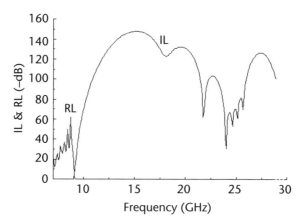

Figure 5.19 Frequency response of corrugated waveguide harmonic reject low-pass filter.

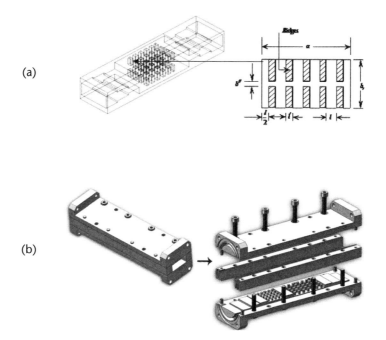

Figure 5.20 (a) Igut waffle-iron filter, and (b) mechanical assembly of a waffle-iron filter. (Reprinted with permission from Stellenbosch University, SA; courtesy Prof. Petrie Meyer and Ms. Susan Maas.)

multiple longitudinal slots in the corrugated waveguide filter. This process renders each capacitive iris into a section of multiple ridge waveguides, as shown in Figure 5.20(a). A corrugated waveguide filter with mode-suppressing longitudinal slots is known as a waffle-iron filter [12–14]. As can be seen from the figure, the filter has a main center section with slotted and corrugated parts and two multisection stepped impedance transformers at the two ends. Like a corrugated waveguide filter, a waffle-iron filter can be designed using the same procedure. The overall filter response can be optimized using the mode-matching method. Figure 5.20(b) shows the mechanical assembly of a waffle-iron filter.

5.3.3 Quasi-Distributed Element TEM Filters

Besides coaxial line and waveguide technology, planar transmission line technology plays an important role in microwave low-pass filter realization. In many applications, a low-pass filter with very steep skirt selectivity (around −70 dB within above 10% of the cutoff frequency) is a necessity. Typical examples are cellular radio base stations and radar warning receiver multiplexers. Conventional Chebyshev filters are inadequate for these types of applications. As a result, a pole extraction technique using elliptic or generalized Chebyshev responses, described in Chapter 4, is used. Figure 5.21 shows the general lumped element prototype of an elliptic or generalized Chebyshev filter. The element values are obtained by using the method described in Chapter 4 or the NTK-CKTTOOL software [15].

The first step is to scale the prototype values to actual terminating impedances and the cutoff frequency because the prototype shown in Figure 5.21 is valid for 1Ω terminating impedances and 1-radian/sec cutoff frequency. The procedure is as follows:

Impedance scaling
Let $R_g = R_L = R$. Then

1. Replace all inductors by

$$L'_r = L_r R \tag{5.44a}$$

2. Replace all capacitors by

$$C'_r = \frac{C_r}{R} \tag{5.44b}$$

3. Replace all resistors by

$$R'_r = R_r R \tag{5.44c}$$

Frequency scaling
Let the actual cutoff frequency be $\omega = \omega_c$ radians/sec. Then

1. Replace all inductors by

$$L_r^{\omega_c} = \frac{L'_r}{\omega_c} \tag{5.45a}$$

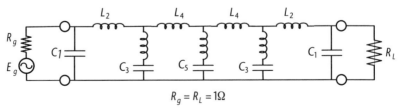

Figure 5.21 Low-pass prototype of elliptic or generalized Chebyshev filter.

2. Replace all capacitors by

$$C_r^{\omega_c} = \frac{C_r'}{\omega_c} \qquad (5.45b)$$

Having performed the impedance and frequency scalings, replace the series inductors by a short section of very high-impedance transmission line (microstrip, stripline, or suspended stripline) of length

$$l_r^L = \frac{1}{\beta_r}\sin^{-1}\left[\frac{\omega_c L_r^{\omega_c}}{Z_r}\right] \qquad (5.46a)$$

where Z_r is an arbitrarily chosen but very high and physically realizable impedance. The shunt capacitances are replaced by a short sections of transmission line of characteristic impedance Z_r and length

$$l_r^C = \frac{1}{\beta_r}\tan^{-1}\left[\frac{Z_r}{\omega_r C_r^{\omega_c}}\right] \qquad (5.46b)$$

Obviously, since transmission lines of different lengths and impedances realize different elements of the filter, the filter is a noncommensurate line filter. Also, the designer should note that β_r and Z_r are mathematically related in case of inhomogeneous transmission lines such as microstrip and suspended stripline. When using homogeneous transmission lines such as stripline or coaxial line, β_r and Z_r are mathematically unrelated. An open-circuit quarter-wavelength stub of characteristic impedance Z_0 replaces the series resonant shunt elements comprising of L and C. Z_0 is given by

$$Z_0 = \frac{4}{\pi}\sqrt{\frac{L}{C}} \qquad (5.46c)$$

Note that all open-circuited stubs must be shortened owing to the effects of open-end capacitance. Also all line lengths should be adjusted to nullify the effects of the T-junction and step junction discontinuities (see Chapter 3).

Shunt series resonant circuits can also be realized using a compound stub. A compound stub is a combination of a UE and an open stub. The transformation of series shunt element into a compound stub is shown in Figure 5.22. The equivalence uses Kuroda identity and the element values are given by

$$Z_{UE} = Z_L + Z_C \qquad (5.46d)$$

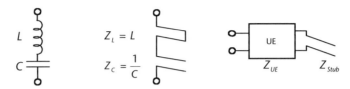

Figure 5.22 Commensurate equivalent of a shunt series resonant circuit.

$$Z_{stub} = \frac{Z_C}{Z_L}(Z_L + Z_C) \qquad (5.46e)$$

In today's world of computer-aided design, it is not necessary to manually perform the above computations in order to design a planar low-pass filter. For example, WAVECON [6] is an excellent tool for this purpose. Table 5.2 shows the WAVECON analysis and design file of a stripline elliptic low-pass filter realized on a 62-mils RT-Duroid™ substrate ($\varepsilon_r = 2.22$). Figure 5.23(a) shows the layout of the filter and Figure 5.23(b) shows the computed frequency response.

The above technique of realizing a low-pass filter can be used with any suitable planar transmission line medium. The basic rule for realization of quasi-lumped elements remains the same for all such transmission lines. One such transmission line that offers the convenience of printed circuit board techniques and the highest Q-factor among most planar transmission lines is the suspended microstrip line. The subject of suspended microstrip line has been described in detail in Chapter 3. The discontinuities involved in suspended microstrip line filters can be approximated by the corresponding homogeneous stripline discontinuities for all practical purposes, provided that the substrate is thin and the dielectric constant is low and

Table 5.2 Wavecon Analysis and Design File for Stripline Low-Pass Filter

Stripline Shunt Stub Low-Pass Filter

5.0000	GHz Cutoff Frequency	5.0000	GHz Bandwidth
0.0430	dB Ripple	5	Poles
0.0400	Inches Ground-plane Spacing	2.2200	Dielectric Constant
0.0005	Inches Conductor Thickness	62.673	Ohms Internal Impedance
50.000	Ohms Input Line Impedance	0.0314	Inches Input Line Width
50.000	Ohms Output Line Impedance	0.0314	Inches Output Line Width

Sect Numb	Element Value	Zshunt Ohms	Zseries Ohms	Lshunt Inches	Lseries Inches	Wdth-Shnt Inches	Width-Ser Inches
1	0.9705	36.373	—	0.1531	—	0.0499	—
2	1.3719	—	107.98	—	0.1894	—	0.0059
3	1.8004	36.373	—	0.2421	—	0.0499	—
4	1.3719	—	107.98	—	0.1894	—	0.0059
5	0.9705	36.373	—	0.1531	—	0.0499	—

5.4 High-Pass Filter Design 233

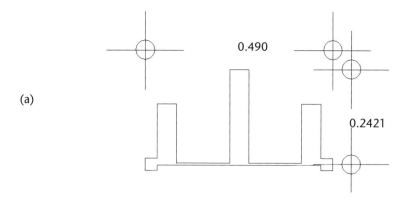

(a)

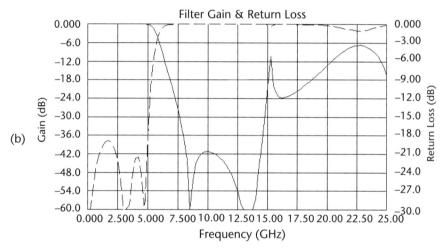

(b)

Figure 5.23 (a) WAVECON layout of an extracted pole stripline low-pass filter, and (b) WAVECON analyzed frequency response of the low-pass filter shown in Figure 5.23(a).

the strips are symmetrically placed between the top and bottom ground planes. However, the overall circuit can be optimized using full-wave analysis [5].

5.4 High-Pass Filter Design

A true high-pass filter is realized in lumped element form only. Any distributed element high-pass filter behaves like a wideband band-pass filter. In microwave technology, the art of high-pass filter realization differs considerably when lumped, TEM element, or waveguide elements are used. However, in many applications a wideband band-pass filter is used instead of a high-pass filter.

5.4.1 Low-Pass to High-Pass Transformation

Consider the frequency response of a low-pass filter with cutoff frequency at ω_c radians/sec, shown in Figure 5.24. From (4.29) the general equation for the transfer function shown in Figure 5.24 can be written as

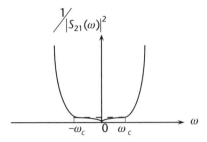

Figure 5.24 General low-pass filter response.

$$|S_{21}(\omega)|^2 = \frac{1}{1 + F^2\left(\dfrac{\omega}{\omega_c}\right)} \tag{5.47}$$

Replacing ω/ω_c by ω_c/ω in (5.47) gives

$$|S_{21}(\omega)|^2 = \frac{1}{1 + F^2\left(\dfrac{\omega_c}{\omega}\right)} \tag{5.48}$$

where F is the approximation function defined in Section 4.2, which is the transfer function of a high-pass filter with cutoff frequency at ω_c. Figure 5.25 shows the graphical representation of (5.48). Let us reconsider Figure 4.6 of a low-pass prototype circuit, as shown in Figure 5.26.

The above networks assume the forms shown in Figure 5.27 when low-pass to high-pass transformation is applied. Figure 5.28 shows the high-pass prototype circuit with impedance inverters.

Derivation of element values in the above networks follows the same procedure as described for low-pass prototypes in Chapter 4. However, such a derivation is preceded by low-pass to high-pass transformation given by (5.47) and (5.48).

5.4.2 Quasi-Lumped Element High-Pass Filter

At RF and microwave frequencies, a high-pass filter is easily realized using a quasi-lumped element approach. In that approach, the series capacitors are realized in lumped element form and the shunt inductors are realized by a section of short-circuited transmission line. A coaxial line version of the filter is shown in Figure 5.29. The structure can be modeled and optimized using full-electromagnetic analysis. However, the initial design should use series gap discontinuity capacitance models for thick striplines for the capacitances (see (3.75) to (3.79)). In case the gap capacitance is inadequate, one can use a disk capacitor (circular or rectangular, as shown in Figure 5.30) in place of the thick stripline gap. The capacitance of a circular disk capacitor is given by

5.4 High-Pass Filter Design

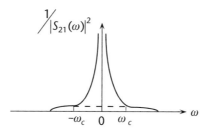

Figure 5.25 Low-pass to high-pass transformation.

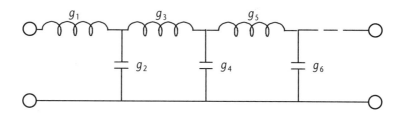

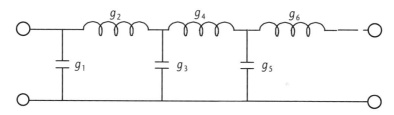

Figure 5.26 Low-pass prototype with unity cutoff and terminations.

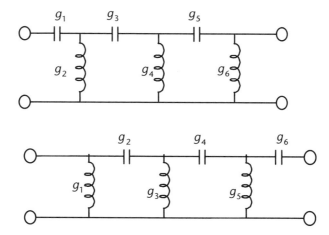

Figure 5.27 Low-pass prototype transformed to high-pass.

$$C = \varepsilon_0 \varepsilon_r \frac{\pi r^2}{d} \, pf \qquad (5.49)$$

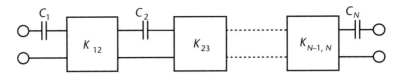

K-inverter couple highpass prototype

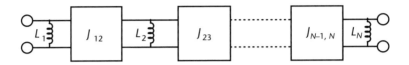

J-inverter couple highpass prototype

Figure 5.28 High-pass prototype involving impedance and admittance inverters.

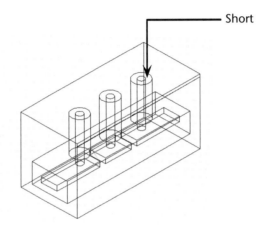

Figure 5.29 Coaxial line high-pass filter.

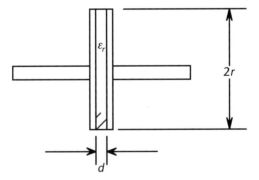

Figure 5.30 Circular disk capacitor.

where $\varepsilon_0 = 0.225$ pf/inch is the permittivity of free space. The length of a shorted coaxial line shunt inductor is obtained from

$$L = 0.085 Z_0 l \quad \text{Henrys} \tag{5.50}$$

5.4 High-Pass Filter Design

All dimensions in the above equations are in inches. Although the capacitors are lumped, the inductors can be either shorted stubs or very short lumped inductors. The passband of the filter may not be very wide if stubs are used.

5.4.3 Levy's Procedure for Planar High-Pass Filter Design

In 1982 Peter LaTourrette [16] proposed a method for the design of high-pass filters that accommodated the physical length of the series capacitors and thereby guaranteed more accurate prediction of passband return loss of the filter. His work presented several configurations that facilitated the practical realization of the filters particularly in suspended stripline. Based on Peter LaTourrette's work, Ralph Levy presented a unified powerful procedure for the realization of planar high-pass filters [17] using suspended stripline. The design procedure is based on the homogeneous distributed prototype filter, shown in Figure 5.31. The transfer function of the filter may be written using the transformed variable theory of Robert Wenzel [18]. The variable W is defined in terms of Richard's variable S via the conformal transformation

$$W^2 = 1 + \frac{S^2}{\Omega_c^2} \tag{5.51}$$

where

$$S = j\tan\theta \tag{5.52}$$

and θ is the commensurate electrical length, and the transformed cutoff frequency Ω_c is given by

$$\Omega_c = j\tan\theta_c \tag{5.53}$$

where θ_c is the electrical length at the equiripple passband edge. Figure 5.31 shows the Nth- (odd N) order prototype with $(N-1)/2$ shunt series resonant elements and transmission zero producing elements and $(N+1)/2$ UEs. All series capacitors produce a single-order pole at dc. Each shunt resonant element produces a double-order transmission zero at a finite frequency $S = j\Omega_p$. Each UE produces a single-order transmission zero at $S = j1$. The corresponding Chebyshev approximation problem is solved by forming the function

$$f(W) = \prod_{i=1}^{N} \sqrt{\frac{W_i + W}{W_i - W}} \tag{5.54a}$$

where

$$\prod_{i=1}^{N}(W_i + W) = (1+W)\left\{\sqrt{1-\frac{\Omega_p^2}{\Omega_c^2}}+W\right\}^{\frac{(N-1)}{2}}\left\{\sqrt{1+\frac{1}{\Omega_c^2}}+W\right\}^{\frac{(N+1)}{2}} \tag{5.54b}$$

Using the above equations, the equiripple transfer function becomes

$$\frac{P_0}{P_L} = 1 + \varepsilon^2 \left| \frac{f(W) + f(-W)}{2} \right|^2 \quad (5.55)$$

Figure 5.31 shows that each UE is followed and preceded by two identical capacitors. Therefore, the synthesis of the network from the transfer function has to be done accordingly. It starts with the extraction of a UE, followed by a partial extraction of a series capacitor of a value such that a shunt series resonant circuit can be extracted next. Using the Kuroda transform, a part of the series capacitor is transferred to the other side of the UE so that the two capacitors on both sides of the UE are equal. Then the series shunt resonators are extracted, and the extraction and Kuroda transformation cycle is repeated to form the entire network. Peter LaTourette [16] showed how to use a commercial circuit solver [19] to accomplish the synthesis. Figure 5.32 shows the evolution and physical realization of such high-pass filters is best done in suspended stripline or microstripline because the series capacitor values are usually too high to be realized by gap capacitors. Let us consider UEs with identical series capacitors on both sides, as shown in Figure 5.31. Figure 5.32 shows in a step-by-step fashion how a coupled line section with extra identical transmission lines on either side can represent the UE capacitor combination.

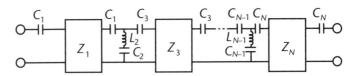

Figure 5.31 Homogeneous distributed high-pass filter.

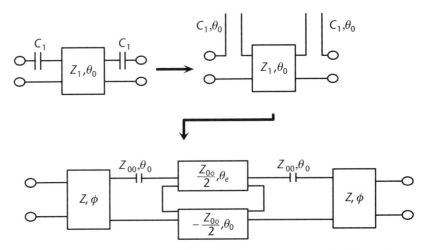

Figure 5.32 Homogeneous prototype section and inhomogeneous coupled line section.

Following the arguments in [17], the above network can be further simplified for all practical purposes as shown in Figure 5.33. In the approximate equivalent circuit, the following relations hold:

$$Z_{0e} = \frac{1}{C_1} \tag{5.56a}$$

$$\frac{Z_{0e} - Z_{0o}}{2} = Z_1 \tag{5.56b}$$

Solving the above equation, Z_{0e} and Z_{0o} are obtained. Knowing these values, direct synthesis equations (3.136) to (3.141) can be used for the dimensions if a homogeneous broadside-coupled suspended stripline, shown in Figure 3.29, is used. If the broadside-coupled suspended microstripline shown in Figure 3.39 is used, then an optimization-based synthesis routine should be used. Such a routine may use the models in (3.267) to (3.278). The quarter-wavelength frequency for all commensurate line lengths, including the series ones, is chosen on the basis of the physical layout or performance. However, it must be as high as the midpoint of the passband.

The shunt-series resonant circuits can be realized as compound stubs shown in Figure 5.22, using (5.46). The overall layout is shown in Figure 5.34. The design method will guarantee first-pass success if the junction parasitics of the transmission line are correctly taken into consideration using accurate analysis of the designed circuit. However, using a full EM solver, the initial design based on the theory can be optimized before actual fabrication.

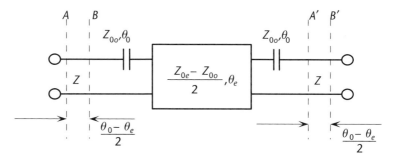

Figure 5.33 Approximate equivalent circuit of an inhomogeneous coupled line section.

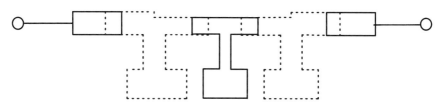

Figure 5.34 Pseudoelliptic planar high-pass filter. Solid and dotted lines represent conductor patterns on either side of the broadside-coupled line.

Besides Levy's systematic approach one can also begin with the basic lumped element prototype and follow Peter LaTourrette's procedure to arrive at the final form of the circuit. Let us consider a 6 to 18 GHz high-pass filter that has greater than 60 dB rejection at a frequency 15% below the cutoff frequency. Figure 5.35 shows the step-by-step procedure, starting from a lumped element prototype and then following Peter LaTourrette's procedure to arrive at the final form of the circuit.

Levy designed the filter for the same specification but used five UEs instead of four. Figure 5.36 shows Levy's prototype equivalent circuit and the layout of the circuit board in suspended stripline.

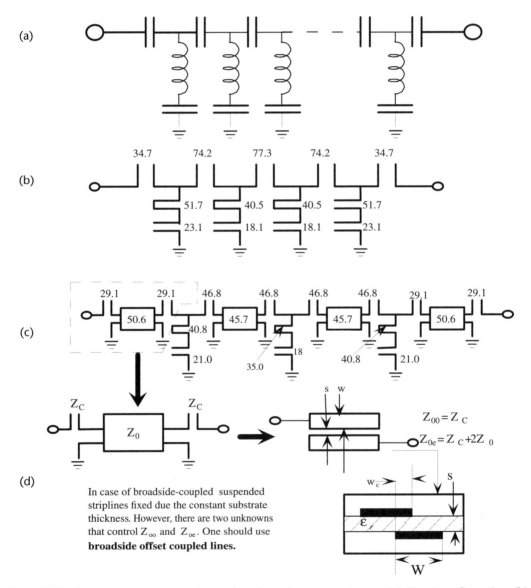

Figure 5.35 Step-by-step procedure starting from basic low-pass prototype. (a) Circuit configuration, (b) distributed equivalent, (c) distributed equivalent using UEs and Szentimei's [19] software S/FILSYN, and (d) parallel-coupled line equivalence of a UE with identical stubs on both sides and offset broadside-coupled line realization.

5.4 High-Pass Filter Design

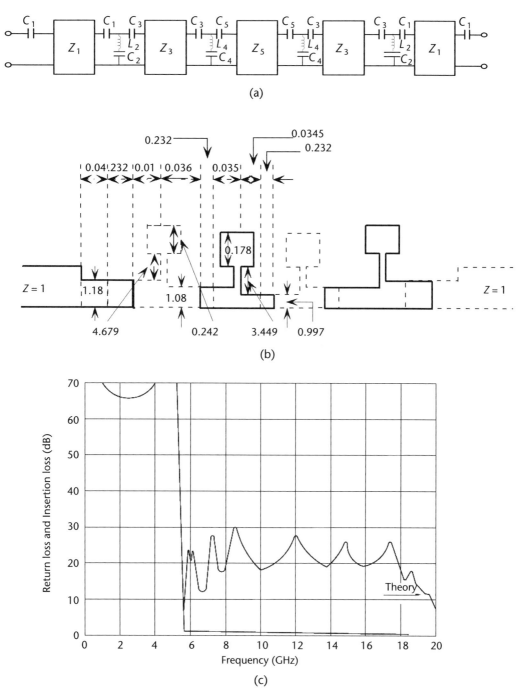

Figure 5.36 (a) Equivalent circuit of Levy's five-unit element design. C1 = 11,088, C2 = 4.3539, C3 = 4.7867, C4 = 5.9081, C5 = 4.6494, Z1 = 1.0801, L2 = 4.4497, Z3 = 1.0640, L4 = 3.2791, Z5 = 1.0629. (b) Layout of the circuit board. Substrate thickness = 0.015 inch, ground plane spacing = 0.100 inch, substrate εr = 2.22. (c) Measured frequency response of a 6 to 18 GHz filter. (Courtesy of Dr. Ralph Levy.)

It is worthwhile to mention that RF and microwave high-pass filters can be realized in microstrip form very conveniently by using planar shunt stub inductors as well as resonant circuits and commercially available high-Q surface-mountable lumped chip capacitors.

5.4.4 Waveguide High-Pass Filter Design

Waveguide high-pass filters are best designed by using a section of below-cutoff waveguide. Since the waveguide operates in the evanescent mode, it acts like an attenuator where the attenuation depends on the length of the section. The propagation constant of a waveguide is given by

$$\gamma = \sqrt{\left(\frac{2\pi}{\lambda_0}\right)^2 - \left(\frac{2\pi}{\lambda_c}\right)^2} = \alpha \quad \text{Nepers/unit length} \quad (5.57)$$

where λ_0 is the operating and λ_c is the fundamental mode cutoff wavelength of the waveguide. When the operating frequency is below the cutoff frequency, then $\lambda_c > \lambda_0$. Consequently α is a positive real number. If the physical length of the below-cutoff waveguide is l, then the wave will undergo an attenuation of $8.686\,\alpha l$ dB. Exactly at the cutoff frequency of the waveguide the waveguide will switch to a propagating mode and the wave will propagate unattenuated above the cutoff. Therefore, if the width of the waveguide is so chosen as to have the same fundamental mode cutoff frequency as the desired cutoff frequency of the high-pass filter, then the structure will behave as the desired high-pass filter and the stopband attenuation can be controlled by controlling the length of the waveguide. The guide-width at either end of the filter can be matched with standard propagating waveguides using taper transitions, as shown in Figure 5.37. The overall filter can be optimized using mode,matching or any other suitable method. As an example, let us consider the design of an X-band high-pass filter with cutoff at 8.33 GHz. The width of a rectangular waveguide with fundamental mode cutoff at 8.33 GHz is 18 mm ($\lambda_c = 36.00$ mm). Figure 5.38 shows the frequency response of a 60-mm long section of the waveguide. The simulation and optimization of the filter was obtained by using the mode-matching method based software WASP-NET from Microwave Innovation Group in Germany [25].

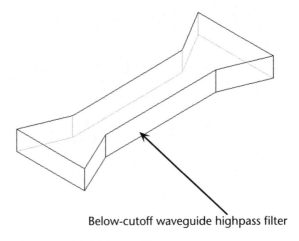

Figure 5.37 Rectangular waveguide high-pass filter.

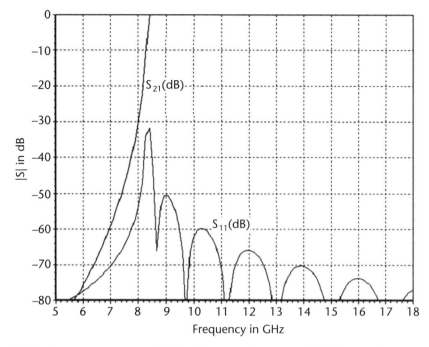

Figure 5.38 Frequency response of X-band rectangular waveguide high-pass filter (WASP-NET analysis).

5.5 Band-Stop Filter Design

Rejection of an undesired band of frequencies in a microwave system is, in most cases, accomplished by the use of a low-pass, high-pass, or band-pass filter. However, in certain cases, the rejection of a relatively narrow band of frequencies with very high rejection levels cannot be achieved without the use of a band reject or band-stop filter. By definition [20], a band-stop filter must pass signals below and above a pair of specified passband edge frequencies (ω_1, ω_2) with a specified return loss and it must reject or attenuate signals between stopband edge frequencies (ω_3, ω_4) by specified values of minimum loss where $\omega_1 < \omega_3, < \omega_4 < \omega_2$. The band-stop filter designer will obtain these values from the imposed requirements for the filter, modifying them to include margins for practical tolerances, tuning capability environmental conditions, and any other rule of thumb based on experience.

5.5.1 Low-Pass to Band-Stop Transformation

Let us reconsider the basic low-pass prototype as shown in Figure 5.39.

Using the frequency transformation [21]

$$\frac{1}{\omega'} = \frac{1}{\varpi \omega_1'} \left(\frac{\omega}{\omega_0} - \frac{\omega_0}{\omega} \right) \tag{5.58a}$$

where ω' is the radian frequency of the low-pass prototype filter and ω is that of the band-stop filter. The rest of the parameters are defined in Figure 5.40, which is the

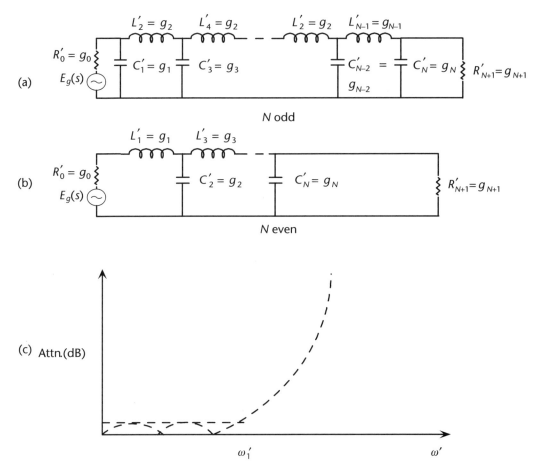

Figure 5.39 Basic low-pass prototype network and the response.

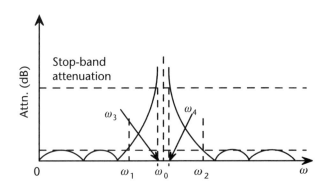

Figure 5.40 Frequency response of band-stop prototype.

response of the prototype band-stop filter obtained from using the above frequency transformation in the low-pass response in Figure 5.39.

$$\omega_0 = \sqrt{\omega_1 \omega_2} \tag{5.58b}$$

5.5 Band-Stop Filter Design

$$\varpi = \frac{\omega_2 - \omega_1}{\omega_0} \tag{5.58c}$$

Table 5.3 summarizes the above mapping. Figure 5.41 shows the corresponding band-stop filter networks

The element values of the band-stop filter prototype are given by

$$x_i = \omega_0 L_i = \frac{1}{\omega_0 C_i} = \frac{1}{\varpi \omega_1' g_i} \tag{5.59}$$

for series branches

$$b_j = \omega_0 C_j = \frac{1}{\omega_0 L_j} = \frac{1}{\varpi \omega_1' g_j} \tag{5.60}$$

Table 5.3 Low-Pass to Band-Stop Mapping Table

ω'	ω
	0 or ∞
	$\omega_{1,2} = \omega_\infty \left\{ \left(1 + \frac{\varpi^2}{4}\right)^{\frac{1}{2}} \pm \frac{\varpi}{2} \right\} \approx \omega_0 \left(1 \pm \frac{\varpi}{2}\right)$ when $\pi \ll 1$
	ω_0

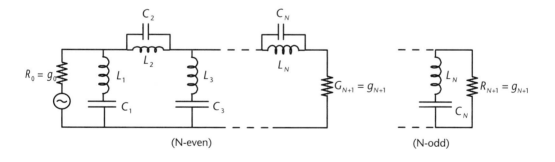

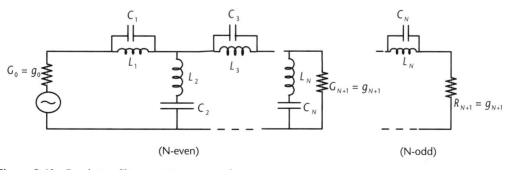

Figure 5.41 Band-stop filter prototype networks.

for shunt branches.

The reactance slope parameter x of a reactance $X = \omega L - 1/\omega C$ at ω_0 is

$$x = \left.\frac{\omega_0}{2}\frac{dX}{d\omega}\right|_{\omega=\omega_0} = \omega_0 L = \frac{1}{\omega_0 C} \qquad (5.61)$$

and the susceptance slope parameter b of a susceptance $B = \omega_0 C - 1/\omega_0 L$ at ω_0 is

$$b = \left.\frac{\omega_0}{2}\frac{dB}{d\omega}\right|_{\omega=\omega_0} = \omega_0 C = \frac{1}{\omega_0 L} \qquad (5.62)$$

The above results are included in (5.59) and (5.60). The networks shown in Figure 5.41 have the same impedance levels as the corresponding low-pass prototypes in Figure 5.39. However, when the impedance level is different, then each L and R has to be multiplied by the new impedance and each G and C should be divided by the new impedance. Once again, prototypes involving inverters (impedance or admittance) are most advantageous in microwave band-stop filter realization. Such networks are shown in Figure 5.42. Out of these four prototypes, the first two, (a) and (b), involving shunt series resonant circuits are suitable for transmission line implementation using shunt stubs. The other two, series parallel resonant circuits, are suitable for rectangular waveguide implementation of band-stop filters.

In the circuit shown in Figure 5.42(a), the input and output impedances are set to Z_0. In case of a Butterworth or Chebyshev filter with N odd, the entire line of impedance inverters can be uniform, with 90° lines, all having impedances $Z_1 = Z_0$. For N even,

$$Z_1 = \frac{Z_0}{\sqrt{g_0 g_{N+1}}} \qquad (5.63)$$

When the slope parameters determined by Figure 5.42(a) are either too small or too large to realize physically, they may be adjusted up or down, respectively, by controlling the impedances of the K-inverters. Formulas given in Figure 5.42(b) are general. It should be mentioned that if Z_i are chosen unequal, then greater reflections result somewhere in the passband than would occur with an impedance level $Z_i = Z_0$. Figure 5.42(c) and (d) show the duals of the networks with series branches. Such duals are useful in waveguide band-stop filter realization. Figure 5.43 shows the schematic of a narrowband band-stop filter that can be realized using any form of planar transmission line or a coaxial line. The circuit is based on the network shown in Figure 5.42(a), where the inductors are realized using short-circuited stubs of length below 90° and the capacitors are realized using gap coupling of the stubs to the main line that realize the inverters. The equivalent circuit of the short-circuited stub connection is shown in Figure 5.44. The figure also shows various other stub configurations for realization of narrow band band-stop filters. The required design equations can be derived in the following way:

5.5 Band-Stop Filter Design

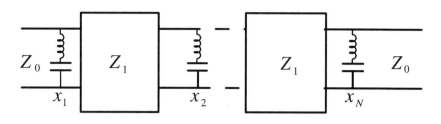

(a)

$$\frac{x_1}{Z_0} = \frac{1}{\omega_1' g_0 g_1 \overline{\omega}}$$

$$\frac{x_i}{Z_0} = \left(\frac{Z_1}{Z_0}\right)^2 \frac{g_0}{\omega_1' \overline{\omega} g_i}; (N = even)$$

$$\frac{x_i}{Z_0} = \frac{1}{\omega_1' \overline{\omega} g_0 g_i}; (N = odd)$$

$$\frac{Z_1}{Z_0} = \frac{1}{\sqrt{g_0 g_{N+1}}}$$

$$Z_0 = Z_1; (N = odd)$$

(b)

$$\frac{x_1}{Z_0} = \frac{1}{\omega_1' g_0 g_1 \overline{\omega}}$$

$$\frac{x_i}{Z_0} = \left(\frac{Z_1 Z_3 \cdots Z_{i-1}}{Z_0 Z_2 \cdots Z_{i-2}}\right)^2 \frac{g_0}{\omega_1' \overline{\omega} g_i}; (N = even)$$

$$\frac{x_i}{Z_0} = \left(\frac{Z_2 Z_4 \cdots Z_{i-1}}{Z_1 Z_3 \cdots Z_{i-2}}\right)^2 \frac{1}{\omega_1' \overline{\omega} g_0 g_i}; (N = odd)$$

$$\left(\frac{Z_0}{Z_N}\right)^2 = \left(\frac{Z_0 Z_2 \cdots Z_{N-2}}{Z_1 Z_3 \cdots Z_{N-1}}\right)^2 \frac{1}{g_0 g_{N+1}}; (N = even)$$

$$\left(\frac{Z_0}{Z_N}\right)^2 = \left(\frac{Z_1 Z_3 \cdots Z_{N-2}}{Z_2 Z_4 \cdots Z_{N-1}}\right)^2 \frac{1}{g_0 g_{N+1}}; (N = odd)$$

Figure 5.42 Inverter-coupled band-stop filter prototypes.

Let the electrical length of the stub at the center of the stopband be ϕ_0. Therefore, at resonance

$$Z_b \tan\phi_0 = \frac{1}{\omega_0 C_b} \tag{5.64}$$

Since φ is linearly proportional to ω, using

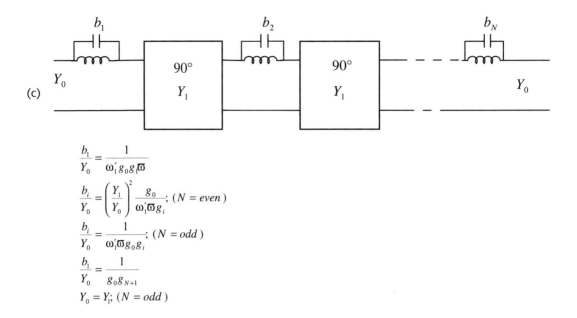

$$\frac{b_1}{Y_0} = \frac{1}{\omega_1' g_0 g_1 \varpi}$$

$$\frac{b_i}{Y_0} = \left(\frac{Y_1}{Y_0}\right)^2 \frac{g_0}{\omega_1' \varpi g_i}; \ (N = even)$$

$$\frac{b_i}{Y_0} = \frac{1}{\omega_1' \varpi g_0 g_i}; \ (N = odd)$$

$$\frac{b_1}{Y_0} = \frac{1}{g_0 g_{N+1}}$$

$$Y_0 = Y_1; \ (N = odd)$$

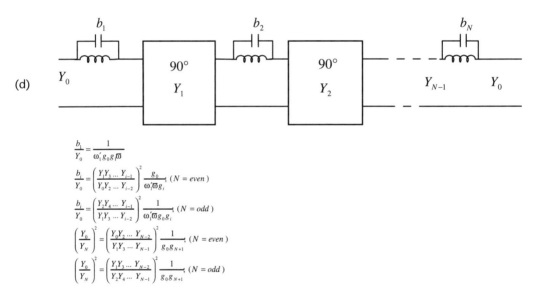

$$\frac{b_1}{Y_0} = \frac{1}{\omega_1' g_0 g_1 \varpi}$$

$$\frac{b_i}{Y_0} = \left(\frac{Y_1 Y_3 \ldots Y_{i-1}}{Y_0 Y_2 \ldots Y_{i-2}}\right)^2 \frac{g_0}{\omega_1' \varpi g_i}; \ (N = even)$$

$$\frac{b_i}{Y_0} = \left(\frac{Y_2 Y_4 \ldots Y_{i-1}}{Y_1 Y_3 \ldots Y_{i-2}}\right)^2 \frac{1}{\omega_1' \varpi g_0 g_i}; \ (N = odd)$$

$$\left(\frac{Y_0}{Y_N}\right)^2 = \left(\frac{Y_0 Y_2 \ldots Y_{N-2}}{Y_1 Y_3 \ldots Y_{N-1}}\right)^2 \frac{1}{g_0 g_{N+1}}; \ (N = even)$$

$$\left(\frac{Y_0}{Y_N}\right)^2 = \left(\frac{Y_1 Y_3 \ldots Y_{N-2}}{Y_2 Y_4 \ldots Y_{N-1}}\right)^2 \frac{1}{g_0 g_{N+1}}; \ (N = odd)$$

Figure 5.42 (continued)

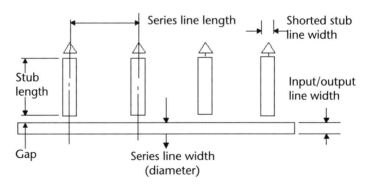

Figure 5.43 Configuration of a capacitor stub-coupled narrowband band-stop filter. (From [6].)

5.5 Band-Stop Filter Design

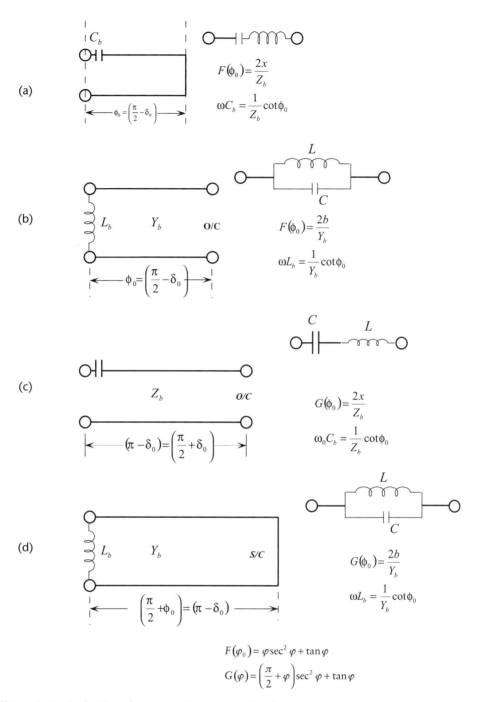

Figure 5.44 Realization of resonant circuits using distributed elements. (a) and (c) are suitable for planar and coaxial lines, and (b) and (d) for waveguides.

$$\frac{d\omega}{\omega} = \frac{d\phi}{\phi} \qquad (5.65)$$

the reactance slope parameter is determined from

$$x = \frac{\omega_0}{2} \frac{d}{d\omega}\left(Z_b \tan\phi - \frac{1}{\omega C_b}\right)\bigg|_{\phi=\phi_0, \omega=\omega_0} = \frac{Z_b}{2} F(\phi_0) \quad (5.66)$$

where

$$F(\phi) = \phi \sec^2 \phi + \tan \phi \quad (5.67)$$

Out of the three unknowns, Z_b, C_b, and ϕ_0, one is selected arbitrarily. In most cases it is Z_b. The required slope parameter is determined from the equations in Figure 5.42. The required electrical length ϕ_0 is obtained by numerically solving (5.67) and using the assumed value of Z_b. Once Z_b and ϕ_0 are known, C_b is obtained from (5.64). The above operations can be performed using WAVECON software very accurately. Table 5.4 shows the input/output files for the design of a 5.705-GHz stripline band-stop filter. See Figure 5.43 for definitions of geometrical parameters of the filter. The computed frequency response of the filter is shown in Figure 5.45.

For a given set of Y_b and ϕ_0, the susceptance slope parameter of a near-π radian line is twice as great as the reactance slope parameter of a near-π/2 radian line. For the near-π radian line one should use [20]

$$G(\phi) = \left(\frac{\pi}{2} + \phi\right) \sec^2 \phi + \tan \phi \quad (5.68)$$

instead of $F(\phi)$, which is used in near-π/2 radian line cases (cases (a) and (b) in Figure 5.44). It can be shown that

Table 5.4 WAVECON Input/Output Files for Stripline Band-Stop Filter Design

Stripline Shunt Stub Narrowband Bandstop Filter   05/07/2006   01:49:16

5.7050	GHz Center Frequency	0.0060	GHz Bandwidth
0.0200	Db Ripple	5	Poles
0.1240	Inches Groundplane Spacing	2.2000	Dielectric Constant
0.0007	Inches Conductor Thickness		
50.000	Ohms Input Line Impedance	0.1004	Inches Input Line Width
50.000	Ohms Output Line Impedance	0.1004	Inches Output Line Width

Sect Numb	Element Value	Z-Shnt Ohms	Z-Ser Ohms	Zshnt-L Inches	Zser-L Inches	Zshnt-W Inches	Zser-W Inches	Gap Inches	Cap Pf
1	g_1 0.8471	74.418		0.3548		0.0500		0.0778	0.0089
			50.034		0.3487		0.1003		
2	g_2 1.3448	74.418		0.3564		0.0500		0.0807	0.0083
			50.020		0.3487		0.1003		
3	g_3 1.6748	74.418		0.3451		0.0500		0.0614	0.0140
			50.020		0.3487		0.1003		
4	g_4 1.3448	74.418		0.3562		0.0500		0.0802	0.0083
			50.034		0.3487		0.1003		
5	g_5 0.8471	74.418		0.3549		0.0500		0.0781	0.0089

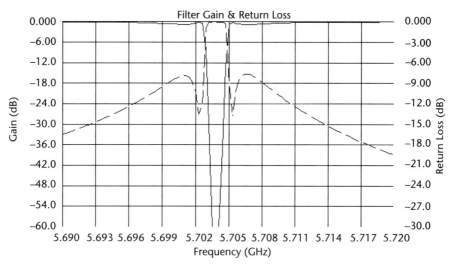

Figure 5.45 Computed frequency response of 5.705-GHz stripline band-stop filter. The above design can be perfected using a full-wave simulator. (From [5].)

$$G(\phi) = 2F(\phi) + \frac{2\delta - \sin(2\delta)}{\pi} \quad (5.69)$$

where

$$\delta = \frac{\pi}{2} - \phi \quad (5.70)$$

which is the amount by which the stub length falls short of $\pi/2$ or π radians. For all practical purposes, the last term in (5.69) can be neglected compared to the first term.

5.5.2 Determination of Fractional 3-dB Bandwidth of a Single Branch

The resonant electrical length ϕ_0 and the stub impedance Z_b can be translated accurately into physical dimensions when a planar transmission line like strip, microstrip, or coaxial line is used. However, determination of the capacitive gap for C_b or the inductive coupling L_b is a cumbersome job. Before the advent of modern CAD procedures, it was done using experimental means. Consequently, a considerable amount of bench tuning of the overall band-stop filter was needed. Today the conventional experimental adjustment can be achieved using simulation tools. Each stub length and its capacitive or inductive coupling with the main line is adjusted using an analysis software so that the 3-dB bandwidth of the overall two-port network offers the same value as the stop bandwidth of the band-stop filter at the center frequency of the filter. While a particular stub is adjusted, all other stubs are either removed or detuned.

5.5.3 Effects of Dissipation Loss in Band-Stop Filters

There are three effects of dissipation loss or finite unloaded resonator Q on the frequency response of a band-stop filter [21]:

1. *The peak attenuation inside the stop band is not infinite, but remains limited.* Let us consider the equivalent circuit of a band-stop filter with finite unloaded Q of the resonators, as shown in Figure 5.46. Analyzing the circuit gives the minimum attenuation in the passband as [21]

$$L_A = 10\log_{10}\left[\left\{\frac{E^2}{4R_0}\right\}\frac{1}{V^2 G_{N+1}}\right] = 10\log_{10}\left\{\frac{(R_0 G_1 R_2 G_3 \cdots R_N G_{N+1})^2}{4R_0 G_{N+1}}\right\} \quad (5.71)$$

2. *The reflection coefficient does not reach unity anywhere inside the stopband, resulting in an imperfect short circuit and loss of power on reflection.* Lack of perfect short or open circuit in the passband due to the completely resistive nature of the circuit causes imperfect return loss in the passband.

3. *There is additional loss in the passband of the filter.* The additional loss in the passband, in the vicinity of $\Delta\omega$ around ω_0, is given by

$$L_A = \frac{4.343}{R_0}\left[\frac{1}{G_1 X_1^2} + \frac{1}{R_2 (R_0 B_2)^2} + \cdots\right] \text{ dB} \quad (5.72)$$

where

$$\left.\begin{array}{l} X_1 \approx 2x_1 \dfrac{\Delta\omega}{\omega_0} \\[6pt] B_2 \approx 2b_2 \dfrac{\Delta\omega}{\omega_0} \\[6pt] \text{etc.} \end{array}\right\} \quad (5.73)$$

Figure 5.47 shows the effect of finite unloaded Q on the response of the stripline filter in Table 5.4.

The response shown in Figure 5.45 assumes an infinite unloaded Q. The response shown in Figure 5.47 assumes finite dielectric loss and pure copper metallization. Therefore, we conclude that the desired frequency response cannot be physically realized using a duroid-substrate stripline configuration.

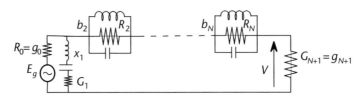

Figure 5.46 Equivalent circuit of a band-stop filter with finite unloaded Q.

5.5 Band-Stop Filter Design

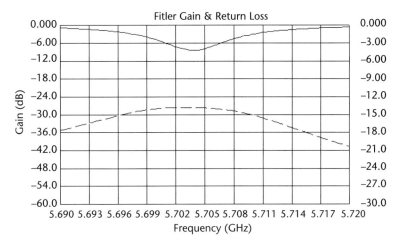

Figure 5.47 Band-stop filter response for finite conductivity.

5.5.4 Stub-Loaded Band-Stop Filter

Capacitive stub-loaded band-stop filters are well suited for narrow bandwidth filters or filters with a narrow stopband. Alternatively, when the bandwidth is moderately narrow or wide, the direct stub-loaded band-stop filter configuration is preferable. The configuration of a direct stub-loaded transmission line band-stop filter is shown in Figure 5.48.

Table 5.5 shows the design parameters for stub-loaded band-stop filters. To use Table 5.5, the following parameters are relevant:

1. Left-hand terminating impedance Z_A;
2. Low-pass prototype values $g_j (j = 0,1,2 \ldots, N + 1)$;
5. Center frequency ω_0 of stop-band;
6. The bandwidth parameter a.

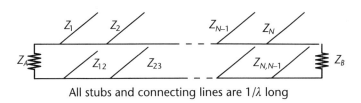

All stubs and connecting lines are $1/\lambda$ long

Figure 5.48 Stub-loaded band-stop filter network.

Table 5.5 Exact Equations for Band-Stop Filter with $\lambda/4$ Spacings between Stubs

N = number of stubs.

Z_A, Z_B = terminating impedances.

Z_j (j =1 to N)= Impedances of open-circuit shunt stubs.

$Z_{j-1, j}$ (j =2 to N) = connecting line impedances.

g_j= Low-pass prototype element values.

$\Lambda = \omega'_1 a$ where ω'_1 is the cutoff frequency of the low-pass prototype and a is the bandwidth parameter defined in (5.74). The terminating impedance Z_A is arbitrary.

The bandwidth parameter a is given by

$$a = \cot\left(\frac{\pi}{2}\frac{\omega_1}{\omega_0}\right) \qquad (5.74)$$

where the frequency parameters ω_0 and ω_1 are defined in Figure 5.40. Figure 5.49 shows the general layout of the filter [21].

Case $N = 1$

$$Z_1 = \frac{Z_A}{\Lambda g_1 g_0}, \quad Z_B = \frac{Z_A g_2}{g_0}$$

Case $N = 2$

$$Z_1 = Z_A\left(1 + \frac{1}{\Lambda g_0 g_1}\right), \quad Z_{12} = Z_A(1 + \Lambda g_0 g_1)$$

$$Z_2 = \frac{Z_A g_0}{\Lambda g_2}, \quad Z_B = Z_A g_0 g_2$$

Case $N = 3$

Z_1^*, Z_2, and Z_{12} are same as in case $N = 2$

$$Z_3 = \frac{Z_A g_0}{g_4}\left(1 + \frac{1}{\Lambda g_4 g_5}\right), \quad Z_{23} = \frac{Z_A g_0}{g_4}(1 + \Lambda g_3 g_4), \quad Z_B = \frac{Z_A g_0}{g_4}$$

Case $N = 4$

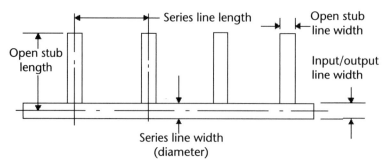

Figure 5.49 Wide stopband shunt stub band-stop filter layout.

5.5 Band-Stop Filter Design

$$Z_1 = Z_A\left(2 + \frac{1}{\Lambda g_0 g_1}\right), \quad Z_{12} = Z_A\left(\frac{1 + 2\Lambda g_0 g_1}{1 + \Lambda g_0 g_1}\right),$$

$$Z_2 = Z_A\left(\frac{1}{1 + \Lambda g_0 g_1} + \frac{g_0}{\Lambda g_2 (1 + \Lambda g_0 g_1)^2}\right), \quad Z_{23} = \frac{Z_A}{g_0}\left(\Lambda g_2 + \frac{g_0}{1 + \Lambda g_0 g_1}\right), \quad Z_3 = \frac{Z_A}{\Lambda g_0 g_2},$$

$$Z_{34} = \frac{Z_A}{g_0 g_5}(1 + \Lambda g_4 g_5), \quad Z_4 = \frac{Z_A}{g_0 g_5}\left(1 + \frac{1}{\Lambda g_4 g_5}\right), \quad Z_B = \frac{Z_A}{g_0 g_5}$$

Case $N = 5$

Z_1, Z_2, Z_3, Z_{12}, and Z_{23} are same as in case $N = 4$.

$$Z_4 = \frac{Z_A}{g_0}\left(\frac{1}{1 + \Lambda g_5 g_6} + \frac{g_6}{\Lambda g_4 (1 + \Lambda g_5 g_4)^2}\right), \quad Z_{34} = \frac{Z_A}{g_0}\left(\Lambda g_4 + \frac{g_6}{1 + \Lambda g_5 g_6}\right),$$

$$Z_5 = \frac{Z_A g_6}{g_0}\left(2 + \frac{1}{\Lambda g_5 g_6}\right), \quad Z_{45} = \frac{Z_A g_6}{g_0}\left(\frac{1 + 2\Lambda g_5 g_6}{1 + \Lambda g_5 g_6}\right),$$

$$Z_B = \frac{Z_A g_6}{g_0}$$

The equations shown in Table 5.5 can be used in the design of a stripline, microstrip, or coaxial line band-stop filter. Figure 5.50 shows a typical WAVECON design file for a stripline band-stop filter. Table 5.6 shows the output file.

Figure 5.51 shows the stripline band-stop filter response and Figure 5.52 shows the layout of the filter.

Shunt stub-loaded bandstop filters cannot be realized easily using waveguide resonators. Consequently, the dual form of shunt-stub loading using series-stub loading shown in Figure 5.53 is used. If the bandwidth of the filter is very narrow, the impedances of the series stubs will become very low. Therefore, it is necessary to replace them with half-wavelength cavity resonators with coupling irises (see Figure 5.44(d)). However, in reality the distance between adjacent series stubs gives rise to undesired interactions and degrades the filter response. To circumvent this

Transmission Media: (M,S,R,B,SM,SS)	S-stripline
Filter Center Frequency (GHz)	3.0000
Filter Bandwidth (GHz)	2.7000
Ripple	0.1000
Number of Filter Sections (Zeros)	5
Input Impedance (Ohms)	50.000
Output Impedance (Ohms)	50.000
Relative Dielectric Constant	2.0000
1/2 Groundplane Spacing or Board Height (Inches)	0.0320
Strip or Bar Thickness (Inches)	0.0014
Distance above Mstrip Board to Top Shield (Inches)	0.0000
Distance below Mstrip Board to Bot Shield (Inches)	0.0000

Figure 5.50 WAVECON input file for stripline band-stop filter design.

Table 5.6 Wavecon Output File for Wideband Stripline Band-Stop Filter

Stripline Shunt Stub Wideband Band-Stop Filter

3.0000	GHz Center Frequency	2.7000	GHz Bandwidth
0.1000	dB Ripple	5	Poles
0.0640	Inches Groundplane Spacing	2.0000	Dielectric Constant
0.0014	Inches Conductor Thickness		
50.000	Ohms Input Line Impedance	0.0531	Inches Input Line Width
50.000	Ohms Output Line Impedance	0.0531	Inches Output Line Width

Sect Numb	Element Value	Zshunt Ohms	Zseries Ohms	Lshunt Inches	Lseries Inches	Wdth-Shnt Inches	Width-Ser Inches
1	1.1468	141.46		0.6941		0.0035	
			81.417		0.7387		0.0216
2	1.3712	28.437		0.7008		0.1158	
			96.844		0.8019		0.0140
3	1.9750	21.889		0.7391		0.1595	
			96.844		0.8019		0.0140
4	1.3712	28.437		0.7008		0.1158	
			81.417		0.7387		0.0216
5	1.1468	141.46		0.6941		0.0035	

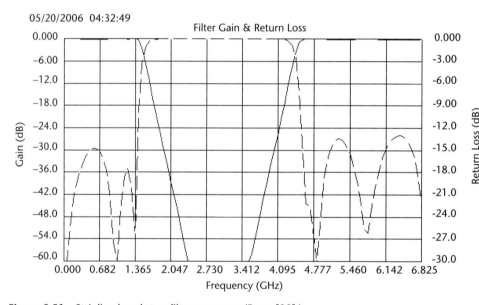

Figure 5.51 Stripline band-stop filter response. (From [22].)

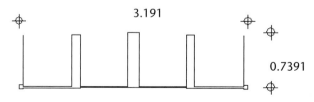

Figure 5.52 Stripline band-stop filter layout. (From [22].)

5.5 Band-Stop Filter Design

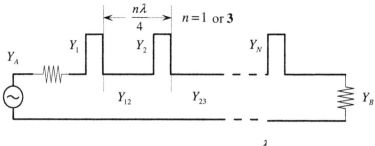

Figure 5.53 Series stub-loaded band-stop prototype.

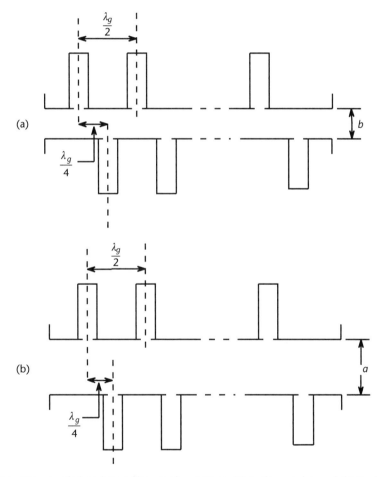

Figure 5.54 Waveguide band-stop filter configurations: (a) E-plane stubs, and (b) H-plane stubs.

problem, the distance between adjacent stubs is increased. The iris dimensions are so chosen to offer the required reactance slope parameters.

The first step toward waveguide band-stop filter design is frequency normalization because a waveguide is a dispersive transmission line. The normalization is accomplished using reciprocal guide wavelength, λ_{g0}/λ_g. For a lumped element or a transmission line filter supporting the TEM mode, the fractional bandwidth is given by

$$w = \frac{\omega_2}{\omega_0} - \frac{\omega_0}{\omega_1} \qquad (5.75)$$

The corresponding equation for a waveguide band-stop filter is

$$w = \frac{\lambda_{g0}}{\lambda_{g2}} - \frac{\lambda_{g1}}{\lambda_{g0}} \qquad (5.76)$$

where λ_{g0}, λ_{g1}, and λ_{g2} are the guide wavelength at the midstopband, lower cutoff, and upper cutoff frequencies, respectively, of the filter. Due to such scaling, the actual bandwidth of a waveguide filter will be $(\lambda_{g0}/\lambda_0)^2$ times the bandwidth of a stripline filter that has been designed from the same prototype. As mentioned above, the iris coupled and approximately $\lambda_g/2$ stub-coupled waveguide band-stop filters have two configurations shown in Figure 5.54. Although the figure shows $\lambda_g/4$ spacing between adjacent stubs, the distance between the stubs may be $3\lambda_g/4$ for the reasons mentioned above. Each stub is slightly below half a wavelength long. The equivalent network of an E-stub-loaded filter is shown in Figure 5.42(c) and (d), while Figures 5.42(a) and (b) correspond to the H-stubloaded case. Another viable structure for a waveguide band-stop filter realization is shown in Figure 5.55. It consists of a ridged waveguide. The ridge height is approximately $\lambda_g/4$.

5.5.4 Design Steps for Waveguide Band-Stop Filters

Let us consider the design of an E-plane stub resonator and iris-coupled waveguide band-stop filter, shown in Figure 5.54. The normalized susceptance B/Y_b of each coupling iris can be determined from the analytical formula

$$\frac{B}{Y_b} = -\frac{\lambda_g a b'}{4\pi M_1'} \qquad (5.77)$$

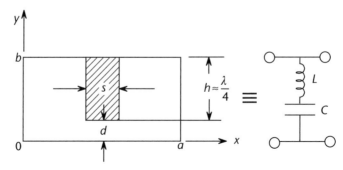

Figure 5.55 Ridged waveguide resonator elements for waveguide band-stop filter.

where the magnetic polarizability of the slot iris is given by [23]

$$M_1' = \frac{M_1}{1-\left(\frac{2L}{\lambda}\right)^2} \times 10^{-\left(1.36\frac{t}{L}\right)\sqrt{1-\left(\frac{2L}{\lambda}\right)^2}} \tag{5.78}$$

$$\frac{M_1}{L^3} = 0.1096\left(\frac{w}{L}\right)^3 - 0.2312\left(\frac{w}{L}\right)^2 + 0.3559\left(\frac{w}{L}\right) + 0.0244 \tag{5.79}$$

where a and b' are the width and the height, respectively, of the stub-resonator-forming waveguide. Equation (5.77) is valid for the iris cut in walls of infinitesimal thickness t and having a length l, and l is assumed to be much less than one-half free space wavelength. Equation (5.79) is based on measured data by Cohn [23] for zero wall thickness of the slot (see Figure 5.55) and equation (5.78) is the correction for the finite wall thickness [24].

Equation (5.79) can be inverted for synthesis of a band-stop filter as

$$\frac{w}{L} = -35.693\left(\frac{M_1}{L^3}\right)^3 + 21.793\left(\frac{M_1}{L^3}\right)^2 + 0.622\left(\frac{M_1}{L^3}\right) + 0.0001 \tag{5.80}$$

5.5.4.1 Design Example of a Three-Resonator Waveguide Band-Stop Filter

Assume that the filter has the following specifications:

1. Center frequency $f_0 = 10$ GHz
2. Fractional stopband width $\varpi = 0.05$
3. Waveguide is WR90

The susceptance slope parameters for the two outer resonators are determined using the equations in Figure 5.42(c). We obtain

$$\frac{b_1}{Y_0} = \frac{b_3}{Y_0} = 12.528 \tag{5.81a}$$

$$\frac{b_2}{Y_0} = 18.232 \tag{5.81b}$$

We choose the characteristic admittance of the stub waveguides to be the same as that of the main waveguide. Therefore

$$G(\phi_{01}) = G(\phi_{03}) = \frac{2b_1}{Y_0} = 25.056 \quad (5.82a)$$

$$G(\phi_{02}) = 36.434 \quad (5.82b)$$

Using the relation [21]

$$G(\phi) \approx 2F(\phi) = 2\left[\phi \sec^2 \phi + \tan \phi\right] \quad (5.83)$$

we obtain

$$\phi_{01} + 90° = \phi_{03} + 90° = 159.5° \quad (5.84a)$$

$$\phi_{02} + 90° = 163° \quad (5.84b)$$

The guided wavelength λ_{g0} in a WR90 waveguide at 10 GHz is 1.5631 inches. Therefore, the stub lengths are

$$L_1 = L_3 = \frac{\lambda_{g0} 159.5°}{360°} = 0.693 \text{ inch} \quad (5.85a)$$

$$L_2 = \frac{\lambda_{g0} 163°}{360°} = 0.709 \text{ inch} \quad (5.85b)$$

In the next step, the magnetic polarizabilities of the slots are obtained from (5.77) as

$$M'_{11} = M'_{13} = 0.0167 \text{ inch}^3 \quad (5.86a)$$

$$M'_{12} = 0.0137 \text{ inch}^3 \quad (5.86b)$$

From (5.78) we obtain

$$M_{1k} = \frac{M'_{1k}\left[1 - \left(\frac{2L}{\lambda}\right)^2\right]}{10^{-\left(\frac{1.36t}{L}\right)\sqrt{1-\left(\frac{2L}{\lambda}\right)^2}}}, \quad (k = 1, 2, 3) \quad (5.87)$$

Let us assume the wall thickness $t = 0.5$ mm and the slot length $L = 10$ mm. At 10 GHz, $\lambda = 30$ mm. This gives

5.5 Band-Stop Filter Design

$$M_{1k} = 0.6244 M'_{1k} \qquad (5.88)$$

Combining (5.86) and (5.88) gives

$$M_{11} = M_{13} = 0.0104 \text{ inch}^3 \qquad (5.89a)$$

$$M_{12} = 0.0086 \text{ inch}^3 \qquad (5.89b)$$

Using (5.80) gives the slot widths as

$$W_{1,3} = 0.2200 \text{ inches} \qquad (5.90a)$$

$$W_2 = 0.1658 \text{ inches} \qquad (5.90b)$$

and the design is complete. However, an analysis of the structure using the full-wave approach shows that the frequency response is considerably different from the desired one, as shown in Figure 5.58. The difference is due to the fact that the above approximate design method ignores the phase contributions from the abrupt step junctions involved with the coupling slots in the main guide as well as in the guides forming the stubs. This problem can be solved either by using tuning screws or by rigorous analysis and optimization. Figure 5.59 shows the response of the filter

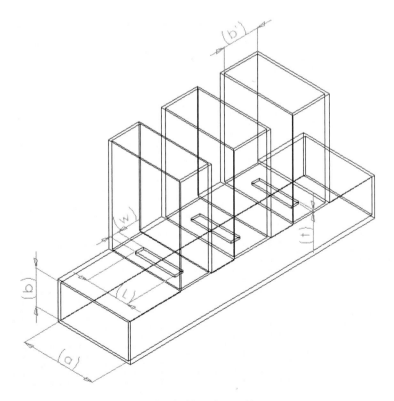

Figure 5.56 E-plane iris-coupled stub-loaded band-stop filter.

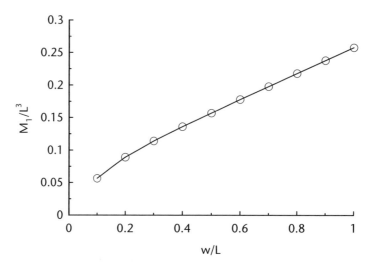

Figure 5.57 Graphical representation of (5.79).

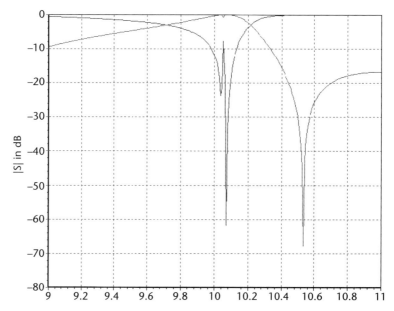

Figure 5.58 Computed response of waveguide band-stop filter before optimization. Frequency is in gigahertz. (From [24].)

after computer optimization [25] using the above design dimensions as the starting guess. The optimized values are

$$L_1 = L_3 = 0.7012 \text{ inches} \tag{5.91a}$$

$$L_2 = 0.7047 \text{ inches} \tag{5.91b}$$

$$W_{1,3} = 0.223 \text{ inches} \tag{5.91c}$$

5.5 Band-Stop Filter Design

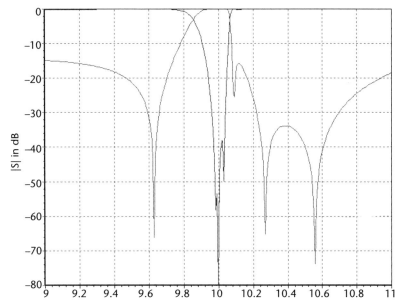

Figure 5.59 Optimized response of waveguide band-stop filter. (From [24].)

$$W_2 = 0.167 \text{ inches} \tag{5.91d}$$

Spacing between adjacent slots = 0.342 inches

References

[1] Kuroda, K., "Derivation Methods of Distributed Constant Filters from Lumped Constant Filters," Lectures at Joint Meeting of Kousi Branch of Institute of Elec. Commun. of Elec. and of Illimn. Engrs. of Japan, p. 32 (in Japanese), October 1952.

[2] Carlin, H. J., "Distributed Circuit Design with Transmission Line Elements," *Proc. IEEE*, Vol. 59, No. 7, July 1971, pp. 1059–1081.

[3] Meadley, M. W., *Microwave and RF Circuits: Analysis, Synthesis, and Design,* Norwood, MA: Artech House, 1993.

[4] Rhodes, J. D., *Theory of Electrical Filters,* London: John Wiley and Sons, 1976.

[5] EM3DS—Full-Wave Electromagnetic Simulation Software, MEMRESEARCH, www.memresearch.com, Italy, 2006.

[6] WAVECON-Procap, Microwave Circuit Design and Analysis software, Version 1, Escondido, CA, 2001.

[7] Rhodes, J. D., *Theory of Electrical Filters,* London: John Wiley and Sons, 1976.

[8] Levy, R., "A Generalized Design Technique for Practical Distributed Reciprocal Networks," *IEEE Trans. Microw. Theory Techn.*, Vol. 21, August, 1973, pp. 519–524.

[9] WAVEFIL—Waveguide Filter Design Software, Polar Waves Consulting, Windham, NH.

[10] Levy, R., "Tapered Corrugated Waveguide Lowpass Filters," *IEEE Trans. Microw. Theory Techn.*, Vol. 21, August 1973, pp. 525–532.

[11] Balasubramnaian, R., "Computer Aided Design of Waveguide Filters Using the Mode Matching Method," M.Sc. thesis, Department of Electrical Engineering, University of Saskatchewan, Saskatoon, 1977.

[12] Sharp, E., "A High-Power Wide-Band Waffle Iron Filter," *IEEE Trans. Microw. Theory Techn.*, Vol. 11, No. 2, March 1953, pp. 111–116.

[13] Young, L., and B. M. Schiffman, "New Improved Types of Waffle Iron Filters," *Proc. IEE (London)*-110, July 1963, pp. 1191–1198.

[14] Young, L., "Postcript to Two Papers on Waffle Iron Filters," *IEEE Trans. Microw. Theory Techn.*, November 1963, pp. 555–557.

[15] Network Tool Kit, Version 1.2.x, NorthEastern Microwave Inc., 2004.

[16] LaTourette, P., "High Pass Filter Design," *12th European Microwave Conference Proceedings*, 1982, pp. 233–238.

[17] Levy, R., "New Equivalent Circuits for Inhomogeneous Coupled Lines with Synthesis Applications," *IEEE Trans. Microw. Theory Techn.*, Vol. 36, No. 6, June 1988, pp. 1087–1094.

[18] Wenzel, R., "Synthesis of Combline and Capacitively Loaded Interdigital Filters of Arbitrary Bandwidth," *IEEE Trans. Microw. Theory Techn.*, Vol. 19, , August 1971, pp. 678–688.

[19] Szentirmai, G., " A New Computer Aid for Microwave Filter Design", *IEEE MTT-S International Microwave Symposium Digest*, 1980, pp. 413–416.

[20] Schiffman, B. M., and G. L. Matthaei, "Exact Design of Bandstop Microwave Filters," *IEEE Trans. Microw. Theory Techn.*, Vol. 12, No. 1, January 1964, pp. 6–15.

[21] Young, L., G. L. Matthaei, and E.M.T. Jones, "Microwave Bandstop Filters with Narrow Stop Bands," *IRE Trans. Microw. Theory Techn.*, Vol. 10, No. 2, November 1962, pp. 416–427.

[22] WAVECON, Filter Design Software, Escondido, CA, 1994.

[23] Cohn, S. B., "Determination of Aperture Parameters by Electrolytic Tank Method," *Proc. IRE,* Vol. 39, November 1951, pp. 1416–1421.

[24] Cohn, S. B., "Microwave Coupling by Large Aperture," *Proc. IRE,* Vol. 40, June 1952, pp. 696–699.

[25] WASPNET, Microwave Innovation Group, GmbH & Co. KG. Bremen, Germany, 2008.

CHAPTER 6

Band-Pass Filters

6.1 Theory of Band-Pass Filters

Like any other electrical filter, a true band-pass filter is realized using lumped elements only. A distributed element band-pass filter exhibits periodic band-pass response in the frequency domain. The basic building block of a band-pass filter is a combination of two coupled resonant networks, as shown in Figure 6.1. The center frequency of the filter is given by

$$f_0 = \frac{1}{2\pi}\sqrt{\frac{L}{C}} \tag{6.1}$$

Adjustment of the input and output coupling capacitors C_0 matches the filter to the external systems. The mutual coupling M between the resonators determines the bandwidth of the filter. However, from a fundamental point of view, the theory of band-pass filter design begins with the transformation of a suitable low-pass prototype filter, which in turn begins with the corresponding frequency and impedance scalings as follows. Consider the low-pass filter prototypes shown in Figure 6.2(a) and (b). As we know from Chapter 4, the frequency response of the network has the form shown in Figure 6.2(c). For $\omega_1' = 1$ and, $g_0 = g_{n+1} = 1$ the network becomes a low-pass prototype with the frequency response shown in Figure 6.3. Using the frequency transformation [1]

$$\omega' = \frac{\omega_0}{\omega_2 - \omega_1}\left(\frac{\omega}{\omega_0} - \frac{\omega_0}{\omega}\right) \tag{6.2}$$

we obtain the band-pass filter prototype frequency response shown in Figure 6.4. The corresponding band-pass filter network is shown in Figure 6.5. The analytical relationship between ω and ω' is given by

$$\omega = \left[\frac{\omega_2 - \omega_1}{2}\omega' \pm \sqrt{\left\{\frac{\omega_2 - \omega_1}{2}\right\}^2 \omega' + \omega_0^2}\right] \tag{6.3}$$

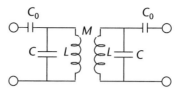

Figure 6.1 Coupled resonators as the basic building blocks of a band-pass filter.

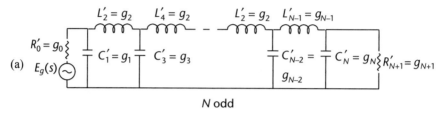

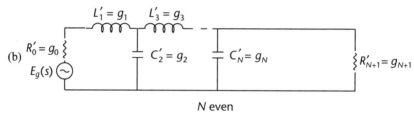

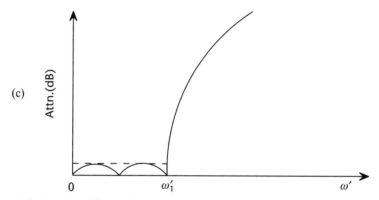

Figure 6.2 Low-pass filter and the frequency response.

and

$$\omega_0 = \sqrt{\omega_1 \omega_2} \qquad (6.4)$$

The element values in the network in Figure 6.5 are given by

$$L'_r C'_r = \omega_0^{-2} = [\omega_1 \omega_2]^{-1} \qquad (6.5)$$

$$L'_r = \frac{g_r}{\omega_2 - \omega_1} \qquad (6.6)$$

6.1 Theory of Band-Pass Filters

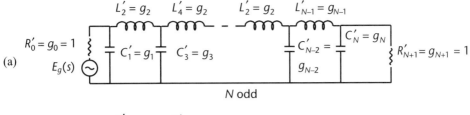

(a) N odd

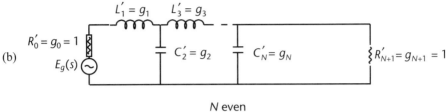

(b) N even

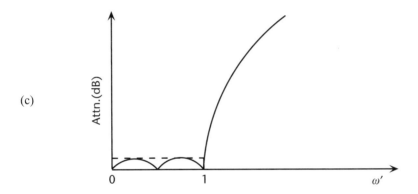

(c)

Figure 6.3 Low-pass prototype and the frequency response.

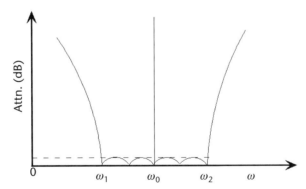

Figure 6.4 Transformation of low-pass filter response to band-pass filter response.

for series elements and

$$L'_r C'_r = \omega_0^{-2} = [\omega_1 \omega_2]^{-1} \tag{6.7}$$

$$C'_r = \frac{g_r}{\omega_2 - \omega_1} \tag{6.8}$$

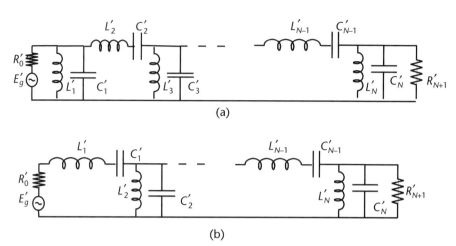

Figure 6.5 Band-pass filter transformation: (a) is for n-even, and (b) is for n-odd.

for shunt elements (g_r were defined in Chapter 4). Once all L'_r and C'_r are obtained, impedance scaling described in Chapter 4 is applied in order to arrive at the final component values. A comparison of Figures 6.2 and 6.5 shows that the series inductors of the low-pass filter have become series resonant circuits and the shunt capacitors in the low-pass filter have become parallel resonant circuits in the band-pass filter. Let us consider an example showing the step-by-step procedure for band-pass filter design from the low-pass prototype. Figure 6.6(a) [2] shows the network of a third-order Chebyshev low-pass prototype that offers 26-dB return loss or -0.01-dB ripple in the passband. Being a low-pass prototype, it has 1Ω terminations at both ends and a 1-rad/sec cutoff frequency. Let us transform the network to a band-pass filter with a bandwidth of 0.0159 GHz/sec centered on 0.318 GHz/sec. Using (6.3) to (6.8) gives us the band-pass filter network shown in Figure 6.6(b). If we change the terminating impedances to 50Ω from 1Ω, the filter network in Figure 6.6c results, where each inductor value of Figure 6.6(b) has been multiplied by 50 and each capacitor value in Figure 6.6(b) has been divided by 50. The computed frequency response is shown in Figure 6.7. Table 6.1 shows the element transformation from low-pass form to band-pass form [3].

We mentioned in Chapter 4 that elliptic and extracted pole low-pass filters involve series resonant shunt elements and parallel resonant series elements. Transformation of such elements in low-pass to band-pass mapping is done as shown in Table 6.2. For the series resonant network:

$$C_1 = \frac{1}{L_1} = \frac{\Delta\omega(v-1)}{2vL_n} \quad (6.9a)$$

$$C_2 = \frac{1}{L_1} = \frac{\Delta\omega(v-1)}{2vL_2} \quad (6.9b)$$

and for a parallel resonant network

6.1 Theory of Band-Pass Filters

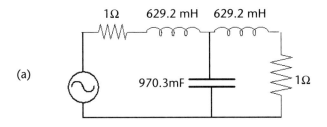

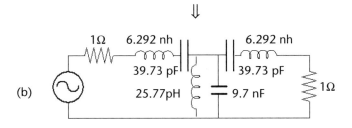

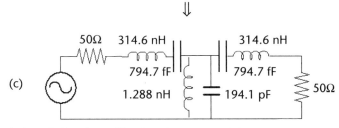

Figure 6.6 Evolution of a band-pass filter from low-pass prototype.

$$L_1 = \frac{1}{C_1} = \frac{\Delta\omega(v-1)}{2vL_n} \tag{6.10a}$$

$$C_2 = \frac{1}{L_1} = \frac{\Delta\omega(v-1)}{2vL_2} \tag{6.10b}$$

where

$$v = \sqrt{\left(\frac{2}{\omega_\infty \Delta\omega}\right)^2 - 1} \tag{6.11}$$

$$\omega_{\infty 1} = \sqrt{\frac{v+1}{v-1}} \tag{6.12a}$$

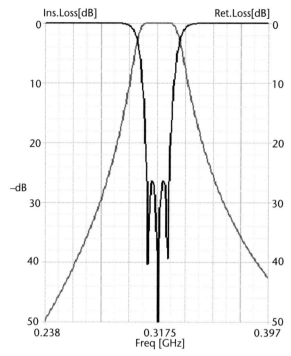

Figure 6.7 Computed frequency response of band-pass filter shown in Figure 6.6 (analysis by Ansoft Designer).

Table 6.1 Low-Pass to Band-Pass Transformation

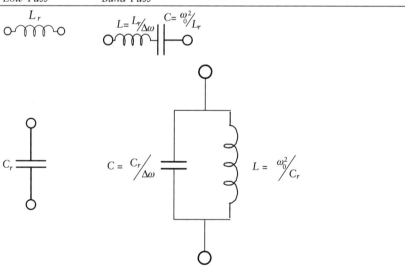

$$\omega_{\infty 2} = \sqrt{\frac{v-1}{v+1}} \qquad (6.12b)$$

6.1 Theory of Band-Pass Filters

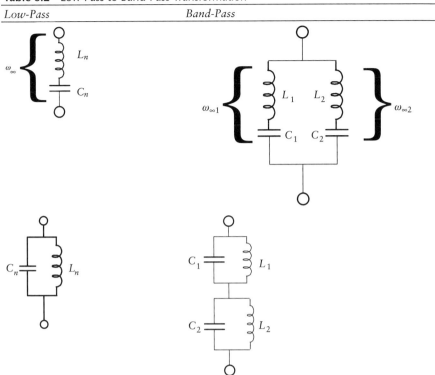

Table 6.2 Low-Pass to Band-Pass Transformation

Figure 6.8 shows the network of a fifth-order extracted pole low-pass filter [2]. In the network the transmission zeros are produced by series resonances in the shunt arms. The cutoff frequency of the filter is at 0.478 GHz and the $f_\infty = 0.525$ GHz. Figure 6.9 shows the computed frequency response of the filter.

Figure 6.10(a) [2] shows the network when it is transformed, using (6.9a) through (6.12b), into a band-pass filter centered at 0.478 GHz/sec and a ripple bandwidth of 0.0478 GHZ/sec. Figure 6.10(b) shows the computed frequency response of the band-pass network for comparison. Figure 6.11 shows the element values when the the terminations are 50 Ω.

Like low-pass filters, band-pass filters can be realized using K- or J-inverters. In fact, use of inverters facilitates realization of distributed element band-pass filters. Figure 6.12 shows the transformation of a low-pass ladder into a band-pass filter, where ω_0 is the center frequency in radians/sec and $\Delta\omega$ is the bandwidth of the filter in radians/sec [4]. The inverter values are obtained from (4.140) to (4.150) depending on the desired frequency response.

The K-inverters in Figure 6.12 can be realized using the Π-network shown in Figure 6.13.

The negative shunt capacitances of the inverter are absorbed in the capacitances of the adjoining resonant networks. This eventually leads to the form of the filter network shown in Figure 6.14. The element values are given by [5]

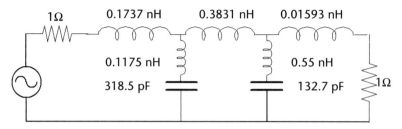

Figure 6.8 Extracted pole low-pass filter (cutoff is at 0.478 GHz/sec and $f_\infty = 0.525$ GHz).

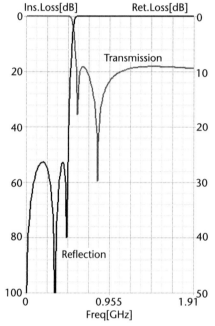

Figure 6.9 Extracted pole low-pass filter response (cutoff is at 0.478 GHz/sec and $\omega_\infty = 0.525$ GHz/sec). (From [2].)

$$C_{01} = C_{N,N+1} = \frac{1}{\omega_0 \sqrt{v-1}} \qquad (6.13a)$$

$$C_{k,k+1} = \frac{K_{k,k+1}}{v\omega_0} \quad (k = 1, \cdots, N-1) \qquad (6.13b)$$

$$C_{11} = \frac{C_1}{\omega_0} - \frac{\sqrt{v-1}}{v\omega_0} - C_{12} \qquad (6.13c)$$

6.1 Theory of Band-Pass Filters

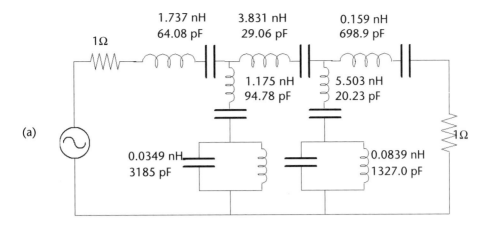

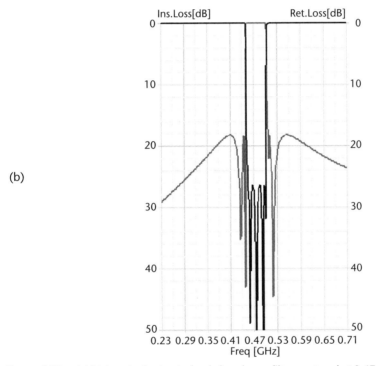

Figure 6.10 (a) Network of extracted pole band-pass filter centered at 0.478 GHz and 0.0478 GHz bandwidth.

$$C_{NN} = \frac{C_N}{\omega_0} - \frac{\sqrt{\nu-1}}{\nu\omega_0} - C_{N-1,N} \qquad (6.13d)$$

$$C_{kk} = \frac{C_k}{\omega_0} - C_{k-1,k} - C_{k,k+1} \qquad (6.13e)$$

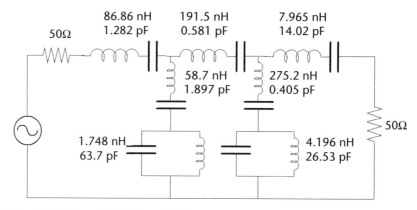

Figure 6.11 Final design with 50Ω terminations.

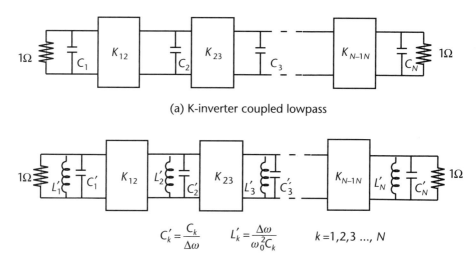

(a) K-inverter coupled lowpass

$$C'_k = \frac{C_k}{\Delta\omega} \qquad L'_k = \frac{\Delta\omega}{\omega_0^2 C_k} \qquad k=1,2,3\ldots,N$$

(b) K-inverter coupled bandpass

Figure 6.12 K-inverter-coupled band-pass prototype.

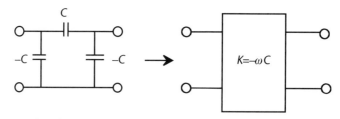

Figure 6.13 Narrowband lumped element inverter using a capacitive Π-network

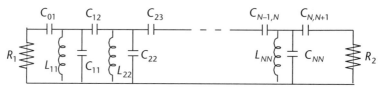

Figure 6.14 Capacitively coupled LE band-pass filter.

$$v = \frac{\omega_o}{\Delta\omega} \qquad (6.13f)$$

As the frequency rises above the upper cutoff frequency, all series capacitors offer less and less reactance and eventually become a series short circuit between the load and the generator. At the same time, all shunt inductors become an open circuit, and due to the shunt capacitors, we get a single shunt capacitor at a very high frequency. This results in 2N–1 transmission zeros at dc and a single transmission zero at infinite frequency. Consequently, the response is more selective below the lower cutoff than the upper cutoff. Following the same line of argument we find that inductive couplings instead of capacitive couplings can reverse the asymmetry. Use of alternate capacitive and inductive couplings will result in a symmetrical response.

6.2 Distributed Transmission Line Form of Capacitively Coupled Band-Pass Filter

6.2.1 Gap-Coupled Transmission Line Band-Pass Filters

Following the basic transmission line theory we see that a section of transmission line of length equal to a multiple of half wavelengths has the equivalent circuit of a shunt-parallel-resonant circuit, as shown in Figure 6.15. Figure 6.16 shows the equivalent circuit of a series gap in a transmission line. Therefore, one can realize the distributed line equivalent of a capacitive Π-network using a series gap discontinuity in a transmission line [6].

We can, therefore, use a series gap discontinuity as the impedance inverter and a half-wavelength-long section of transmission line as the shunt resonator in a capacitively coupled band-pass filter. The configuration of such a filter is shown in Figure 6.17 [7].

The transmission line can be the center conductor of a coaxial line, stripline, microstrip line, or suspended microstrip line. The design equations are as follows [8]:

$$w = \frac{\Delta f}{f_0} \qquad (6.14a)$$

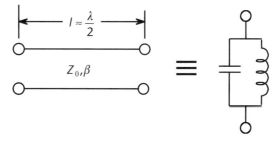

Figure 6.15 Equivalent circuit of a half-wavelength transmission line.

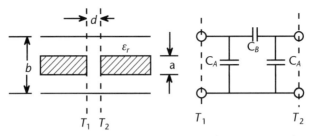

Figure 6.16 Equivalent circuit of a series gap discontinuity of a transmission line.

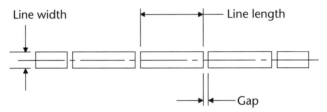

Figure 6.17 Layout of a gap-coupled band-pass filter.

$$\frac{K_{01}}{Z_0} = \sqrt{\frac{\pi}{2} \frac{\varpi}{g_0 g_1 \omega_1'}} \qquad (6.14\text{b})$$

$$\left.\frac{K_{k,k+1}}{Z_0}\right|_{k=1\cdots N-1} = \frac{\pi \varpi}{2\omega_1'} \sqrt{\frac{1}{g_k g_{k+1}}} \qquad (6.14\text{c})$$

$$\frac{K_{N,N+1}}{Z_0} = \sqrt{\frac{\pi}{2} \frac{\varpi}{g_N g_{N+1} \omega_1'}} \qquad (6.14\text{d})$$

where the prototype cutoff frequency ω_1' has been defined in Figure 6.2. Once the K-inverter values have been obtained from the above equations, the gaps between adjacent resonators are adjusted so that the scattering parameters of a gap are related to the corresponding K-inverter as

$$K = \left| \tan\left(\frac{\phi}{2}\right) + \tan^{-1} x_s \right| \qquad (6.15\text{a})$$

$$\phi = -\tan^{-1}(2x_p + x_s) - \tan^{-1} x_s \qquad (6.15\text{b})$$

$$jx_s = \frac{1 - S_{12} + S_{11}}{1 - S_{11} + S_{12}} \tag{6.15c}$$

$$jx_p = \frac{2S_{12}}{(1 - S_{11})^2 - S_{12}^2} \tag{6.15d}$$

The edge-to-edge line length l_j of the jth resonator of the filter is given by [8]

$$l_j = \frac{\lambda_{g0}}{2\pi}\left[\pi + \frac{1}{2}(\phi_j + \phi_{j+1})\right] \tag{6.15e}$$

The scattering matrix of the K-inverter forming a gap between resonators can be determined from the equivalent network of the gap (see Chapter 3 for certain transmission lines like striplines or coaxial lines). Otherwise, for any other form of transmission line it can be obtained using an electromagnetic simulator. In a simulator-based approach, the gap is analyzed using an electromagnetic simulator and the two-port scattering matrix, with the reference planes at the strip edges forming the gap, obtained at the center frequency of the filter. Once the S-matrix is known, (6.15a) through (6.15d) are used to calculate the values of the K-inverters. The above procedure has been programmed in a commercial software called WAVECON [7]. WAVECON synthesis and analysis results for a 14-GHz stripline gap-coupled band-pass filter are shown below. Figure 6.18(a) shows the input file, Figure 6.18(b) shows the filter dimensions, and Figure 6.18(c) shows the computed response. The substrate is Rogers RT-Duroid 5880. The main disadvantage of a half-wave gap-coupled filter is the appearance of spurious passband at twice the center frequency of the filter. As well, the filter is very long in size. In addition to that, the gaps become too small to be fabricated in the case of wideband filters. The last problem is often solved for planar filters using broadside coupling or interdigital or multilayer capacitor coupling leading to a more complicated structure.

6.2.2 Edge Parallel-Coupled Band-Pass Filters

A more compact form of a band-pass filter is a half-wave resonator quarter-wave coupled band-pass filter. Let us consider the equivalent network of an edge parallel-coupled transmission lines of length θ and the even- and the odd-mode impedances Z_{oe} and Z_{oo}, respectively, as shown in Figure 6.19.

Cascading several such edge parallel-coupled sections of length $\theta = 90°$ in tandem leads to a band-pass filter. Figure 6.20 shows the layout of the filter and the equivalent network.

For the first coupled pair of lines [9]:

$$\frac{Z_0}{K_{01}} = \sqrt{\frac{\pi\varpi}{2g_0 g_1}} \tag{6.16a}$$

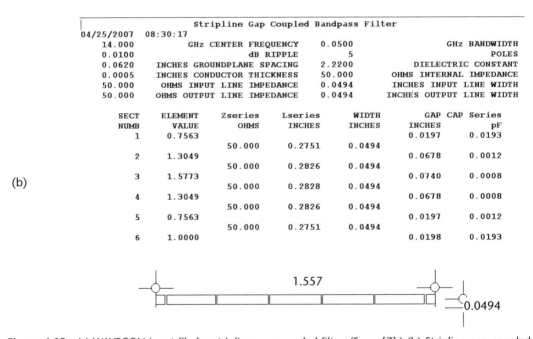

Figure 6.18 (a) WAVECON input file for stripline gap-coupled filter. (From [7].) (b) Stripline gap-coupled filter dimensions and layout. (From [7].) (c) Computed frequency response of stripline gap-coupled band-pass filter.

For the intermediate coupled pair of lines:

$$\frac{Z_0}{K_{k,k+1}} = \frac{\pi\varpi}{2\omega'_1\sqrt{g_0 g_1}} \qquad (k = 1,\cdots,N-1) \tag{6.16b}$$

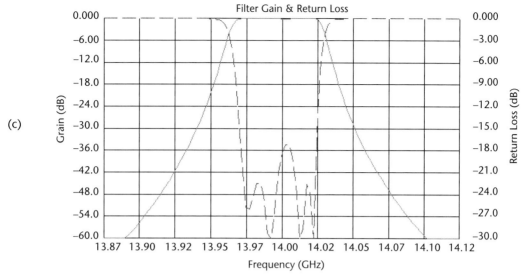

Figure 6.18 (continued)

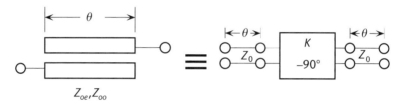

Figure 6.19 Edge parallel-coupled transmission lines and the equivalent K-inverter.

and for the outermost coupled pair of lines:

$$\frac{Z_0}{K_{N,N+1}} = \sqrt{\frac{\pi w}{2 g_N g_{N+1}}} \tag{6.16c}$$

Once the reciprocals of the normalized K-inverter values are known from the equations, the even- and odd-mode characteristic impedances of the corresponding coupled pair of lines are obtained from

$$(Z_{oe})_{k,k+1} = Z_0 \left[1 + \frac{Z_0}{K_{k,k+1}} + \left(\frac{Z_0}{K_{k,k+1}}\right)^2 \right] \tag{6.17a}$$

$$(Z_{oo})_{k,k+1} = Z_0 \left[1 - \frac{Z_0}{K_{k,k+1}} + \left(\frac{Z_0}{K_{k,k+1}}\right)^2 \right] \tag{6.17b}$$

where $k = 0$ to N.

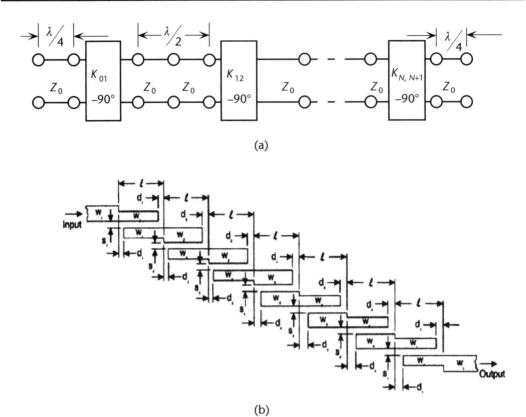

Figure 6.20 Edge parallel-coupled band-pass filter and the equivalent network.

In Figure 6.20 the length of each coupled section is $\lambda/4$. However, each open end of a line is shortened by an amount d_k to account for the fringing fields at the open ends. For a stripline filter, the foreshortening is given by (3.85) in Chapter 3. For any other transmission line the open-end correction is best done by using an electromagnetic simulator. Knowing the even- and odd-mode impedances of a coupled pair of lines, the corresponding line width and separation are obtained by using a suitable synthesis scheme or equation. For example, in the case of a stripline, (3.123) and (3.124) in Chapter 3 are used. WAVECON [7] synthesis and analysis results for a 10-GHz stripline edge parallel-coupled band-pass filter are shown below. Figure 6.21(a) shows the input file, Figure 6.21(b) shows the filter dimensions, Figure 6.21(c) shows the layout, and Figure 6.21(d) shows the computed response. The substrate is Rogers RT-Duroid 5880.

Planar edge parallel-coupled filters are usually enclosed within a box-type enclosure, of which the lateral dimensions are kept as small as possible so that no waveguide mode propagates through the structure up to the highest frequency of operation of the filter. That is, if the desired highest frequency of operation is f_H, then the width D of the enclosure is kept less than $c/(2f_H \sqrt{\varepsilon_r})$, where c is the velocity of electromagnetic wave in free space and ε_r is the dielectric constant of the medium filling the space between the strips and the ground planes. In order to achieve this objective, the layout is tilted at an angle to keep D as small as possible, as shown in Figure 6.22. If waveguide modes of the enclosure of an edge parallel-coupled band-pass filter are properly suppressed, the first spurious pass band of the filter

6.2 Distributed Transmission Line Form of Capacitively Coupled Band-Pass Filter

(a)

Transmission Media: (M,S,R,B,SM,SS)	S
Filter Center Frequency (GHz)	10.000
Filter Bandwidth (GHz)	1.0000
Ripple	0.0100
Number of Filter Sections (Poles)	6
Out-of-Band Freq where Attenuation needed (GHz)	6.0000
Attenuation at that Frequency (dB)	40.000
Input Impedance (Ohms)	50.000
Output Impedance (Ohms)	50.000
Internal Impedance (Ohms)	50.000
Relative Dielectric Constant	2.2200
Strip or Bar Thickness (Inches)	0.0005
1/2 Groundplane Spacing or Board Height (Inches)	0.0310
Distance above Mstrip Board to Top Shield (Inches)	0.0000
Distance below Mstrip Board to Bot Shield (Inches)	0.0000

(b)

```
     10.000          GHz CENTER FREQUENCY      1.0000           GHz BANDWIDTH
     0.0100                    dB RIPPLE            6                  POLES
     0.0620       INCHES GROUNDPLANE SPACING    2.2200    DIELECTRIC CONSTANT
     0.0005       INCHES CONDUCTOR THICKNESS   50.000      OHMS INTERNAL IMPEDANCE
    50.000        OHMS INPUT LINE IMPEDANCE    0.0494     INCHES INPUT LINE WIDTH
    50.000        OHMS OUTPUT LINE IMPEDANCE   0.0494    INCHES OUTPUT LINE WIDTH

    SECT     ELEMENT       ZOE       ZOO     LENGTH    WIDTH     GAP
    NUMB      VALUE        OHMS      OHMS    INCHES    INCHES   INCHES
      1      0.7813       82.847    37.824   0.1913    0.0314   0.0027
      2      1.3600       58.907    43.531   0.1909    0.0461   0.0133
      3      1.6896       55.798    45.303   0.1909    0.0477   0.0195
      4      1.5350       55.427    45.548   0.1909    0.0480   0.0206
      5      1.4970       55.798    45.303   0.1909    0.0477   0.0195
      6      0.7098       58.907    43.531   0.1909    0.0461   0.0133
      7      1.1007       82.847    37.824   0.1913    0.0314   0.0027
```

Figure 6.21 (a) Parfile input file for edge parallel-coupled stripline band-pass filter, (b) Parfile output file for edge parallel-coupled stripline band-pass filter, (c) edge parallel-coupled stripline band-pass filter layout, and (d) computed frequency response of edge parallel-coupled stripline band-pass filter. (From [7].)

appears centered at twice the frequency of the fundamental passband. There are several methods for suppressing the spurious passband. One very popular method is using a stepped impedance resonator, as shown in Figure 6.23. The resonance condition is given by [10]

$$\theta_0 = \tan^{-1} \sqrt{K} \tag{6.17c}$$

The first spurious passband of the filter occurs around

(c)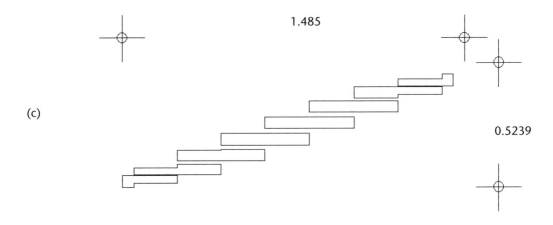

(d)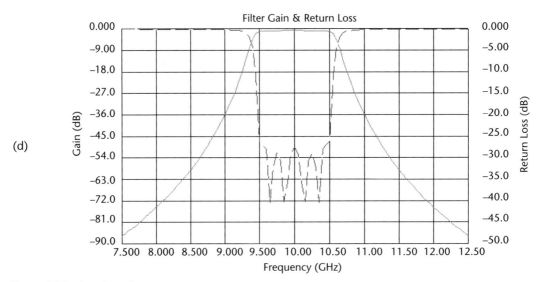

Figure 6.21 (continued)

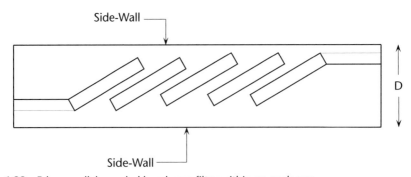

Figure 6.22 Edge parallel-coupled band-pass filter within an enclosure.

6.2 Distributed Transmission Line Form of Capacitively Coupled Band-Pass Filter

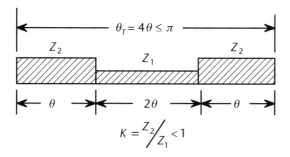

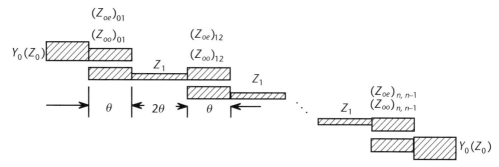

Figure 6.23 Stepped impedance edge-coupled resonator filter layout.

$$f_{s1} = f_0 \frac{\pi}{2\tan^{-1}\sqrt{K}} \quad (6.17d)$$

The design equations are

$$\left.\begin{aligned} J_{01} &= Y_0\sqrt{\frac{2w\theta_0}{g_0 g_1}} \\ J_{j,j+1} &= Y_0 \frac{2w\theta_0}{\sqrt{g_j g_{j+1}}} \dots (j = 1 \cdots n-1) \\ J_{n,n+1} &= Y_0\sqrt{\frac{2w\theta_0}{g_n g_{n+1}}} \end{aligned}\right\} \quad (6.17e)$$

$$\left(\frac{Z_{oe}}{Z_0}\right)_{k,k+1} = \frac{1 - \left(\frac{J_{k,k+1}}{Y_0}\right)\cosec\theta + \left(\frac{J_{k,k+1}}{Y_0}\right)^2}{1 - \left(\frac{J_{k,k+1}}{Y_0}\right)^2 \cot^2\theta} \quad (6.17f)$$

$$\left(\frac{Z_{oe}}{Z_0}\right)_{k,k+1} = \frac{1-\left(\frac{J_{k,k+1}}{Y_0}\right)\cosec\theta + \left(\frac{J_{k,k+1}}{Y_0}\right)^2}{1-\left(\frac{J_{k,k+1}}{Y_0}\right)^2\cot^2\theta} \qquad (6.17g)$$

6.2.3 Hairpin-Line Filter

A major drawback of half-wave edge parallel-coupled band-pass filter is its size. A compact form of the filter is the hairpin-line filter shown in Figure 6.24 [11]. This is a folded form of the edge parallel-coupled filter shown in Figure 6.21(c).

When each inner resonator of the filter in Figure 6.20(b) is folded in the form of a hairpin, a hairpin-line filter results. Obviously, the compactness of such a structure depends on the spacing between the two arms of a hairpin-line resonator. Smaller spacing results in more undesired coupling between the two arms of a resonator and less desired coupling between adjacent resonators. Hence, the self- or undesired coupling in each resonator becomes an important factor in filter design. Since the self-coupling in each resonator is an arbitrary factor that depends on the value of the spacing chosen by the designer, the design begins with a bandwidth correction factor given by [11]

$$\Delta f_c = \frac{\Delta f_d}{1-\Lambda} \qquad (6.18a)$$

$$\Lambda = 0.001|CP|^3 + 0.0012|CP|^2 - 0.0735|CP| + 0.7284 \qquad (6.18b)$$

where CP is the coupling between the arms of a single hairpin resonator in decibels (dB), Δf_c is the corrected bandwidth, and Δf_d is the desired bandwidth of the filter. Figure 6.24 defines the parameters for a hairpin-line filter [11] and design equations are presented below.

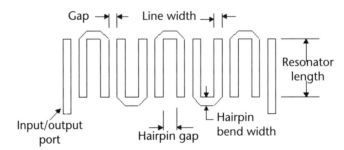

Figure 6.24 Hairpin-line filter.

6.2.3.1 Design Equations for Hairpin-Line

N Order of low-pass prototype filter

g_i Low-pass prototype filter element values ($i = 0, 1, 2, ..., N+1$)

ω_1' Low-pass prototype radian frequency

ϖ fractional bandwidth of the filter $= \dfrac{\Delta f_c}{f_0}$

$$\theta_1 = \frac{\pi}{2}\left(1 - \frac{\varpi}{2}\right)$$

$$G_i = \frac{1}{\sqrt{\omega_1' g_{i-1} g_i}} \text{ for } i = 1 \text{ and } N+1$$

$$G_i = \frac{1}{\omega_1' \sqrt{g_{i-1} g_i}} \text{ for } i = 2, 3, ..., N$$

h Arbitrary positive dimensionless parameter usually less than unity. It controls the internal immittance level of the filter.

$k = 10^{-\frac{|CP|}{10}}$ CP is the self-coupling in decibels.

for $i = 1$ and $N+1$	for $i = 1$ and $N+1$
$A_{11}^i = 1$	$A_{11}^i = h\tau$
$A_{12}^i = \sqrt{h}G_i$	$A_{12}^i = hG_i \sin\theta_i$
$A_{22}^i = h\left[G_i^2 + 1\right]$	$\tau = \dfrac{1}{2}\tan\theta_1$

$L_{p,i} = \dfrac{A_{22}^i + A_{11}^{i+1}}{2(1-k)}$	$p = 2i\,(i = 1, 2, ..., N)$
$L_{p+1,p+1} = L_{pp}$	
$L_{11} = A_{11}^i$	
$L_{2N+2,2N+2} = A_{11}^{N+1}$	
$L_{p,p+1} = A_{12}^i$	$p = 2i - 1\,(i = 1, 2, ..., N+1)$
$L_{p,p+1} = kL_{pp}$	$p = 2i\,(i = 1, 2, ..., N)$

$$[l_{ij}/Z_A v^{-1}] = [L] = \begin{bmatrix} L_{11} & L_{12} & 0 & 0 & 0 & \cdots & 0 \\ L_{12} & L_{22} & L_{23} & 0 & 0 & \cdots & 0 \\ 0 & L_{23} & L_{33} & L_{34} & 0 & \cdots & 0 \\ 0 & 0 & & & & & 0 \\ 0 & 0 & & \ddots & & & \vdots \\ \vdots & \vdots & & & \ddots & & L_{2N,2N+1} \\ 0 & 0 & 0 & \cdots & 0 & L_{2N+1,2N} & L_{2N+1,2N+1} \end{bmatrix}$$

where

v is the velocity of light in the medium
l_{ij} are the self or mutual inductance per unit length of the ith and jth conductors
Z_A is the source and load impedance

The normalized capacitance matrix of the parallel-coupled conductors is given by

$$[C] = [L]^{-1}$$

where

$$C_{ij} = \frac{c_{ij} Z_A}{v^{-1}}$$

c_{ij} are the self- or mutual capacitance per unit length of the ith and jth conductors.

Knowing the self- and the mutual capacitances of each conductor, the corresponding widths and spacings between the conductors are obtained by using a synthesis routine based on (3.102) to (3.117). Figure 6.25(a) shows a typical WAVECON [7] input file for a hairpin-line filter design. Figure 6.25(b) shows the output file and the filter layout. The computed frequency response is shown in Figure 6.25(c).

The input and output connections to a hairpin-line filter or an edge parallel-coupled filter are realized by capacitive coupling between the two outermost resonators and open-ended straight transmission lines. A major drawback of such a technique is that the coupling gap becomes unrealistically small when the bandwidth of the filter is large. This obstacle is removed by eliminating the input- and output-coupled lines and directly tapping the two outermost resonators, as shown in Figure 6.26. The location of the tapping point on the resonators is given by [12]

$$l = \frac{2L}{\pi} \sin^{-1} \left[\sqrt{\frac{\pi Z_0}{2 Z_r Q_e}} \right] \qquad (6.19a)$$

where Z_0 is the system impedance (usually equal to 50Ω), Z_r is the resonator impedance and Q_e, the external quality factor, is given by

6.2 Distributed Transmission Line Form of Capacitively Coupled Band-Pass Filter

(a)

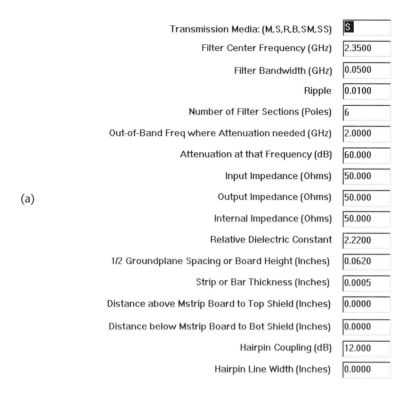

(b)

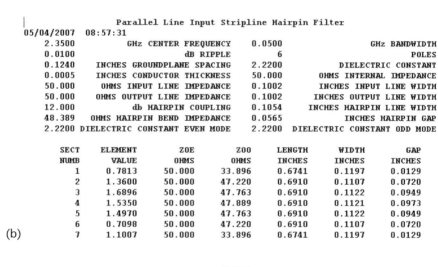

Figure 6.25 (a) WAVECON input file for a 4.35-GHz hairpin-line filter, (b) output file and the layout for a hairpin-line filter, and (c) computed frequency response of a hairpin-line filter, (From [7].)

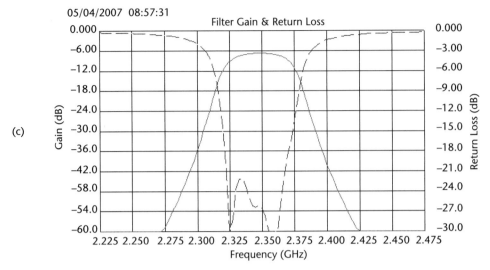

Figure 6.25 (continued)

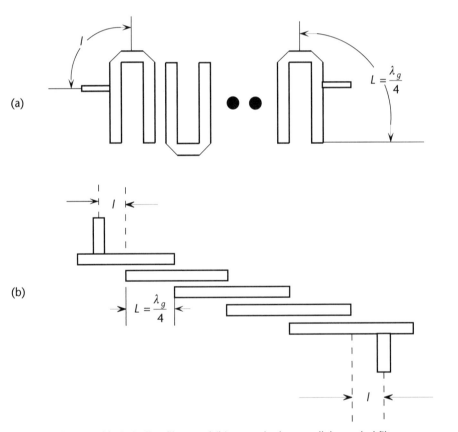

Figure 6.26 (a) Tapped hairpin-line filter, and (b) tapped edge parallel-coupled filter.

$$Q_e = \frac{g_0 g_1 f_0}{\Delta f} \tag{6.19b}$$

In the case of coupled resonators in a homogeneous medium, the guide wavelength is given by

$$\lambda_g = \frac{c}{f_0 \sqrt{\varepsilon_r}} \quad (6.20)$$

where c is the velocity of light in free space and ε_r is the dielectric constant of the medium. However, in the case of coupled resonators embedded in an inhomogeneous medium, the guide wavelength is given by

$$\lambda_g = \frac{c}{f_0 \sqrt{\varepsilon_{effr}}} \quad (6.21a)$$

where ε_{effr} is the Dell-Imagine [13] weighted average of effective dielectric constant of the coupled transmission line. It is given by

$$\varepsilon_{effr} = \frac{Z_{oe}\sqrt{\varepsilon_{effe}} + Z_{oo}\sqrt{\varepsilon_{effo}}}{Z_{oe} + Z_{oo}} \quad (6.21b)$$

It is obvious from the above equation that different resonators will have different lengths in an edge parallel-coupled and a hairpin-line filter because the even- and odd-mode parameters will be different for different coupled sections based on the coupling values. That is not a severe problem for narrow and moderate bandwidth filters. However, for wideband filters, a considerable amount of postproduction tuning is needed to achieve the desired center frequency and passband return loss. In addition to that, tapped line filters often show degraded passband return loss due to the parasitic discontinuity effect of the T-junctions. Equation (6.19) assumes a perfect T-junction with zero parasitic effect. In reality, however, the effect of the T-junction discontinuity off-tunes the two outermost resonators and the filter becomes nonsynchronous. Fortunately, the modern simulator-based approach alleviates these design problems considerably and in most cases a first-pass success is guaranteed. The approach will be discussed in Section 6.2.6.

6.2.4 Interdigital Filters

An interdigital filter [14] is a very popular device in microwave systems. It offers a very good compromise between size and resonator Q-factor. In fact, it is a modified form of a hairpin-line filter. Assume that an infinite coupling exists between the arms of a hairpin-line resonator. As a result, the gap between the arms becomes zero. Also, since the entire length of the hairpin is equal to one half of a wavelength, a virtual short circuit exists at the middle of the resonator. Figure 6.27(a) shows how, in the limit, a hairpin-line resonator degenerates into a single quarter-wavelength resonator with one end grounded. As a result, a hairpin-line filter becomes what is known as an interdigital filter, which is shown in Figure 6.27(b). The cor-

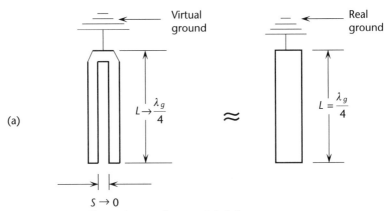
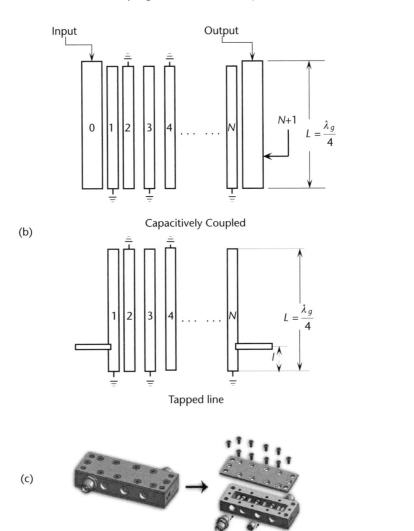

Figure 6.27 (a) Degeneration of a hairpin-line resonator to a quarter-wavelength resonator, (b) capacitively coupled and tapped line interdigital filters, and (c) mechanical assembly of an interdigital filter. (Reprinted with permission from Stellenbosch University, SA, courtesy of Prof. Petrie Meyer and Ms. Susan Maas.)

responding equivalent circuits are shown in Figure 6.28 [15]. Figure 6.27(c) shows the mechanical assembly of a typical interdigital filter.

The above interdigital filter configurations are the traditional forms used from the invention of interdigital filter in 1960s to the end of the 1990s. Design equations for such forms are available in [16]. In the modern approach, most interdigital filters are realized with all resonators having nearly equal impedances. This approach leads to simpler design equations and easy realizability, especially when round rods are used. Also, in most cases tapped resonators are used because the tap-point location can be adjusted with relative ease during postproduction tuning. The equations, from Caspi and Adelman [17], are as follows:

$$\left. \begin{array}{l} \theta = \dfrac{\pi}{2}\left(1 - \dfrac{\Delta f}{2 f_0}\right) \\[6pt] Y = \dfrac{Y_1}{\tan \theta} \\[6pt] J_{i,i+1} = \dfrac{Y}{\sqrt{g_i g_{i+1}}}\bigg|_{i=1\cdots N-1} \\[6pt] Y_{i,i+1} = J_{i,i+1} \sin \theta \big|_{i=1\cdots N-1} \\[6pt] C_1 = C_N = \dfrac{Y_1 - Y_{12}}{v} \\[6pt] C_i = \dfrac{Y_1 - Y_{i-1} - Y_{i+1}}{v}\bigg|_{i=2,\cdots N-1} \\[6pt] C_{i,i+1} = \dfrac{Y_{i,i+1}}{v} \\[6pt] Y_T = Y_1 - \dfrac{Y_{12}^2}{Y_1} \\[6pt] \theta_T = \dfrac{\sin^{-1}\left(\sqrt{\dfrac{Y \sin^2 \theta}{Y_0 g_0 g_1}}\right)}{1 - \dfrac{\Delta f}{2 f_0}} \\[6pt] C_T = \dfrac{\cos \theta_T \sin^3 \theta_T}{\omega_0 Y_T \left(\dfrac{1}{Y_0^2} + \dfrac{\cos^2 \theta_T \sin^2 \theta_T}{Y_T^2}\right)} \end{array} \right\} \quad (6.22)$$

In the above equations θ_T is the electrical angle of the tap-point location on the outermost resonators from the ground, v is velocity of light in medium F propagation ($1.18 \times 10^{10}/\sqrt{\varepsilon_r}$ inches/second). The physical location of the tap point is given by

$$l = \dfrac{\theta_T}{2\pi} \lambda_g = \dfrac{\theta_T}{2\pi}\left(\dfrac{v}{f_0}\right) = \dfrac{\theta_T}{2\pi}\left(\dfrac{c}{f_0 \sqrt{\varepsilon_{effr}}}\right) \qquad (6.23)$$

where c is the velocity of light in free space, v is the velocity of light in the filter cavity, and ε_{effr} is the effective dielectric constant of the medium. For a homogeneous

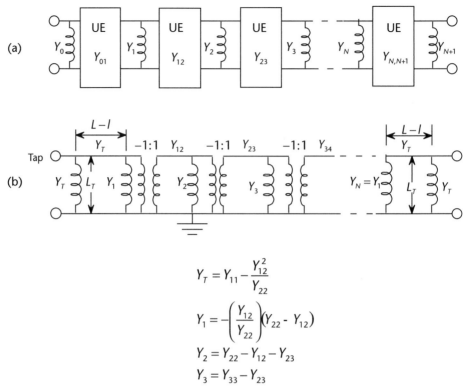

Figure 6.28 (a) Equivalent circuit of a capacitively coupled interdigital filter (Figure 6.27(a)), and (b) equivalent circuit of a tapped interdigital filter (Figure 6.27(b)). (From [15].)

medium such as a stripline or slabline, it is the dielectric constant of the space completely filling the entire filter. For an inhomogeneous medium such as a microstrip or suspended microstrip, the designer should use the Dell-Imagine equation [13] (see (6.21b)). Due to the loading effect of the tapping, the resonant frequency of the two outermost resonators changes from f_0, so the resonators are tuned back to f_0 by adding compensating capacitors of value C_T at the open ends. Once the self- and mutual capacitances per unit length of each resonator are obtained from the above equations, the width of each resonator and the spacings between them are obtained by using the synthesis and analysis equations in Chapter 3. Figure 6.29 shows the WAVECON [7] input file, output file, and the computed frequency response of a slabline interdigital filter.

6.2.5 Capacitively Loaded Interdigital Filters

The advantage of a capacitively loaded interdigital filter, shown in Figure 6.30, is that the resonators in it can be shorter than a quarter wavelength, leading to a more compact filter. However, the compactness is achieved at the cost of lower unloaded resonator Q-factor and hence resulting in higher passband insertion loss and band-edge rounding in the frequency response.

A viable approach for designing the capacitively loaded interdigital filter is as follows. Use (6.22) for 90° long resonators and obtain an initial set of values for line widths and spacings. Then, if a resonator electrical length of θ_0 is assumed, the

6.2 Distributed Transmission Line Form of Capacitively Coupled Band-Pass Filter

(a)

Transmission Media: (M,S,R,RS,B,SM,SS)	R
Filter Center Frequency (GHz)	2.3180
Filter Bandwidth (GHz)	0.0500
Ripple	0.0430
Number of Filter Sections (Poles)	5
Out-of-Band Freq where Attenuation needed (GHz)	2.3880
Attenuation at that Frequency (dB)	46.000
Input Impedance (Ohms)	50.000
Output Impedance (Ohms)	50.000
Internal Impedance (Ohms)	50.000
Relative Dielectric Constant	1.0000
1/2 Groundplane Spacing or Board Height (Inches)	0.3110
Strip or Bar Thickness (Inches)	0.0000
Distance above Mstrip Board to Top Shield (Inches)	0.0000
Distance below Mstrip Board to Bot Shield (Inches)	0.0000
Rod Filter - Rod Diameter (Inches)	0.2500
Rod Filt - Distance Rod Edge to Side Wall (Inches)	0.3000
Rod Filt - Distance Rod End to End Wall (Inches)	0.1800
Rod Filter Input Line Diameter (Inches)	0.0500

(b)

```
                Tapped Round Rod Interdigital Filter
05/09/2007  10:55:03
   2.3180        GHz CENTER FREQUENCY      0.0500           GHz BANDWIDTH
   0.0430                      dB RIPPLE        5                   POLES
   0.6220   INCHES GROUNDPLANE SPACING      1.0000      DIELECTRIC CONSTRANT
                                           69.182     OHMS INTERNAL IMPEDANCE
  50.000      OHMS INPUT LINE IMPEDANCE    0.3441     INCHES INPUT LINE WIDTH
  50.000     OHMS OUTPUT LINE IMPEDANCE    0.3441    INCHES OUTPUT LINE WIDTH
   0.1864      INCHES INPUT TAP TO SHORT   0.1864    INCHES OUTPUT TAP TO SHORT
  -0.006      INCHES INPUT ADDED LENGTH   -0.006     INCHES OUTPUT ADDED LENGTH
```

SECT NUMB	ELEMENT VALUE	ZOE OHMS	ZOO OHMS	LENGTH INCHES	WIDTH INCHES	GAP INCHES
1	0.9705	70.238		1.1427	0.2507	
			68.363			0.7046
2	1.3719	70.244		1.1430	0.2500	
			68.730			0.7482
3	1.8004	70.246		1.1347	0.2499	
			68.730			0.7480
4	1.3719	70.244		1.1430	0.2500	
			68.363			0.7046
5	0.9705	70.238		1.1427	0.2507	

Figure 6.29 (a) WAVECON input file for a slabline interdigital filter, and (b) WAVECON output file corresponding to the input file in Figure 6.29(a). (From [7].) (c) Computed frequency response of slabline interdigital filter.

loading capacitance connected between the ground and the open end of a resonator is given by

$$C_{sk}\big|_{k=1,2\cdots N} = \frac{\cot\theta_0}{2\pi f_0 Z_r} \quad (6.24)$$

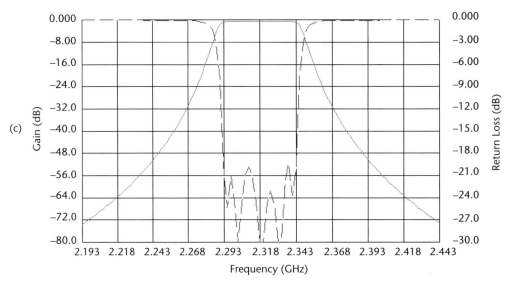

Figure 6.29 (continued)

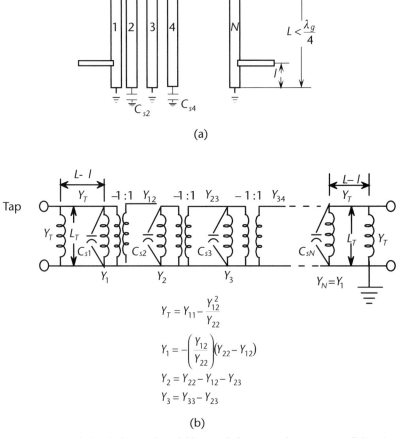

Figure 6.30 Capacitively loaded interdigital filter and the equivalent circuit. All line lengths not shown are equal to L.

where Z_r, the resonator impedance, is assumed to be the same for all resonators. Use a repeated-analysis-based computer optimization routine to obtain the desired frequency response. The optimization parameters are the line widths, line spacings, loading capacitances, and the tap-point location. The analysis routine may be based on transmission line and lumped element theory or full electromagnetic modeling. The latter is the most realistic and effective one for several reasons. One main reason is the physical form of the loading capacitance. Figure 6.31 shows several such forms used in practice.

Although reasonably accurate analytical models exist for the forms of capacitors shown in Figure 6.31, the best way to design capacitively loaded interdigital filters involving such forms is 3-D electromagnetic simulation. The design approach is based on interresonator couplings. It is worthwhile to explain the theory behind the interresonator-coupling-based approach of filter design.

6.2.6 Band-Pass Filter Design Based on Coupling Matrix

Consider the generalized network of the multiple-coupled resonator filter shown in Figure 6.32 [18].

The corresponding network equation is

$$[e] = [Z][i] \tag{6.25a}$$

In expanded form

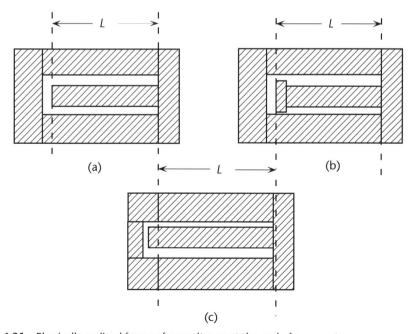

Figure 6.31 Physically realized forms of capacitance at the end of a resonator.

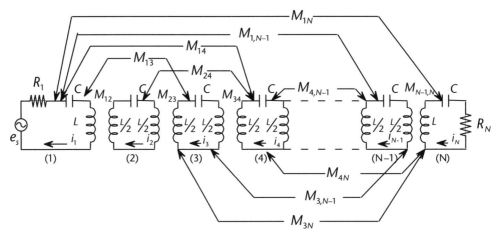

Figure 6.32 Generalized network of a multiple-coupled resonator filter.

$$\begin{bmatrix} e_s \\ 0 \\ 0 \\ 0 \\ \bullet \\ \bullet \\ \bullet \\ 0 \end{bmatrix} = \begin{bmatrix} S+R_1 & -j\omega M_{12} & -j\omega M_{13} & \bullet & \bullet & \bullet & \bullet & -j\omega M_{1N} \\ -j\omega M_{12} & S & -j\omega M_{23} & \bullet & \bullet & \bullet & \bullet & \bullet \\ -j\omega M_{13} & -j\omega M_{23} & S & \bullet & \bullet & \bullet & \bullet & \bullet \\ \bullet & \bullet & \bullet & \bullet & \bullet & \bullet & \bullet & \bullet \\ \bullet & \bullet & \bullet & \bullet & \bullet & \bullet & \bullet & \bullet \\ \bullet & \bullet & \bullet & \bullet & \bullet & \bullet & \bullet & \bullet \\ \bullet & \bullet & \bullet & \bullet & \bullet & \bullet & \bullet & \bullet \\ -j\omega M_{1N} & \bullet & \bullet & \bullet & \bullet & -j\omega M_{N-1,N} & S+R_N \end{bmatrix} \begin{bmatrix} i_1 \\ i_2 \\ i_3 \\ \bullet \\ \bullet \\ \bullet \\ \bullet \\ i_N \end{bmatrix} \quad (6.25b)$$

where

$$S = j\left(\omega L - \frac{1}{\omega C}\right) \quad (6.26a)$$

$$\omega_0 = \frac{1}{\sqrt{LC}} \quad (6.26b)$$

The impedance matrix of the network is expressed as

$$Z = ([S][I] + [M_R]) \quad (6.27)$$

where I is the identity matrix of order N, $[S]$ is a matrix with all elements zero except elements (1,1), and (N,N).

$$[M_R] = [R] + j[M] \quad (6.28)$$

The scattering parameters of the filter are given by

6.2 Distributed Transmission Line Form of Capacitively Coupled Band-Pass Filter

$$S_{21} = \frac{2\sqrt{R_1 R_N} f_0}{\omega_0 L \Delta f} [Z']_{N1}^{-1} = \frac{2}{\sqrt{q_{e1} q_{eN}}} [Z']_{N1}^{-1} \qquad (6.29\text{a})$$

$$S_{11} = 1 - \frac{2 R_1 f_0}{\omega_0 L \Delta f} [Z']_{11}^{-1} = 1 - \frac{2}{q_{e1}} [Z']_{11}^{-1} \qquad (6.29\text{b})$$

where the normalized external quality factors q_{e1} and q_{eN} are defined by

$$q_{ei} = \frac{\Delta f}{f_0} \frac{\omega_0 L}{R_i} = \frac{\Delta f}{f_0} Q_{ei} \qquad \text{for } i = 1, N \qquad (6.29\text{c})$$

and

$$[Z'] = \frac{f_0}{\Delta f \omega_0 L} [Z] \qquad (6.29\text{d})$$

The coupling coefficient k_M between the ith and the jth resonators is the ratio of the mutual inductance M_{ij} between them and the self-inductance L

$$k_M = \frac{M_{ij}}{L} \qquad (6.30)$$

From the above analysis, the design of a synchronously tuned band-pass filter depends on the determination of the parameters M_{ij}, L, C, and the external Q-factors Q_{e1}, Q_{eN}. Keep in mind that for a physically symmetric band-pass filter, the designer needs to determine only half the coupling parameters. The design procedure can be described with the help of an example. The example filter is a special type of microstrip filter having the following specifications:

1. Ripple bandwidth $\Delta f = 0.160$ GHz
2. Center frequency $f_0 = 2.20$ GHz
3. Passband return loss $RL = -26$ dB
4. Filter order $N = 4$
5. Transmission line = microstrip
6. Substrate thickness = 50 mils
7. Substrate dielectric constant $\varepsilon_r = 10.5$
8. Resonator impedance $Z_r = 50$ ohms
9. System impedance 50 ohms
10. Special requirement: All resonators should be grounded on the same side. The basic filter configuration is shown in Figure 6.33.

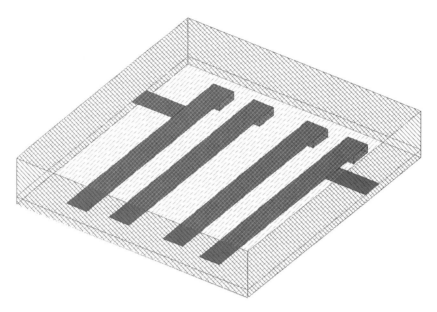

Figure 6.33 Microstrip interdigital filter.

6.2.6.1 Determination of Resonator Spacings

From the specifications it is obvious that it is not a conventional interdigital filter where alternate resonators are grounded on the same side. As a first step toward the design we construct a coupled pair of microstrip resonators grounded on the same side using via holes. The via hole configurations are the same as the ones we intend to use in the actual physical realization of the filter. The structure is shown [19] in Figure 6.34(a) and the frequency response of the structure, obtained experimentally or by a simulator, is shown in Figure 6.34(b). The resonator widths are chosen so as to offer the desired Z_r and the length L is chosen so that together with the effects of the open end and the via hole, the resonators resonate at the center frequency f_0 of the filter. Due to the coupling between the resonators we see two resonant peaks, one below f_0 at f_1 and the other above f_0 at f_2. The coupling coefficient between the resonators is given by [20]

$$k_m = \frac{f_2^2 - f_1^2}{f_2^2 + f_1^2} \tag{6.31}$$

The magnitude of k_m is a function of the separation S between the resonators. Therefore, by simulating the structure for different S we can obtain k_m as a function of S, as shown in Figure 6.35. An analytical equation for interresonator spacing S as function of coupling k_m can be obtained by using a suitable curve-fitting software.

For the filter to be designed, we find that the required low-pass prototype has the parameters

6.2 Distributed Transmission Line Form of Capacitively Coupled Band-Pass Filter 299

(a)

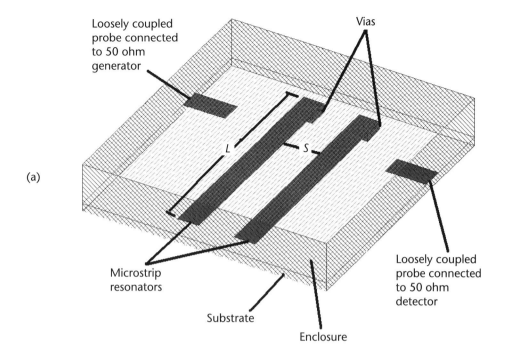

(b)

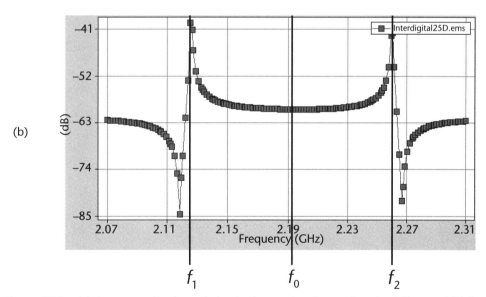

Figure 6.34 (a) Arrangements of coupled pair of resonators in coupling simulation, and (b) frequency response of circuit arrangement shown in Figure 6.34(a).

$$\left.\begin{array}{l} g_1 = 0.7128 \\ g_2 = 1.2003 \\ g_3 = 1.3212 \\ g_4 = 0.6476 \end{array}\right\} \qquad (6.32)$$

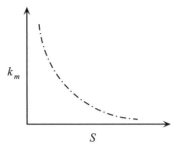

Figure 6.35 Coupling as a function of resonator spacing.

It can be shown that the corresponding coupling coefficients are

$$k_{m12} = k_{m34} = \frac{\Delta f}{f_0\sqrt{g_1 g_2}} = 0.0786 \tag{6.33}$$

$$k_{m12} = \frac{\Delta f}{f_0\sqrt{g_2 g_3}} = 0.0577 \tag{6.34}$$

Using the procedure described above, the corresponding resonator spacings are obtained as

$S_{12} = 50$ mils
$S_{23} = 84$ mils

6.2.6.2 Determination of Tap-Point Locations

The tap-point locations depend on the external quality factors Q_{e1} and Q_{e4}. For the required values, these parameters can be shown to be

$$Q_{e14} \approx Q_{e1} = \frac{g_1 f_0}{\Delta f} = 9.80 \tag{6.35}$$

We construct a microstrip configuration shown in Figure 6.36(a) where a resonator is attached to a 50-ohm finite length tapping line on one side and a very loosely coupled tapping line on the other side. The tapping line is connected to a generator with 50-ohm output impedance and the loosely coupled line is connected to a 50-ohm detector. The network is swept around the center frequency of the filter. Figure 6.36(b) shows the frequency response of the filter.

From Figure 6.36(b) the external Q-factor is obtained as

$$Q_e = \frac{f_0}{\Delta f_{3dB}} \tag{6.36}$$

6.2 Distributed Transmission Line Form of Capacitively Coupled Band-Pass Filter

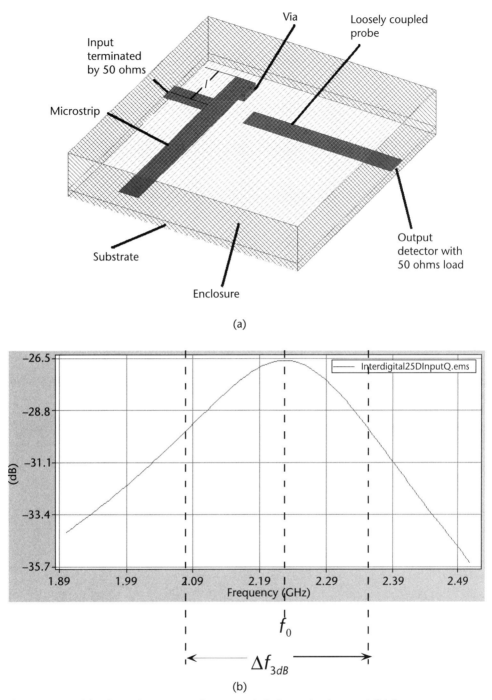

Figure 6.36 (a) Microstrip structure for external Q determination, and (b) frequency response network shown in Figure 6.36(a).

where Δf_{3dB} is the −3 dB bandwidth of the response curve in Figure 3.36(b). The tap-point location l is varied and for each location the structure is analyzed. Corresponding loaded Q, Q_e, is calculated using (6.36) and recorded. The required tap-point location l is obtained when Q_e equals Q_{e1}. For the present filter we obtained

$l = 125$ mils. The final layout of the designed filter is shown in Figure 6.37. The computed frequency response of the filter, based on full EM-wave analysis using EM3DS [19] software, is shown in Figure 6.38.

Having successfully finished the design using a generalized simulator-based method, we can arrive at a few conclusions. Those are:

1. The method is general and it uses the most realistic form of the coupling mechanism, grounding structure, vias, and the parasitics due to tap-point tee-junction discontinuity and the open ends.
2. Irrespective of the type and shape of the cavity and transmission line, the method is the same for all types of band-pass filters (e.g., waveguide, coaxial line, dielectric resonators, and planar transmission lines).

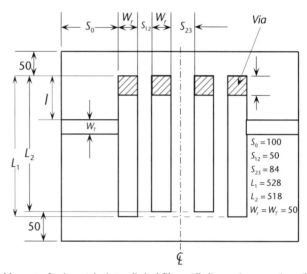

Figure 6.37 Final layout of microstrip interdigital filter. All dimensions are in mils.

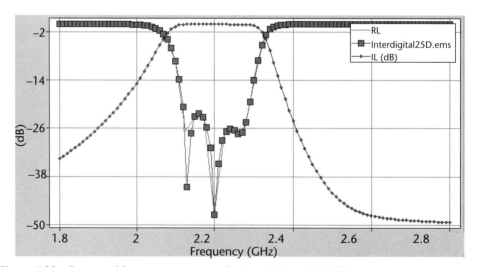

Figure 6.38 Computed frequency response of microstrip interdigital filter.

3. Accuracy of the method depends on how exactly the structure is represented in the simulations and the accuracy of the simulator used.

In a nutshell, the simulator-based method is based on Dishal's method [21] of band-pass filter design using virtual reality. The biggest advantage of the method is that no hardware is necessary to realize the design goal, which is cost effective.

6.2.7 Coaxial Cavity Band-Pass Filter Design

Figure 6.39 shows structures needed in designing an iris coupled coaxial cavity filter using the simulator-based method. Such structures can be created and analyzed using a standard 3-D EM simulator.

6.2.8 Combline Filters

A combline filter [22] is a very popular component used in microwave systems. It is compact in size and offers very wide stopband width depending on the length of the coupled transmission lines used. However, the advantages of a combline filter are obtained at the cost of increased passband insertion loss. A capacitively loaded interdigital filter and a combline filter have almost identical properties. But unlike in an interdigital filter, all transmission lines are grounded on the same side, as shown in Figure 6.40, which is convenient from a manufacturing standpoint. Figure 6.41(a) shows the practical schematic and Figure 6.41(b) shows 3-D view of a fourth-order combline filter with top cover removed.

6.2.8.1 Design Equations for a Combline Filter [17]

Choose:

- N Order of the filter (number of resonators)
- g_i Low-pass prototype filter element values $i = 0$ to $N+1$
- $\varpi = \dfrac{\Delta f}{f_0}$ Fractional bandwidth of the filter
- Y_A Source and load admittances
- Y_r Resonator admittance (same for all resonators)
- θ_0 Electrical length of the resonators at center frequency f_0

Compute:

$$\left.\begin{aligned}
b &= \left(Y_r/2\right)\left[\frac{\theta_0}{\sin^2 \theta_0} + \cot \theta_0\right] \\
J_{i,i+1}\Big|_{i=1\cdots N-1} &= \frac{\varpi b}{\sqrt{g_i g_{i+1}}} \\
y_{i,i+1}\Big|_{i=1,\cdots N-1} &= J_{i,i+1} \tan \theta_0 \\
l &= \frac{\lambda_g}{2\pi}\sin^{-1}\left\{\sqrt{Y_r \varpi \left(\cos \theta_0 \sin \theta_0 + \theta_0\right)/2g_0 g_1 Y_r}\right\}
\end{aligned}\right\} \quad (6.37)$$

Per unit length capacitances

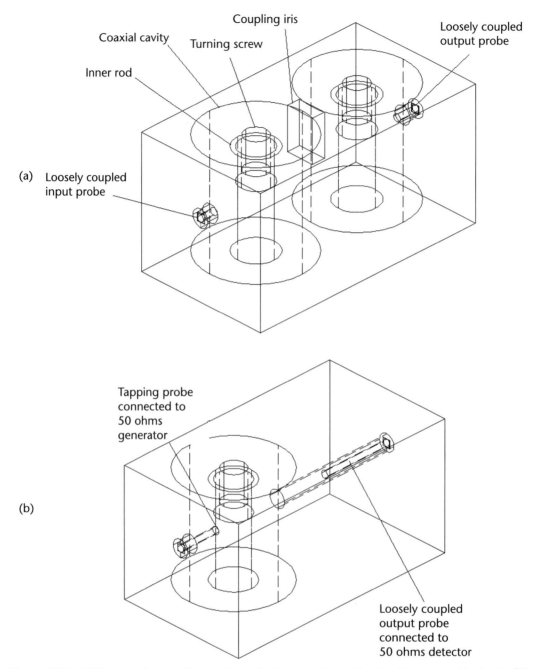

Figure 6.39 (a) Structure for coupling simulation in iris-coupled coaxial cavity filter (created by Empire™), (b) structure for external Q simulation of iris-coupled coaxial cavity filter (created by Empire™), and (c) iris-coupled coaxial cavity filter, final form (created by Empire™).

6.2 Distributed Transmission Line Form of Capacitively Coupled Band-Pass Filter

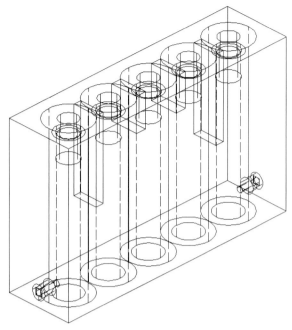

Figure 6.39 (continued)

$$\left.\begin{array}{l}\dfrac{C_{i,i+1}}{\varepsilon}\bigg|_{i=1,\cdots N-1} = \dfrac{\eta_0}{\sqrt{\varepsilon_r}} y_{i,i+1} \\[6pt] \dfrac{C_i}{\varepsilon}\bigg|_{i=1,\cdots N-1} = \dfrac{\eta_0}{\sqrt{\varepsilon_r}}\left[Y_r - y_{12}\right] \\[6pt] \dfrac{C_N}{\varepsilon}\bigg|_{i=1,\cdots N-1} = \dfrac{\eta_0}{\sqrt{\varepsilon_r}}\left[Y_r - y_{N-1,N}\right] \\[6pt] \dfrac{C_i}{\varepsilon}\bigg|_{i=2,\cdots N-1} = \dfrac{\eta_0}{\sqrt{\varepsilon_r}}\left[Y_r - y_{i-1,i} - y_{i,i+1}\right]\end{array}\right\} \quad (6.38)$$

The terminating capacitors are

$$C_{Si}\big|_{i=1,\cdots N} = \frac{Y_r \cot\theta_0}{2\pi f_0} \quad (6.39)$$

Compensating capacitance for tapping on the outermost resonators l and N

$$C_c^S = \frac{Y_r^2 Z_0 \sin(\theta_0 - \Phi_0)}{\left\{\left[\left(\dfrac{\sin\theta_0}{\sin\Phi_0}\right)^3 + \left(\dfrac{\sin\theta_0}{\sin\Phi_0}\right)Y_r^2 Z_0^2 \sin^2(\theta_0 - \Phi_0)\right]2\pi f_0\right\}} \quad (6.40)$$

Therefore, the total capacitance on resonators 1 and N is

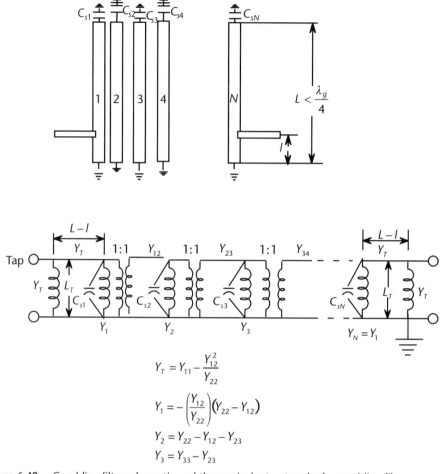

Figure 6.40 Combline filter schematic and the equivalent network of a combline filter.

$$C_1^{Total} = C_c^s + C_{S1} \tag{6.41}$$

The above equations are derived on the basis of lumped element circuit theory, transmission line theory, and assumption of ideal grounding of resonators. In reality, such assumptions are not completely valid due to via inductance, proximity of the enclosure walls, and particularly the way the capacitances are realized. Consequently, the dimensions obtained by using the above design equations need to be refined using full EM simulation or actual postproduction tuning of the filter. In fact, the addition of the compensating capacitance C_c^s amounts to a few extra turns of the tuning screw in postproduction tuning. Besides frequency tuning, it may be necessary to adjust the interresonator coupling as well. Figure 6.41(c) shows how a set of $N-1$ screws is inserted between the resonators to achieve this goal.

As mentioned above, the combline filter is compact in size and offers a wide stopband. In an interdigital filter the first spurious passband occurs at $3f_0$ with a $3\Delta f$ wide passband. For a combline filter the first spurious passband is centered around

6.2 Distributed Transmission Line Form of Capacitively Coupled Band-Pass Filter

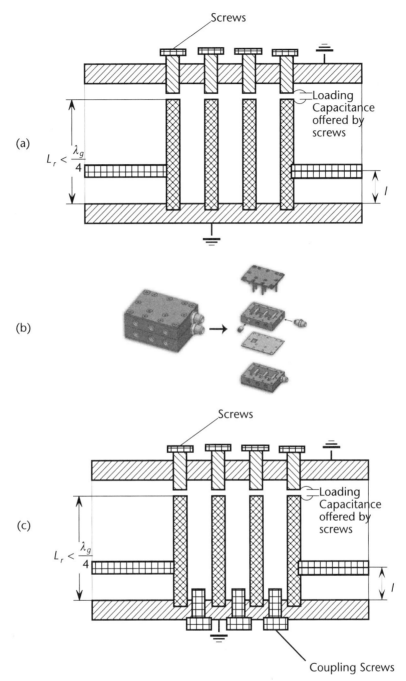

Figure 6.41 (a) Practical schematic and (b) 3-D view of a 3-pole combline filter assembly. Reprinted with permission from Stellenbosch University, SA; courtesy of Prof. Petrie Meyer and Ms. Susan Maas.) (c) Combline filter with tuning screws.

$$f_{Sr} = \frac{180}{\theta_0} f_0 \tag{6.42}$$

Also, shorter than a quarter-wavelength resonator has a lower unloaded Q-factor. If the quarter-wavelength resonator has an unloaded Q of $Q_{\lambda/4}$, and a resonator of electrical length θ_0 has the unloaded Q_{θ_0}, then

$$Q_{\theta_0} = \left(Q_{\lambda/4}\right)\sin^2\theta_0 \tag{6.43a}$$

However, the overall Q_{Total} of the resonator is a combination of the resonator $Q(Q_{\theta_0})$ and the Q-factor Q_c of the lumped capacitor forming the resonator. It is given by

$$\frac{1}{Q_{Total}} = \frac{1}{Q_{\theta_0}} + \frac{1}{Q_c} \tag{6.43b}$$

6.2.8.2 Combline Filter Design Example

We have a set of customer specifications for a square-rod combline filter as follows:

1. Passband: 7.40–8.1 GHz (ripple bandwidth Δf_r = 700 MHz)
2. Insertion loss: 1.00 dB maximum at band edges
3. Rejection (IL): –10 dB at 7.2 and 8.3 GHz
   (Isolation bandwidth Δf_i = 1100 MHz)
4. Passband return Loss (RL): –15.00 dB
5. Operating temperature: –30°C to +70°C
6. Approximate inner dimensions: 1.2" × 0.25" × 0.25"

A good approximation for the initial dimensions of the rods and the spacings between consecutive rods can be obtained from (6.37) to (6.41). However, before proceeding with the initial design, we should analyze the temperature stability of the filter, which will in turn determine the required order N of the filter. The overall temperature range of the filter is 100°C. The coefficient of linear thermal expansion of brass is α_l = 0.000019 length/unit length/°C. Therefore, the shift in the center frequency of the filter due to temperature change is $\Delta f_{sT} = \alpha_l \Delta T f_0 \approx 15$ MHz. Consequently, we need to expand the ripple bandwidth (Δf_r = 700 MHz) by 15 MHz and reduce the isolation bandwidth (Δf_i = 1100 MHz) by the same amount (15 MHz) for temperature stability. The new values for those parameters are Δf_r^{new} = 715 MHz and Δf_i^{new} = 1085 MHz. We design the filter for a 20-dB return loss instead of a 15 dB return loss. The stop-band to pass-band ratio

$$\gamma = \Delta f_i^{new} \big/ \Delta f_r^{new} = 1.5175.$$

And, assuming a Chebyshev response, the required filter order

$$N \geq (IL + RL + 6)\big/20\log\left(\gamma + \sqrt{\gamma^2 - 1}\right) = 4.238$$

Hence, the required order is $N = 5$. According to (6.42), a sixty-degree resonator length will ensure a spurious free stopband up to and beyond 22 GHz. Also, a circuit simulation shows that the required unloaded resonator Q-factor for less than 1 dB insertion loss at the band edges should be more than 600. The corresponding quarter-wavelength resonator Q is obtained from (6.43) is 800. Using (3.66) and a 66 Ω stripline (84 mils × 84 mils square rod between 240 mils ground-plane spacing), we find an unloaded Q-factor of 2609. Considering only 65% of that unloaded Q we have an available Q of 1696, which is more than twice the required Q. Figure 6.41(a) shows the dimensions of the filter as obtained from (6.37) to (6.42a) via WAVECON software.

The analyzed frequency response of the filter is shown in Figure 6.42(b). However, the analysis assumes that the resonators are terminated by lumped capacitors. From Figure 6.42(a), we notice that the resonator rods do not have the same width. For the sake of manufacturing convenience, we use an average value of 84 mils. The open end of each rod and the opposite wall form the terminating capacitance. The value of that capacitance, in case of round rods, is shown in Figure 6.42(c).

For a square rod of dimension 84 mils × 84 mils and a required terminating capacitance of 0.150 pf, $d \approx 10$ mils. Using the computed values of d and the dimensions of the filter in the table, we now analyze and optimize the response of the filter using a full 3-D EM simulator. The optimized dimensions are shown in Figure 6.42(d).

Figure 6A.1 in Appendix 6A shows a very commonly used slot-coupled coaxial-resonator-based combline filter configuration and the associated design equations.

Usually in all cases of combline filter design, when the normalized ground plane separation b/λ becomes more than 0.12, two effects become more and more prominent: (1) the measured bandwidth becomes more than the bandwidth the filter is designed for, and (2) the resonators exhibit larger Q-factors than what is expected from a combline resonator.

When b/λ becomes larger and larger, the supported mode in the structure gradually transitions from pure TEM to a ridged waveguide mode and the filter becomes a ridged waveguide filter. In general, the Q-factor of a combline resonator is given by[1].

$$Q = Kb\sqrt{f} \quad (6.43c)$$

where the constant $K = 1600$, b is in inches and f in gigahertz. Figure 6.42(e) shows the increase in Q as a function of $b\sqrt{f}$. The expansion in bandwidth, however, depends on the specific combline resonator configuration. Therefore, it is wise to design such filters, which Levy, Yao and Zaki[2] called transitional combline/eva-

---

1. See Chapter 5, reference [24].
2. Ralph Levy, Hui-Wen Yao and Kawthar Zaki, "Transitional Combline/Evanescent-Mode Microwave Filters", *IEEE Trans. On MTT.* Vol. 45, No. 12, December 1997.

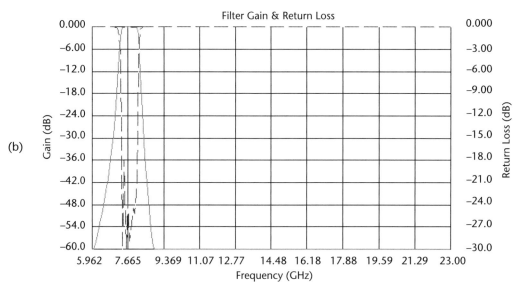

Figure 6.42 (a) Dimensions of the designed combline filter obtained from WAVECON software [7]. (b) computed frequency response of combline filter. (c) Terminating capacitance of a combline filter. (d) optimized dimensions of combline filter. Dimensions are in mils. (e) measured variation of experimental Q with b/λ (Courtesy of Dr. Ralph Levy).

nescent mode filters, using the full-wave simulators and optimization described in Section 6.2.6.

6.2.9 Waveguide Band-Pass Filter Design

Most waveguide band-pass filters consist of a series of half-wavelength-long cavities coupled by inverters formed by waveguide discontinuities. A few such typical discontinuities are shown in Figure 6.43.

6.2 Distributed Transmission Line Form of Capacitively Coupled Band-Pass Filter 311

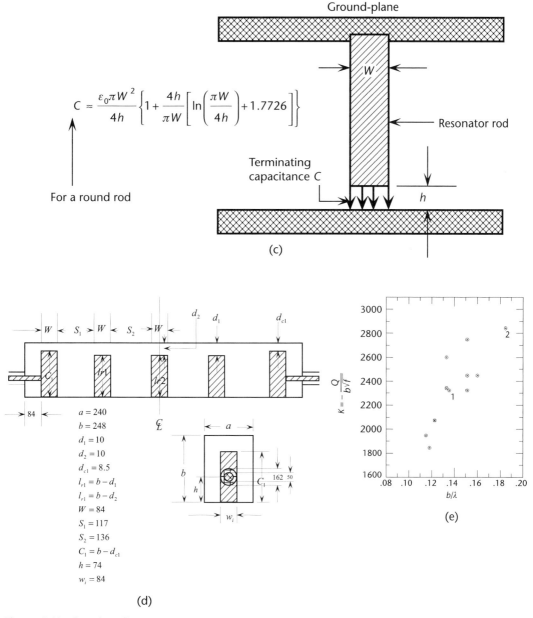

Figure 6.42 (continued)

Irrespective of the type of the discontinuity forming the coupling inverter, the equivalent network of a propagating mode waveguide filter is shown in Figure 6.44. The design procedure is based on a formulation proposed by Rhodes [23] for a distributed stepped impedance low-pass prototype. When the passband ripple ε, the lower and the upper cutoff frequencies f_L and f_H, and the filter order are specified, the procedure is as follows:

1. Determine the midband guide wavelength λ_g by solving

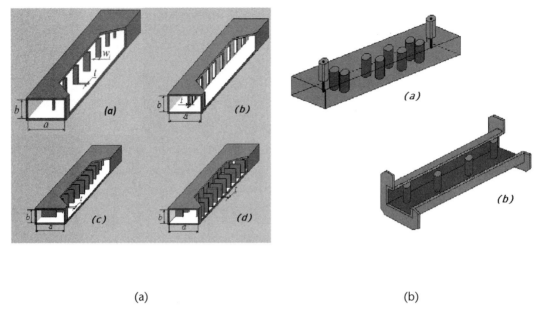

(a)   (b)

Figure 6.43 Iris-coupled waveguide band-pass filter structure: (a) double iris, (b) septum, (c) asymmetric iris, and (d) symmetric iris. (b) Shows post-coupled waveguide band-pass filters, (a) shows dual-post and coaxial interface, and (b) shows a single-post with waveguide interface.

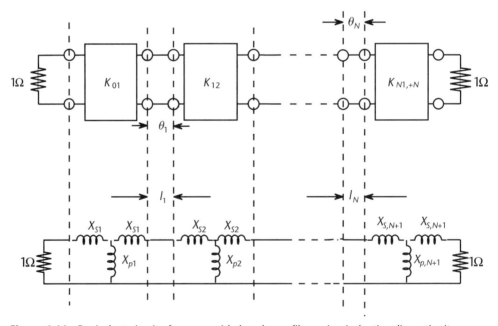

Figure 6.44 Equivalent circuit of a waveguide band-pass filter using inductive discontinuity.

$$\lambda_{gL} \sin\left(\frac{\pi \lambda_{gL}}{\lambda_{g_0}}\right) = -\lambda_{gH} \sin\left(\frac{\lambda_{go}}{\lambda_{gH}}\right) \quad (6.44)$$

For narrowband filters

6.2 Distributed Transmission Line Form of Capacitively Coupled Band-Pass Filter

$$\lambda_{go} \approx \frac{\lambda_{gL} + \lambda_{gH}}{2} \tag{6.45}$$

2. Determine the scaling factor α from

$$\alpha = \frac{\lambda_{go}}{\lambda_{gL} \sin\left(\dfrac{\pi \lambda_{go}}{\lambda_{gL}}\right)} \tag{6.46}$$

3. Calculate the impedances of the distributed elements and impedance inverter values

$$Z_N = \frac{2\alpha \sin\left[\dfrac{(2N-1)}{2N}\pi\right]}{y} - \frac{1}{4y\alpha} \left\{ \frac{y^2 + \sin^2\left(\dfrac{k\pi}{N}\right)}{\sin\left\{\dfrac{(2N+1)}{2N}\pi\right\}} \right.$$

4.
$$\left. + \frac{y^2 + \sin^2\left[\dfrac{(N-1)}{N}\pi\right]}{\sin\left\{\dfrac{(2N-3)}{2N}\pi\right\}} \right. \tag{6.47}$$

$$k = 1,\ldots,N$$

5.
$$y = \sinh\left[\frac{1}{N}\sinh^{-1}\left(\frac{1}{\varepsilon}\right)\right] \tag{6.48}$$

$$K_{k,k+1} = \frac{\sqrt{y^2 + \sin^2\left(\dfrac{k\pi}{N}\right)}}{y\sqrt{Z_k Z_{k+1}}} \tag{6.49}$$

$$k = 0,\cdots,N$$

with

$$Z_0 = Z_{N+1} = 1 \tag{6.50}$$

In order to realize the value of a K-inverter, analyze the waveguide discontinuity that you have chosen from Figure 6.43. For example, if it is a symmetric iris filter (6.43(a, b, d)) then a symmetric iris is analyzed using a suitable analysis method. Once the two-port S-matrix of the jth iris is obtained, the equivalent T-network in Figure 6.44 is obtained by

$$jX_{Sj} = \frac{1 - S_{j12} + S_{j11}}{1 - S_{j11} + S_{j12}} \tag{6.51a}$$

$$jX_{pj} = \frac{2S_{j12}}{\left(1-S_{j11}\right)^2 - S_{j12}^2} \qquad (6.51b)$$

The equivalent network of the discontinuity is shown in Figure 6.45

$$\phi_k = -\tan^{-1}\left(2X_{pk} + X_{Sk}\right) - \tan^{-1}\left(X_{Sk}\right) \qquad (6.52a)$$

$$K_{k,k+1} = \left|\tan\left(\frac{\phi_k}{2} + \tan^{-1}\left(X_{Sk}\right)\right)\right| \qquad (6.52b)$$

The length of the waveguide section between the kth and the $(k+1)$th discontinuity is given by

$$l_k = \frac{\lambda_{go}}{2\pi}\left[\pi + \frac{1}{2}\left(\phi_k + \phi_{k+1}\right)\right] \qquad (6.53)$$

The desired value of a K-inverter is obtained by adjusting the dimensions of a particular discontinuity. For example, it is the iris opening, for an iris discontinuity, it is the septum width for a septum discontinuity and it is the post diameter if it is a post discontinuity. The design steps can be programmed on a computer. Figure 6.46 shows the input and output files for an X-band double-iris-coupled filter. Notice from the synthesis file that the design frequency of the filter is not exactly the average of the lower and upper cutoff frequencies of the filter. This frequency was calculated using (6.44). Also, the rejection at 9.5 GHz is 8 dB less than the rejection specification of 40 dB. This is due to the dispersive property of the waveguide. A physically realized filter, using the above dimensions, may show a narrower bandwidth and an upwardly shifted center frequency due to the tooling radius at the junctions forming the iris walls and the waveguide side walls.

Figure 6.47 shows the interface of a typical commercial waveguide filter design software. Appendix 6B shows the interface of a typical commercial waveguide filter wizard.

6.2.10 Evanescent-Mode Waveguide Band-Pass Filter Design

The type of waveguide band-pass filter described in the previous sections has two elements. Those are the half-wave resonators consisting of waveguide sections

Figure 6.45 Equivalent network of an inductive discontinuity in waveguide.

6.2 Distributed Transmission Line Form of Capacitively Coupled Band-Pass Filter

(a)

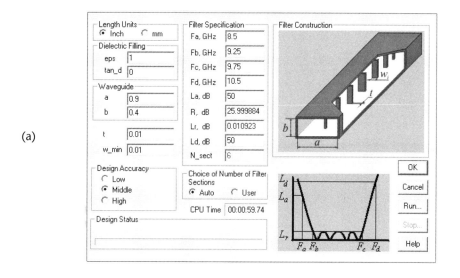

(b)

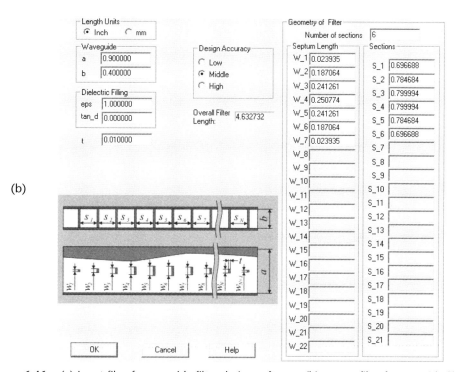

Figure 6.46 (a) Input file of waveguide filter design software, (b) output file of waveguide filter design software, and (c) computed frequency response of a double-iris waveguide filter.

operating in the propagating mode and the inverter forming below cutoff sections containing the discontinuity. We can replace the propagating mode section by a section of ridged waveguide and the inverters forming section can be replaced by a section of below cutoff waveguide without any discontinuity [25]. However, in the process of doing so, each resonator is bounded by two discontinuities between the ridged waveguide and the evanescent mode waveguide. The structure is shown in Figure 6.48 and the complete filter is shown in Figure 6.49. Advantages of an

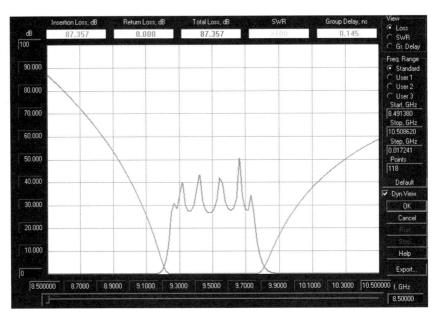

Figure 6.46 (continued)

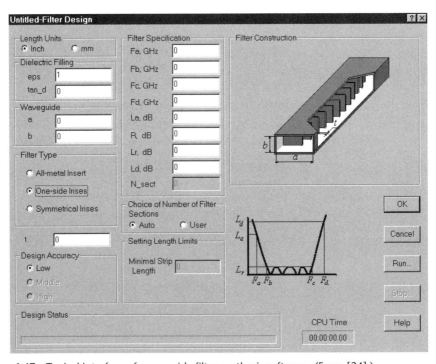

Figure 6.47 Typical interface of waveguide filter synthesis software. (From [24].)

evanescent mode waveguide filter are compact size, steep skirt selectivity, and wide stopband. The disadvantages are lower resonator Q and power-handling capability. Despite the disadvantages, evanescent mode filters are used in many systems where a wide passband width, wide stopband width, and small size are required. In fact, the wide passband width, to some extent, compromises the passband insertion loss.

6.2 Distributed Transmission Line Form of Capacitively Coupled Band-Pass Filter 317

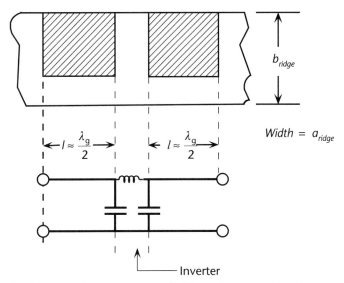

Figure 6.48 Ridged waveguide resonators coupled by an inverter. Dimensions a_{ridge} and b_{ridge} are chosen so that the nonridged section is below cutoff for the fundamental TE_{10} mode of operation.

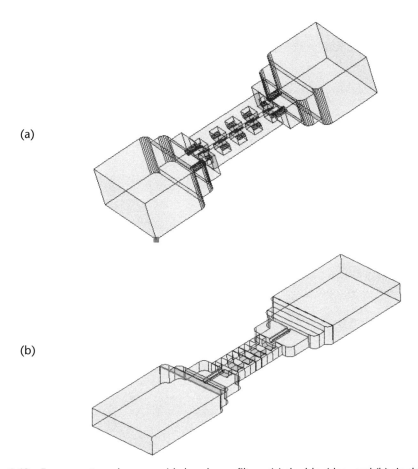

Figure 6.49 Evanescent mode waveguide band-pass filters: (a) double-ridge, and (b) single-ridge.

The equivalent circuit of the filter is the same as that of a propagating mode waveguide band-pass filter (Figure 6.44). The equivalent circuit of the filter is obtained by analyzing the discontinuities between the ridge waveguide and the evanescent mode waveguide using the mode-matching method described in Section 3.10.8. However, other methods like the finite element method or the finite difference time domain method would also be suitable.

Since a_{ridge} and b_{ridge} are chosen to be smaller than a and b, respectively, quarter-wave transformers are used between the evanescent mode filter and the input and output waveguides, as shown in Figure 6.49. Computer optimization is used to obtain the overall frequency response of the filter. WAVEFIL [24] waveguide filter design computer program designs, analyzes, and optimizes evanescent mode band-pass filters. The computed response of an X-band evanescent mode band-pass filter is shown in Figure 6.50.

The stopband width of an evanescent mode waveguide band-pass filter can be improved by interposing inductive septa in the evanescent mode sections of the filter, as shown in Figure 6.51 [25]. The frequency responses of X- and Ku-band evanescent mode filters with added inductive septa are shown in Figure 6.52. The figure shows that stopbands almost up to the third harmonic are achievable using this technique.

With the single-mode approach, the following formula

$$r = \angle S_{11} / \zeta_{10}$$

is used to estimate the initial value of the septum width. ζ_{10} is the propagation constant of the ridged waveguide dominant mode, and $\angle S_{11}$ is the phase angle of the dominant mode from the reflecting septum, as shown in Figure 6.51(e). The design procedure remains the same as that for a conventional evanescent mode filter. The

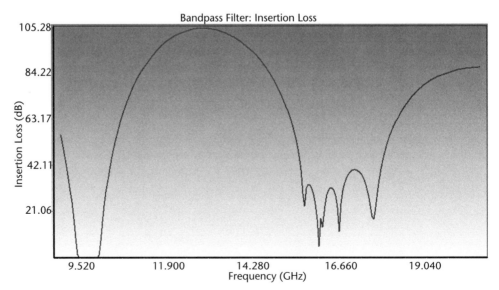

Figure 6.50 Computed frequency response of X-band evanescent mode filter.

6.2 Distributed Transmission Line Form of Capacitively Coupled Band-Pass Filter

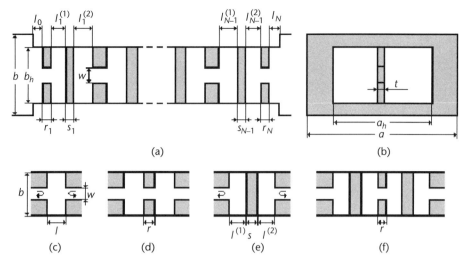

Figure 6.51 Evanescent mode band-pass filters: (c,d) conventional, and (a,e,f) with interposed inductive strips, and (b) end view. (From [25]. Reprinted with permission from the IEEE.)

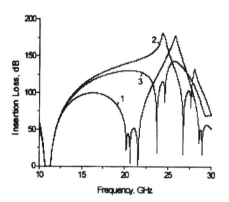 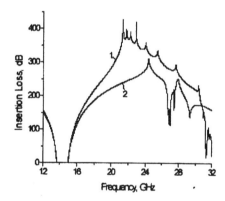

Insertion loss for the 10.5–11.5 GHz five-resonator bandpass filters. Filters' dimensions (in millimeters): $1 - l_0 = l_5 = 2.092$, $l_1 = l_4 = 9.965$, $l_2 = l_3 = 11.035$, $r_1 = r_5 = 1.491$, $r_2 = r_4 = 1.770$, $r_3 = 1.762$; $2 - s = 0.3$, $l_0 = l_5 = 1.391$, $l_1 = l_4 = 1.810$, $l_2 = l_3 = 2.228$, $r_1 = r_5 = 2.756$, $r_2 = r_4 = 4.256$, $r_3 = 3.961$; $3 - s = 0.7$, $l_0 = l_5 = 1.025$, $l_1 = l_4 = 1.179$, $l_2 = l_3 = 1.535$, $r_1 = r_5 = 3.389$, $r_2 = r_4 = 5.523$, $r_3 = 5.118$.

Insertion loss for the 13.7–15.0 GHz eleven-resonator bandpass filters. Filters' dimensions (in millimeters): $1 - s = 0.2$, $l_0 = l_{11} = 2.200$, $l_1 = l_{10} = 1.623$, $l_2 = l_9 = 2.064$, $l_3 = l_8 = 2.331$, $l_4 = l_7 = 2.438$ $l_5 = l_6 = 2.475$, $r_1 = r_{11} = 0.930$, $r_2 = r_{10} = 1.881$, $r_3 = r_9 = 1.521$, $r_4 = r_8 = 1.360$, $r_5 = r_7 = 1.304$, $r_6 = 1.290$; $2 - l_0 = l_{11} = 0.551$, $l_1 = l_{10} = 6.182$, $l_2 = l_9 = 7.310$, $l_3 = l_8 = 7.671$, $l_4 = l_7 = 7.800$, $l_5 = l_6 = 7.846$, $r_1 = r_{11} = 0.518$ $r_2 = r_{10} = 0.791$, $r_3 = r_9 = 0.757$, $r_4 = r_8 = 0.749$, $r_5 = r_7 = 0.747$, $r_6 = 0.746$.

Figure 6.52 Computed frequency response of evanescent mode filter with added inductive strips (From [25]. Reprinted with permission from IEEE.)

only difference is that the calculation of the K-inverter also takes into consideration the reflecting septum.

6.2.11 Cross-Coupled Resonator Filter Design

Let us consider the generalized multiple-coupled resonator filter network in Figure 6.32. The figure shows that coupling may exist among nonadjacent resonators in the filter. So far we have discussed only those filters in which there are no nonadjacent resonator couplings. Nonadjacent resonator couplings are also known as cross couplings. There are several advantages of having cross couplings in a filter [26]. The most important advantage is that for the same filter order, a cross-coupled filter may offer much better selectivity than a Chebyshev filter. Also, using cross couplings of proper sign (negative or positive), considerable flat group delay within the passband can be obtained. However, flatter group delay is obtained at the expense of more delay and reduction in selectivity. Figure 6.53 (a) shows the effect of cross coupling on the frequency response of an otherwise sequentially coupled filter [27]. This circuit is known as a quadruplet. Figure 6.53(b) shows one possible physical realization of quadruplet using coaxial resonators. Figure 6.54 shows a possible realization of a triplet and its frequency response. Figures 6.54(b) shows the frequency response of a triplet for positive and negative cross couplings. Figure 6.55 shows the frequency response of a quadruplet with multiple cross couplings that involve three or more signal paths.

In the first network of Figure 6.55, the outer path 1-2-3 combines with the inner path 1-3 to form one transmission zero. The other transmission zero is formed by the paths 1-3-4 and 1-4. The inductive cross coupling between 1 and 3 generates both transmission zeros on the high frequency side; while capacitive cross coupling between 1 and 3 generates two transmission zeros on the low frequency side.

Figure 6.56 shows two possible quintuplet configurations involving multiple couplings [27]. These are the only two configurations for achieving all three transmission zeros on the same side of the passband. Other combinations exist that produce two transmission zeros above and one below the passband.

The coupling matrix of a cross-coupled filter for a prescribed frequency response and coupling topology can be determined using various methods [27–30]. However, the most versatile one is the computer optimization based method due to Atia, Zaki, and Atia [31]. The method is as follows:

Let us consider the generalized multicoupled filter network in Figure 6.57.

The corresponding two-port representation of the network is shown in Figure 6.58. The transfer function of the network in Figure 6.58 can be written as

$$S_{21} = \frac{1}{1 + \varepsilon^2 \Phi^2(\bar{\lambda})} \quad (6.54)$$

where

$$\Phi(\bar{\lambda}) = \frac{\prod_{i=1}^{n}(\bar{\lambda} - \bar{\lambda}_{zi})}{\prod_{j=1}^{m}(\bar{\lambda} - \bar{\lambda}_{pj})} \quad (6.55)$$

is the characteristic function. The filter is a two-port network driven by a source of internal resistance R_1 and terminated in a load R_2. Also, the normalized frequency is

$$\bar{\lambda} = \frac{f_0}{\Delta f}\left(\frac{f}{f_0} - \frac{f_0}{f}\right) \quad (6.56)$$

6.2 Distributed Transmission Line Form of Capacitively Coupled Band-Pass Filter

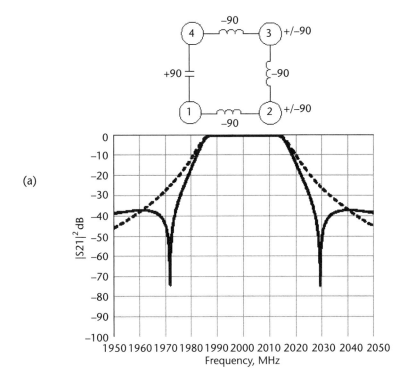

(a)

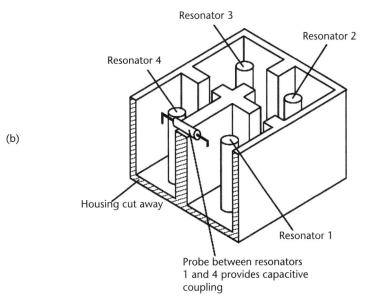

(b)

Figure 6.53 (a) The dashed line is the response of a Chebyshev filter without cross coupling. The solid line is the same filter with negative cross coupling between the first and fourth resonators, and (b) physical realization of a coaxial resonator quadruplet. (From [27]. Reprinted with permission from the IEEE.)

$\bar{\lambda}_{zi}$ and $\bar{\lambda}_{pj}$ are the poles and zeros of the filter transfer function. The scattering parameters of the network are obtained from the impedance matrix using (6.29).

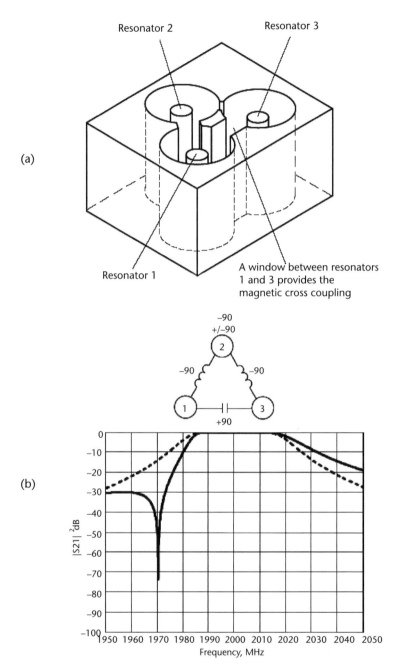

Figure 6.54 (a) Physical structure of coaxial line resonator triplet, and (b) and (c) frequency response of a triplet for different cross couplings. (From [27]. Reprinted with permission from the IEEE.)

In order to obtain the coupling matrix for a given coupling topology, the following penalty function is optimized

$$Errf\left(M_{ij}\right) = \sum_{i=1}^{n}\left|S_{11}\left(\overline{\lambda}_{zi}\right)\right|^{2} + \sum_{i=1}^{n}\left|S_{11}\left(\overline{\lambda}_{zi}\right)\right|^{2} + \left|\varepsilon - \overline{\varepsilon}\right|^{2} \qquad (6.57)$$

6.2 Distributed Transmission Line Form of Capacitively Coupled Band-Pass Filter

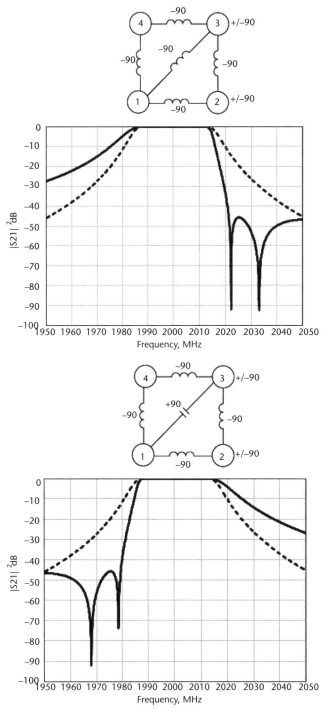

Figure 6.55 Frequency response of a quadruplet with multiple cross couplings. (From [27]. Reprinted with permission from the IEEE.)

The coupling topology matrix has all entries zero, except some of the nondiagonal ones that show the resonator connections. Such entries of the matrix are unity. The topology matrix can be used as the starting guess for the optimization. $\bar{\varepsilon}$ is calculated from the current trial matrix and ε is the desired value of the scale factor

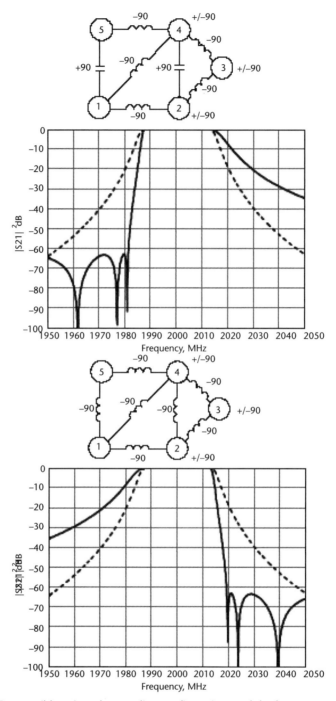

Figure 6.56 Two possible quintuplet-coupling configurations and the frequency responses. (From [27]. Reprinted with permission from the IEEE.)

related to the passband ripple. The method is extremely robust and it converges very rapidly. A computer program can be developed for rapid generation of coupling matrix for a given filter response where the transmission zeros can be conveniently placed (pole-placed) in the frequency domain. Let us consider an example. We need to generate the coupling matrix of a six-pole band-pass filter with two

6.2 Distributed Transmission Line Form of Capacitively Coupled Band-Pass Filter

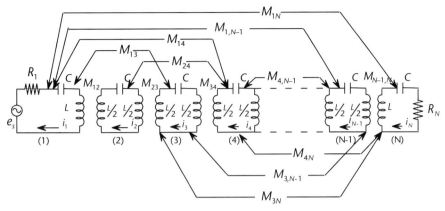

Figure 6.57 General multicoupled resonator filter network.

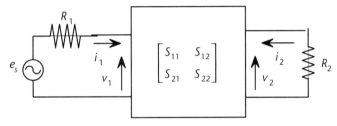

Figure 6.58 Two-port representation of a multicoupled resonator filter.

symmetrically placed transmission zeros above and below the passband. The filter has the following specifications:

1. Lower cutoff frequency = 0.69350 GHz
2. Upper cutoff frequency = 0.69925 GHz
3. Passband ripple = 0.03
4. Filter order = 6

The first transmission zero is at 0.69235 GHz and the second transmission zero is at 0.7004 GHz.

The desired coupling topology is shown in Figure 6.59, and the corresponding topology matrix is shown in (6.58). Using the topology matrix as the initial guess and the above coupling matrix generation scheme, the required filter coupling matrix is generated as shown in (6.59). Generation of the matrix takes less than a second on a personal computer. The computed frequency response of the filter is shown in Figure 6.60.

$$[T] = \begin{bmatrix} 0 & 1 & 0 & 1 & 0 & 0 \\ 1 & 0 & 1 & 0 & 0 & 0 \\ 0 & 1 & 0 & 1 & 0 & 1 \\ 0 & 0 & 1 & 0 & 1 & 0 \\ 0 & 0 & 0 & 1 & 0 & 1 \\ 0 & 0 & 1 & 0 & 1 & 0 \end{bmatrix} \qquad (6.58)$$

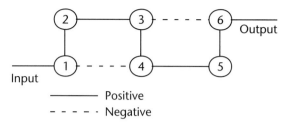

Figure 6.59 Coupling topology of a six-pole band-pass filter.

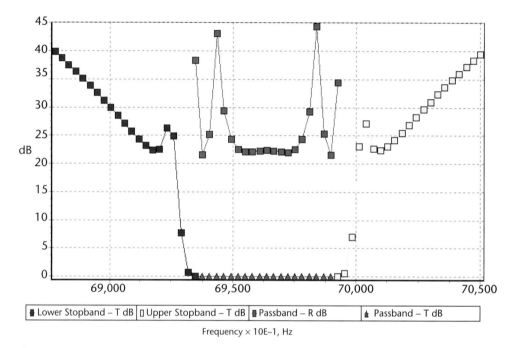

Figure 6.60 Computed frequency response of six-pole cross-coupled band-pass filter.

$$[M] = \begin{bmatrix} 0 & 1.0088 & 0 & -0.2012 & 0 & 0 \\ 1.0088 & 0 & 0.7794 & 0 & 0 & 0 \\ 0 & 0.7794 & 0 & 0.5488 & 0 & -0.2012 \\ -0.2012 & 0 & 0.5488 & 0 & 0.7794 & 0 \\ 0 & 0 & 0 & 0.7794 & 0 & 1.0088 \\ 0 & 0 & -0.2012 & 0 & 1.0088 & 0 \end{bmatrix} \quad (6.59)$$

Having discussed the basic theory on which cross-coupled filters are designed, we are now in a position to describe the design steps of a real-world cavity filter. This design method is based on a tutorial by Mayer and Vogel of Ansoft Corporation [32]. The specifications of the filter are

1. Center frequency $f_0 = 400$ MHz
2. Ripple bandwidth $\Delta f = 15$ MHz
3. Passband ripple = 0.10 dB

4. Filter order $N = 4$
5. Locations of transmission zeros: 0.3876 MHz and 0.4115 MHz.

Figure 6.61 shows the equivalent network of the filter.

Based on the electrical specifications, we use the optimization method described in the previous section to generate the following coupling matrix.

$$[M] = \begin{bmatrix} 0 & 0.7717 & 0 & -0.2513 \\ 0.7717 & 0 & 0.7635 & 0 \\ 0 & 0.7635 & 0 & 0.7717 \\ -0.2513 & 0 & 0.7717 & 0 \end{bmatrix} \quad (6.60)$$

The corresponding coupling values to be realized are shown in the matrix

$$[K] = \begin{bmatrix} Q_L & 0.02894 & 0 & -0.00942 \\ 0.02894 & 0 & 0.02863 & 0 \\ 0 & 0.02863 & 0 & 0.02994 \\ -0.00942 & 0 & 0.02894 & Q_L \end{bmatrix} \quad (6.61)$$

The required value of the external loaded Q, Q_L is 29.69. Figure 6.62 shows the basic resonator. It consists of a coaxial cavity with a rectangular outer conductor and a circular inner conductor. The space inside the cavity is entirely filled with air. The length of the resonator is so chosen as to have a resonance frequency at 400 MHz. In fact, the resonance frequency can be always adjusted by a tuning screw inserted vertically through the top wall of the cavity, as shown in Figure 6.39. In the first step, using an arrangement shown in Figure 6.63 simulates the interresonator couplings. The basic procedure is the same as used in microstrip band-pass filter

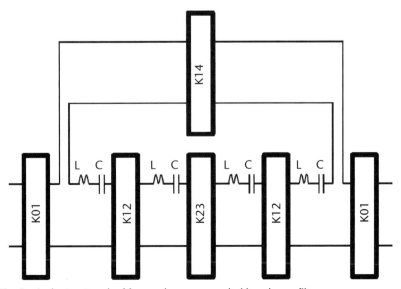

Figure 6.61 Equivalent network of four-pole cross-coupled band-pass filter.

- Cavity 20×20×200 mm
- Metal walls, air inside
- Metal cylinder R=5 mm
- Cylinder stands on cavity floor, does not touch ceiling
- Cylinder length is varied to tune the resonance
- Irises in walls, not shown here, will provide coupling to other resonators

Figure 6.62 Basic resonator-orientation. (From [32]. Reprinted with permission from Ansoft Corp.)

Figure 6.63 Dual resonator-orientation for coupling simulation. (From [32]. Reprinted with permission from Ansoft Corp.)

design, described in Section 6.2.6. Computed dependence on aperture opening is shown in Figure 6.64.

Figure 6.65 shows the locations of irises for different interresonator couplings. Please note that all positive couplings are achieved by irises on the ground of the resonators. An iris near the top of the cavity achieves only negative coupling between the first and fourth resonator. Figure 6.66 shows the circuit arrangement for determination of loaded Q (Q_e).

Keep in mind that while determining the interresonator couplings, we neglected the proximity effects of the nonadjacent resonators. Consequently, when the filter is assembled from the coupling database and the loaded Q database, slightly inaccurate couplings and loaded Q are obtained. Table 6.3 shows the difference between the target values and the achieved ones. Figure 6.67 show the computed

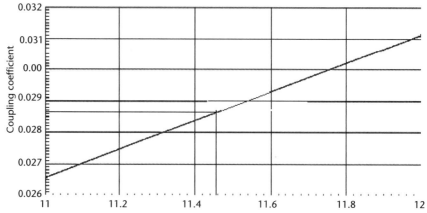

Figure 6.64 Dependence of coupling on iris opening. (From [32]. Reprinted with permission from Ansoft Corp.)

6.2 Distributed Transmission Line Form of Capacitively Coupled Band-Pass Filter

Elliptic filter, basic design

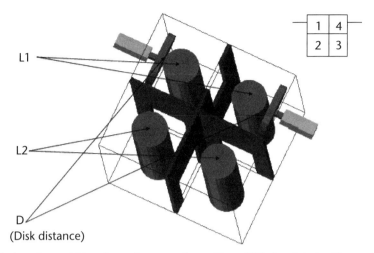

Figure 6.65 Locations of irises for various couplings. (From [32]. Reprinted with permission from Ansoft Corp.)

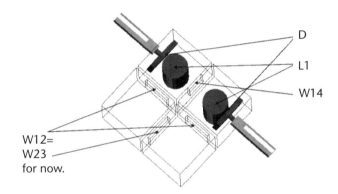

Figure 6.66 Circuit arrangement for determination of loaded Q.

Table 6.3 Curve Fitting

	Curve Fitting Results	Original Targets
f1	= 399.93 MHz	400 MHz
f2	= 400.10 MHz	400 MHz
K12	= 0.03122	0.02894
K23	= 0.02819	0.02863
K14	=−0.00930	−0.00942
Q_L	= 28.43	29.69
d12	=−0.0082	}
d23	=−0.0410	} were zero in circuit
d14	− 0.0363	}

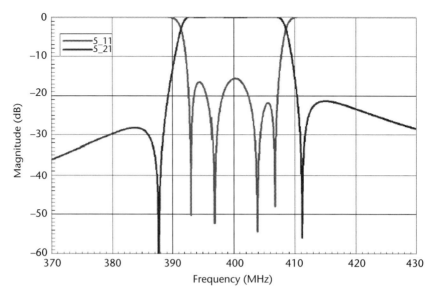

Figure 6.67 Computed frequency response of the filter before adjustment. (From [32]. Reprinted with permission from Ansoft Corp.)

frequency response of the filter. The discrepancy between the target specifications and the achieved ones can be removed in either of two ways: (1) theoretically altering the iris openings and the resonator lengths, and (2) using tuning screws to adjust the resonator lengths and the coupling iris dimensions. The parameters d12, d23, d14 are the required iris adjustment values. However, no matter how accurate the theoretical design is, postproduction bench tuning is invariably needed to obtain the desired frequency response due to mechanical tolerencing during production of the filter. Figure 6.68 shows the response of the tuned filter.

6.2.12 Design of Cross-Coupled Filters Using Dual-Mode Resonators

The operation of a dual-mode filter can be best explained by considering a circular waveguide dual-mode cavity filter [33]. A circular waveguide cavity is capable of holding two dominant degenerate TE_{111} resonant modes having orthogonal transverse field patterns. By the term degenerate, we mean both modes have the same resonant frequencies. However, if there is no discontinuity inside the cavity, both modes exist independently without influencing each other's field pattern or resonance frequency. In other words the modes do not couple. Insertion of a tiny discontinuity at the proper location disturbs the field patterns and mode coupling takes place. Consequently, it is possible to realize a two-pole band-pass filter using only one physical cavity. Figure 6.69 shows a two-cavity, four-pole waveguide filter, which offers an elliptic response. The tilted near square large irises couple the degenerate modes in the same cavity. The amount of coupling depends on the aspect ratio of the iris. However, it does not have any tuning effect on the resonant modes if the tilt is exactly 45 degrees. On the other hand, if the iris is perfectly square and tilted around 45 degrees, then it will only have tuning effects but no mode coupling. In a first step toward the design, a coupling matrix is generated using the procedure

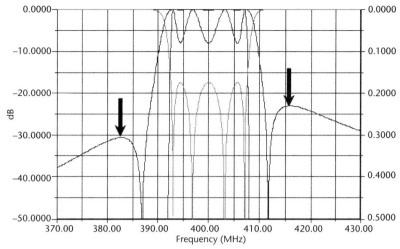

Figure 6.68 Frequency response of tuned filter using HFSS. (From [32]. Reprinted with permission from Ansoft Corp.)

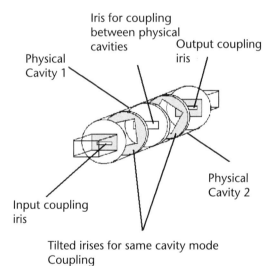

Figure 6.69 Circular cavity four-pole dual-mode band-pass filter.

described in the previous sections. Then the coupling values are transformed into K-inverter values using the following equations [34]

$$K_{01} = K_{N,N+1} = \left(\frac{\lambda}{\lambda_g}\right)\sqrt{\frac{\pi}{2}\frac{\Delta f}{f_0}R_1} \qquad (6.62a)$$

$$K_{pq} = \left(\frac{\lambda}{\lambda_g}\right)^2 \frac{\pi}{2} M_{pq} \qquad (6.62b)$$

where R_1 is the input and the output terminal resistance. In the next step, the intercavity iris dimensions are adjusted by relating the K-inverter to the scattering parameter according to (6.51) and (6.52). The scattering parameters can be obtained by using a suitable EM analysis method. For the jth electrical cavity, the resonance condition is

$$\phi_{j,1} + \beta_j l^i + \phi_{j,2} = \pi \quad j = 2i-1, \cdot 2i \tag{6.63}$$

where β_j is the propagation constant of the jth electrical cavity and $l^i \left(= l_1^i + l_2^i + l_3^i\right)$ is the effective total length of the ith physical cavity, as shown in Figure 6.70. Generally, $\phi_{2i-1,1} + \phi_{2i-1,2}$ may not equal $\phi_{2i,1} + \phi_{2i,2}$ for the ith physical cavity. A slight deviation in the cross-sectional dimensions of the cavity may result is different values for $\beta_{2i-1}l^i$ and $\beta_{2i}l^i$ to compensate for the loading between the $(2i-1)$th and the $(2i)$th electrical cavities. Adjustment of cavity dimensions alters the coupling and vice versa. Therefore, an iterative method is needed. The method is implemented using computer optimization. Figure 6.71 shows the flow diagram for the optimization based iteration scheme.

In order to determine the shape and the tilt of each tilted iris, the procedures described in [35–37] are used. As mentioned above, the shape of the tilted iris depends on the amount of coupling between the degenerate modes that is needed. It amounts to computing the split in resonance frequency when the modes are coupled. The coupling coefficient is given by

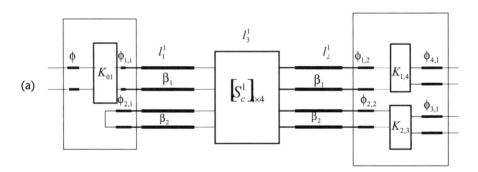

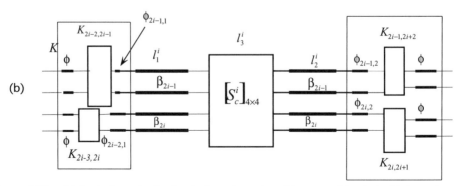

Figure 6.70 (a) Equivalent circuit of a dual-mode cavity with one end loaded with input/output coupling and the other end with intercavity coupling, and (b) equivalent circuit with both ends loaded with intercavity coupling. (From [34]. Reprinted with permission from the IEEE.)

6.2 Distributed Transmission Line Form of Capacitively Coupled Band-Pass Filter 333

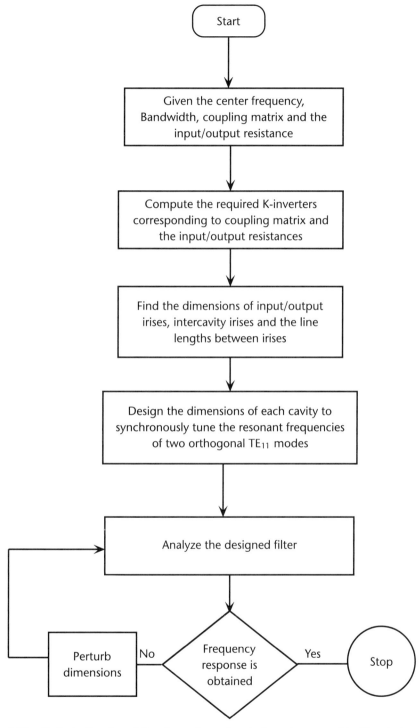

Figure 6.71 Optimization-based iteration scheme for dual-mode waveguide filter design. (From [34]. Reprinted with permission from the IEEE.)

$$k_m = \frac{f_2^2 - f_1^2}{f_2^2 + f_1^2} \qquad (6.64)$$

where f_2 is the higher frequency of the split degenerate resonance frequency and f_1 is the lower one. Equation (6.64) is the same as (6.31). However, in the present case both resonances occur in the same physical cavity. Figure 6.72 shows the implementation of McDonald's method [36] for determination of f_1 and f_2.

Figure 6.73(a) shows the design dimensions of a 600-MHz circular cavity dual-mode six-pole filter and Figure 6.73(b) shows the computed but unoptimized frequency response. The coupling matrix of the filter is given by (6.59). At each step of the design the mode-matching method described in Section 3.10.8 was used.

The entire design process needed approximately 15 minutes on a Pentium III computer with 1-GHz clock frequency and 1 Gb of RAM. Such dual-mode filters can also be realized using square cavities instead of circular ones [38]. Other types of discontinuities like square corner cuts [31] or screws can replace the degenerate mode-coupling tilted irises. The amount of coupling for such types of mode couplers can be predicted with reasonable accuracy using perturbation theory or closed-form equations [37]. The tilted rectangular irises can be replaced by off-circular or elliptic irises [38]. Such irises have special advantages over rectangular ones from a manufacturing point of view because they are devoid of tooling radii at the four corners of the iris. As well, the filters can handle a larger amount of power.

6.2.13 Folded Resonator Cross-Coupled Filters

Waveguide cross-coupled filters can also be realized using a folded resonator configuration [39,40]. Figure 6.74 shows the configuration of a four-pole waveguide iris-coupled cross-coupled filter.

The design steps for folded waveguide cross-coupled filters are as follows:

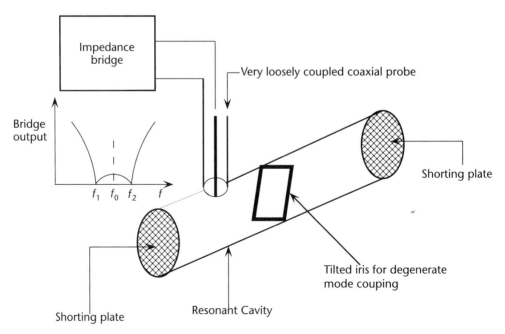

Figure 6.72 Implementation of McDonald's [36] method.

6.2 Distributed Transmission Line Form of Capacitively Coupled Band-Pass Filter 335

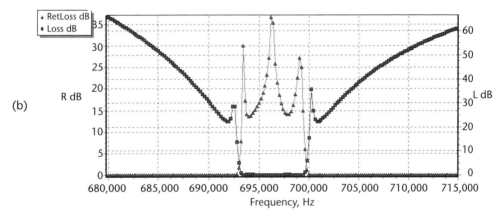

Figure 6.73 (a) Designed dimensions of a six-pole circular cavity dual-mode filter, and (b) computed frequency response of a six-pole circular cavity dual-mode-filter.

1. From the required loaded Q-factor Q_e determine the K-inverter value K_{01} and $K_{N,N+1}$ (assuming the terminations at either end are equal). Then, determine the iris width required to realize the required K-inverter value and the associated phase ϕ_{01}. The method is the same as in realization of sequentially coupled synchronously tuned Chebyshev or Butterworth filter designs described in previous sections.
2. From the computed coupling matrix determine the K-inverter value K_{12}. Repeat the procedure described in step 1 to obtain the iris width and the associated phase ϕ_{12}.
3. Compute the length of the first resonator as

$$l_1 = \frac{\lambda_{go}}{2\pi}\left(\pi - \frac{1}{2}(\phi_{01} + \phi_{12})\right) \tag{6.65}$$

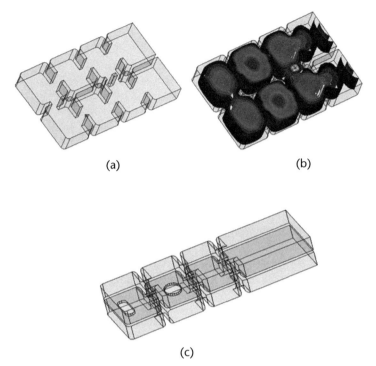

Figure 6.74 (a) H-plane folded cross-coupled waveguide filter, (b) the magnetic field pattern, and (c) E-plane folded cross-coupled filter. (Reprinted with permission from E. Ofli, D.Sc thesis, Swiss Federal Institute of Technology, Zurich.)

4. Repeat step 3 for up to the resonator right before the resonator forming the fold. For example, if it is a six-pole filter, the third and the fourth resonators form the fold. Therefore, we need to repeat step 3 up to the third resonator.

5. Due to symmetry, the dimensions of resonator 6 are the same as that of resonator 1 and the dimensions of resonator 5 are same as those of resonator 2.

6. If it is an E-plane fold, then use a 3-D simulator to analyze the structure shown in Figure 6.75. From the two-port S-matrix compute the K-inverter value using (6.51) and (6.52). Adjust the slot dimensions to realize K_{34} and record the associated phase angle ϕ_{34}. Knowing ϕ_{23} and ϕ_{34}, determine the lengths of the third and the fourth resonators using equation

$$l_3 = \frac{\lambda_{go}}{2\pi}\left(\pi + \frac{1}{2}(\phi_{23} + \phi_{34})\right) \tag{6.66}$$

For H-plane folding, the equation becomes

$$l_3 = \frac{\lambda_{go}}{2\pi}\left(\pi + \frac{1}{2}(\phi_{23} + \phi_{34})\right) + \frac{\lambda_g}{2} \tag{6.67}$$

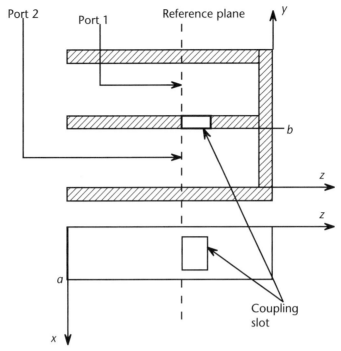

Figure 6.75 E-plane fold with coupling slot on the common broad wall.

7. Add a round hole on the common broad wall between resonators 1 and 6 and analyze and optimize the frequency response of the filter structure using a 3-D simulator.

The entire operation of folded cross-coupled rectangular waveguide band-pass filter design can be performed using standard commercial software like HFSS or WASPNET. Figure 6.76 shows the configuration of an H-plane folded E-plane septum filter with four poles. The computed frequency response of the filter in Figure 6.77 shows the effects of various types of cross couplings on the filter response. Those include the direct cross coupling between the input and the output, also known as source-load coupling [39]. Figure 6.77 shows that source-load coupling has distinct advantages over direct cross coupling as far as skirt selectivity and stopband width are concerned. For an N-pole filter they can implement N transmission zeros in place of $(N–2)$ transmission zeros in an uncoupled source and load filter. The additional two transmission zeros can be properly placed in order to obtain better group delay or an improved passband without affecting the passband ripple, which is controlled by the number of resonators. These advantages, however, come at a price. The filter response is very sensitive to the location of the cross-coupling hole.

The design steps for cross-coupled waveguide filters can be programmed and the entire design process can be made very fast and accurate. Please see Appendix 6B for a description of a commercially available filter wizard that designs cross-coupled waveguide filters.

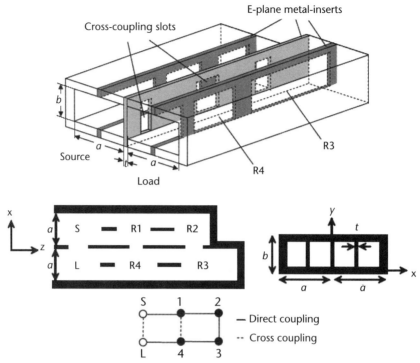

Figure 6.76 Cross-coupled E-plane septum filter with H-plane folding. (From [39]. Reprinted with permission from E. Ofli, D.Sc thesis, Swiss Federal Institute of Technology, Zurich.)

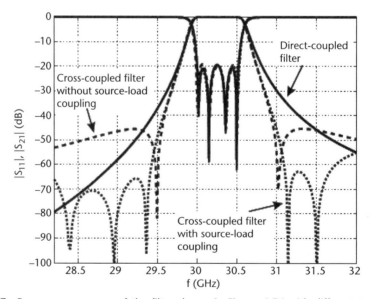

Figure 6.77 Frequency response of the filter shown in Figure 6.76 with different types of cross couplings. (From [39]. Reprinted with permission from E. Ofli, D.Sc thesis, Swiss Federal Institute of Technology, Zurich.)

6.2.14 Cross-Coupled Filters Using Planar Transmission Lines

Cross-coupled filters can also be realized using planar transmission lines such as stripline, suspended stripline, and microstrip line. An excellent account of

6.2 Distributed Transmission Line Form of Capacitively Coupled Band-Pass Filter

realization of such filters can be found in the classic text by Hong and Lancaster [41] Figure 6.78(a) shows the most convenient resonator configurations for cross-coupled planar filters [42]. These structures are modified versions of a basic hairpin resonator. Figure 6.78(b) shows a straight sequentially coupled filter configuration using the resonators shown in Figure 6.78(a). The general forms of cross-coupled planar filters are shown in Figure 6.78(c).

The design method for cross-coupled planar filters is the same as that for cross-coupled waveguide filters. However, the interresonator coupling data and input/output loaded Q databases are generated by using the method shown for the microstrip interdigital filter above. Iliev and Nedelchev [42] have presented the following analytical equations for a folded and capacitively loaded hairpin resonator (see Figure 6.78(b)) cross-coupled filter designs. Inductive, capacitive, or mixed coupling between modified hairpin resonators can be obtained based on the orientations of the resonators. Figure 6.79(a) shows the resonator orientation for magnetic coupling. The corresponding coupling coefficient is given by

$$k_m = \frac{1}{bZ_c}\left\{\frac{Z_c - Z_o \tan\left(\frac{\theta_c}{2}\right)\tan\left(\frac{\theta_s - \theta_c}{2}\right)}{Z_o \tan\left(\frac{\theta_c}{2}\right) - Z_c \tan\left(\frac{\theta_s - \theta_c}{2}\right)} - \frac{Z_c - Z_e \tan\left(\frac{\theta_c}{2}\right)\tan\left(\frac{\theta_s - \theta_c}{2}\right)}{Z_e \tan\left(\frac{\theta_c}{2}\right) - Z_c \tan\left(\frac{\theta_s - \theta_c}{2}\right)}\right\} \quad (6.68)$$

where Z_c is the resonator impedance or the characteristic impedance of the transmission line forming the hairpin. Z_e and Z_o are the even- and odd-mode characteristic impedances, respectively, of the coupled lines, θ_s is the electrical length of the resonator, θ_c is the electrical length of the coupled portion of the resonator, and b is the admittance slope parameter of the resonator. The admittance slope parameter b of the resonator is given by [42]

$$b = -\frac{A+B}{2C} \quad (6.69)$$

where

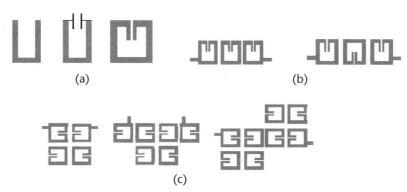

Figure 6.78 (a, b) Modified hairpin resonators, and (c) cross-coupled modified hairpin-line filters. (From [43]. Reprinted with permission from *Microwave Review*.)

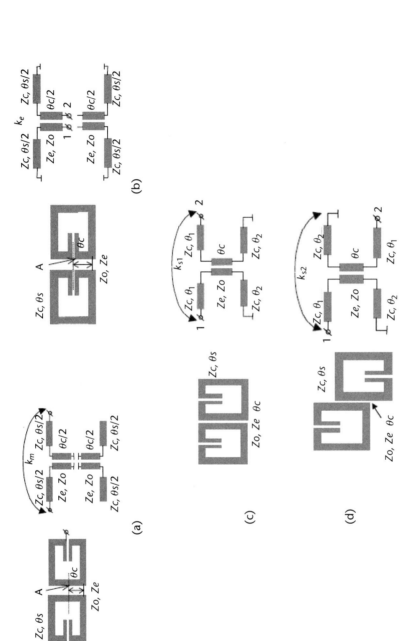

Figure 6.79 (a) Coupled modified hairpin resonators for magnetic coupling, (b) coupled modified hairpin resonators for electric coupling, and (c) mixed coupled modified hairpin resonator configurations. (From [43]. Reprinted with permission from *Microwave Review*.)

6.2 Distributed Transmission Line Form of Capacitively Coupled Band-Pass Filter

$$A = \frac{\theta_p\{(Z_{pe} - Z_{po}) + (Z_{pe} + Z_{po})\cos\theta_s\}}{\sin^2\theta_p} + (Z_{pe} + Z_{po})\theta_s \sin\theta_s \cot\theta_p$$

$$B = \theta_s \cos\theta_s \left\{\frac{Z_{pe}Z_{po}}{Z_c}\cot^2\theta_p - Z_c\right\} + \frac{2Z_{pe}Z_{po}\theta_p \cot\theta_p \sin\theta_s}{Z_c \sin^2\theta_s}$$

$$C = 2Z_{pe}Z_{po}\cot^2\theta_p \cos\theta_s - Z_c(Z_{pe} + Z_{po})\sin\theta_s \cot\theta_p$$

and the electrical length θ_p of the coupled line part of a single resonator is given by (see Figure 6.80).

$$\theta_p = \cot^{-1}\left\{\frac{\sqrt{R^2 + 4Z_c^2 \sin^2\theta_s} - R}{2Z_C \sin\theta_s}\right\}$$

$$R = (Z_{pe} + Z_{po})\cos\theta_s - (Z_{pe} - Z_{po})$$
(6.70)

Figure 6.79(b) shows the orientation of the resonators for electrical coupling between them. The corresponding coupling coefficient is given by

$$k_e = \frac{1}{b}\left(\frac{1}{Z_0}\frac{Z_0 - Z_c \tan\theta_c \tan\left(\frac{\theta_s}{2} - \theta_c\right)}{Z_c \tan\left(\frac{\theta_s}{2} - \theta_c\right) - Z_0 \tan\theta_c} - \frac{1}{Z_e}\frac{Z_e - Z_c \tan\theta_c \tan\left(\frac{\theta_s}{2} - \theta_c\right)}{Z_c \tan\left(\frac{\theta_s}{2} - \theta_c\right) - Z_e \tan\theta_c}\right)$$
(6.71)

Figure 6.79(c) shows the mixed coupling configurations for modified hairpin resonators. The coupling coefficient for configuration (a) is given by

$$k_{s1} = \frac{1}{2b}\left[\frac{1}{|Z_{2o}|} - \frac{1}{|Z_{2e}|}\right]$$
(6.72)

where

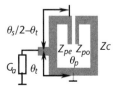

Figure 6.80 Tap-point location in hairpin-line filter. (From [43]. Reprinted with permission from *Microwave Review*.)

$$Z_{2o} = Z_c \frac{Z_{1o} + jZ_c \tan\theta_1}{Z_c + jZ_{1o} \tan\theta_1}$$

$$Z_{2e} = Z_c \frac{Z_{1e} + jZ_c \tan\theta_1}{Z_c + jZ_{1e} \tan\theta_1}$$

$$Z_{1o} = Z_c \frac{Z_l + jZ_o \tan\theta_2}{Z_o + jZ_l \tan\theta_2}$$

$$Z_{1e} = Z_c \frac{Z_l + jZ_e \tan\theta_2}{Z_e + jZ_l \tan\theta_2}$$

$$Z_l = jZ_c \tan\theta_2$$

The coupling coefficient for configuration 6.79c(b) is given by [43]

$$k_{s2} = \frac{1}{2b}\left[\frac{1}{|Z_{2o}|} - \frac{1}{[Z_{2e}]}\right] \tag{6.73}$$

$$Z_{io} = Z_c \frac{Z_{i1} + jZ_c \tan\theta_1}{Z_c + jZ_{i1} \tan\theta_1}$$

$$Z_{ie} = Z_c \frac{Z_{ii} + jZ_c \tan\theta_1}{Z_c + jZ_{ii} \tan\theta_1}$$

$$Z_{ii} = \frac{Z_{i1} Z_{i2}}{Z_{i1} + Z_{i2}}$$

$$Z_{i2} = \frac{1}{2y_{12a}}$$

$$Z_{i1} = \frac{1}{y_{11a} - y_{12a}}$$

$$y_{11a} = j\left[\frac{1}{d}\left\{\frac{\cot\theta_c}{w} + y_c \cot\theta_2\right\}\left\{\frac{\cot\theta_c}{v^2} + \frac{1}{w^2 \sin^2\theta_c}\right\}\right]$$

$$-j\left\{\frac{\cot\theta_c}{w} + \frac{2\cos\theta_c}{dv^2 w \sin^3\theta_c}\right\}$$

$$y_{21a} = \frac{j}{v\sin\theta_c}\left[1 + \frac{1}{d}\left\{\frac{1}{w^2 \sin\theta_c} + \frac{\cot^2\theta_c}{v^2} - 2\left(\frac{\cot\theta_c}{w} + y_c \cot\theta_2\right)\frac{\cot\theta_c}{w}\right\}\right]$$

$$d = -\frac{1}{v^2 \sin\theta_c} + \left(\frac{\cot\theta_c}{w} + y_c \cot\theta_2\right)^2$$

$$v = \left\{\frac{1}{2}\left(\frac{1}{Z_o} - \frac{1}{Z_e}\right)\right\}^{-1}$$

$$w = \left\{\frac{1}{2}\left(\frac{1}{Z_o} + \frac{1}{Z_e}\right)\right\}^{-1}$$

6.2 Distributed Transmission Line Form of Capacitively Coupled Band-Pass Filter

The tapped input and output locations are determined from (see Figure 6.80)

$$\theta_t = \tan^{-1}\sqrt{\frac{G_a}{bQ_{ei}}}, \quad i = 1, \tag{6.74}$$

The above equations are accurate as long as the parasitic effects of the bends and discontinuities in the resonators are negligible at low frequencies. As the frequency rises, one has to take into consideration these parasitic effects. Otherwise, the best way would be the 3-D simulator based approach described for the microstrip interdigital filter. One can always use the generalized coupling matrix for a design. However, in most cases, Levy's [44] low-pass prototype with single transmission zero at real or imaginary frequency may be used as follows.

The relations between the band-pass design parameters and the low-pass elements are [44]

$$Q_{ei} = Q_{eo} = \frac{g_1}{\varpi} \tag{6.75a}$$

$$k_{n,n-1} = k_{N-n, N-n+1} = \frac{\varpi}{\sqrt{g_n g_{n+1}}} \tag{6.75b}$$

$$k_{m,m+1} = \frac{\varpi J_m}{g_m}, \quad \text{for } m = N/2 \tag{6.75c}$$

$$k_{m-1,m+2} = \frac{\varpi J_{m-1}}{g_{m-1}}, \quad \text{for } m = N/2 \tag{6.75d}$$

Levy's [44] low-pass prototype is shown in Figure 6.81.
The design steps are as follows:

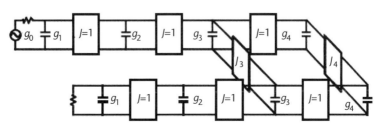

Figure 6.81 Levy's low-pass prototype with a single transmission zero.

1. Choose the electrical length θ_s of the resonator, characteristic impedance Z_c and the even- and odd-mode impedances Z_{pe} and Z_{po} of the coupled port and calculate the line length using (6.70);
2. Calculate the admittance slope parameter using (6.69);
3. Calculate the coupling parameters using (6.68) or (6.71) or (6.72) or (6.73), depending on the type of coupling needed;
4. Determine the tap-point location using (6.74);
5. Determine the geometrical parameters of the filter using a particular substrate;
6. Optimize the frequency response of the filter using an analysis software.

Figure 6.82 shows the frequency responses of a 5% bandwidth four-pole cross-coupled filter centered at 1.44 GHz. The corresponding coupling parameters are shown below.

(a)
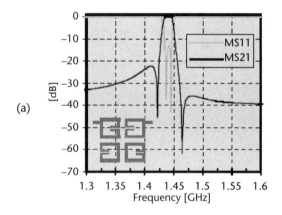

(b)
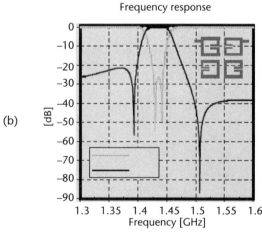

Figure 6.82 (a) A 1% bandwidth cross-coupled filter response, and (b) a 5% bandwidth microstrip cross-coupled filter response. (From [42]. Reprinted with permission from *Microwave Review*.)

6.3 Cross-Coupled Band-Pass Filters with Independently Controlled Transmission Zeros

Whether it is a triplet, quadruplet, or quintuplet, if a single cross-coupling value is altered, the alteration affects all the transmission zeros. This is not convenient from the standpoint of tuning the filter after fabrication. Cascading triplets, quadruplets, quintuplets, and so forth can overcome this drawback of a cross-coupled filter. As an example, let us consider the cross-coupled quadruplets shown in Figure 6.55. Figure 6.83 shows the frequency response of a filter formed by cascading the quadruplets. In the eight-resonator filter, the low-side transmission zeros are produced by the resonators 1–4 and the high-side transmission zeros are produced by resonators 5–8. Design of each N-tuplet can be accomplished by using the method described in Section 6.2.11. The cascaded structure can be further improved by another round of optimization by the method used for the individual N-tuplets. However, Levy has presented closed-form equations for cascaded quadruplets CQ [45]. Cascaded N-tuplets based band-pass filter designs are extremely useful in base station diplexer design for cellular radio. The quadruplets shown in Figure 6.83

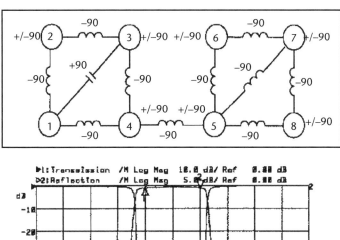
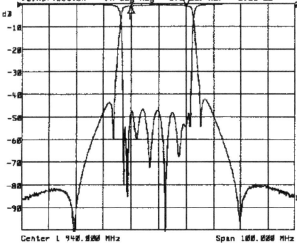

Figure 6.83 Cascaded quadruplets and the frequency response. (From [27]. Reprinted with permission from the IEEE.)

have special significance as basic building blocks in cross-coupled band-pass filter design. They have a diagonal cross-coupling element. The diagonal cross coupling can be eliminated by choosing the complex transmission zero as $s_i = \pm \sigma_i \pm \omega_{p0}$ where ω_{p0} is the center frequency of the passband [46]. In a direct synthesis approach by Sen et al. [46], ω_{p0} depends on the number of transmission zeros at $s = 0$ and $s = \infty$, together with the other transmission zeros. That is, it is not the geometric mean of the upper and the lower cutoff frequencies but somewhere between the geometric and the arithmetic center of the passband. The actual center frequency is found by using an iterative method. Also, all elements of the quadruplet come out to be positive if σ_i and ω_i are chosen in the ranges $\Delta\omega/4 < \sigma_i < \Delta\omega$ and $0.8\omega_{p0} < \omega_i < 1.2\omega_{p0}$ where $\Delta\omega = \omega_{p2} - \omega_{p1}$ is the bandwidth of the filter. In order to achieve a linear phase response, one should select [46]

$$\left. \begin{aligned} \sigma_i &= \frac{\Delta\omega}{2} \\ \omega_i &\approx \omega_{p0} \end{aligned} \right\} \tag{6.76}$$

The above condition also guarantees CQ sections without cross couplings. An excellent account of synthesis of N-tuplet filters can be found in a review article by Yildirim et al. [47]. For CAD of cross-coupled filters and filters involving cascaded N-tuplets, FILPRO [48] is a usefull software. While dealing with multiresonator cross-coupled band-pass filters, the possible topologies are virtually innumerable depending on the number of resonators and the types of coupling. However, in reality only a small fraction of those topologies are used in practice and a large majority are of academic interest only.

6.4 Unified Approach to Tuning Coupled Resonator Filters

In the preceding sections we described the design of coupled resonator band-pass filters. In general, a filter can be designed using approximate circuit theory or transmission line theory. However, in all cases each approximate design can be exactly tuned, before fabrication, in virtual reality with the help of exact electromagnetic simulation. A good designer should consider the effects of manufacturing tolerances in the simulation and computer tuning so that the fabricated prototype will need the minimum or very little tuning. In the case of the expected performance of the filter deviates from the predicted one, postproduction tuning becomes necessary. Fortunately, the art of experimental tuning of multicoupled resonator filter advanced considerably during the absence of modern EM simulators. Therefore, it can be used to some extent to correct any undesirable performance of a band-pass filter. We briefly discuss the most commonly used tuning method for the sake of completeness. The method due to Ness [49] is as follows. It can be shown that [49] the reflection group delay of a band-pass filter contains all the necessary information regarding coupling coefficients among various resonators. One can, therefore, use a vector network analyzer to adjust the couplings. We reconsider the evolution of an inverter coupled band-pass filter from the corresponding low-pass filter in

Figure 6.84 [49]. It can be shown that the reflection group delays of the circuits are related to the filter specifications, as shown in Table 6.4.

The corresponding shapes of the reflection group delay response and the coupling values are shown in Table 6.5. The step-by-step tuning procedure can be illustrated by considering a filter design example [49]. The procedure is applicable to any general filter structure provided that the actual resonators and coupling networks are accurately modeled by LC networks over the frequency band of interest. Suppose we need to design a band-pass filter with the following specifications and a Chebyshev response:

Center frequency $f_0 = 2.300$ GHz
Ripple bandwidth $\Delta f = 26.90$ MHz
Passband ripple $= 0.01$ dB
Number of sections $N = 6$

Table 6.6 shows the required computed reflection group delay and coupling values based on the formulas in Table 6.4.

The filter was realized in combline form [49] using round rods and tapped input and output ports, as shown in Section 6.2.8. The resonance frequency of each resonator was precisely controlled by tuning and coupling screws, as shown in Figure 6.41(c). Using a calibrated vector network analyzer, the reflection group delay for the input and output resonators were set to 18.5 ns at 2.30 GHz. In the next step. resonator 6 is shorted and the tuning process started at resonator 1. The basic steps are as follows [49]:

1. Short all resonators except resonator 1.

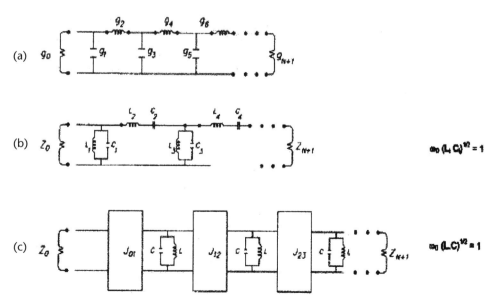

Figure 6.84 Circuit elements for (a) low-pass, (b) band-pass, and (c) inverter-coupled band-pass filter structures.

Table 6.4 Reflection Group Delays for Filters

	Low Pass Prototype	Band Pass Filter	Inverter Coupled Filter
n = 1	$\Gamma_{d1} = \dfrac{4 g_0 g_1}{\Delta\omega}$	$\Gamma_{d1} = 4 C_1 Z_0$	$\Gamma_{d1} = \dfrac{4 Q_E}{\omega_0}$; $Q_E = \dfrac{g_0 g_1 \omega_0}{\Delta\omega}$
n = 2	$\Gamma_{d2} = \dfrac{4 g_2}{g_0 \Delta\omega}$	$\Gamma_{d2} = \dfrac{4 L_2}{Z_0}$	$\Gamma_{d2} = \dfrac{4}{\omega_0 Q_E k_{12}^2}$; $k_{12} = \dfrac{\Delta\omega}{\omega_0 (g_1 g_2)^{1/2}}$
n = 3	$\Gamma_{d3} = \dfrac{4 g_0 (g_1 + g_3)}{\Delta\omega}$	$\Gamma_{d3} = 4(C_1 + C_3) Z_0$	$\Gamma_{d3} = \Gamma_{d1} + \dfrac{4 Q_E k_{12}^2}{\omega_0 k_{23}^2}$; $k_{23} = \dfrac{\Delta\omega}{\omega_0 (g_2 g_3)^{1/2}}$
n = 4	$\Gamma_{d4} = \dfrac{4 (g_2 + g_4)}{g_0 \Delta\omega}$	$\Gamma_{d4} = \dfrac{4 (L_2 + L_4)}{Z_0}$	$\Gamma_{d4} = \Gamma_{d2} + \dfrac{4 k_{23}^2}{\omega_0 Q_E k_{12}^2 k_{34}^2}$; $k_{34} = \dfrac{\Delta\omega}{\omega_0 (g_3 g_4)^{1/2}}$
n = 5	$\Gamma_{d5} = \dfrac{4 g_0 (g_1 + g_3 + g_5)}{\Delta\omega}$	$\Gamma_{d5} = 4(C_1 + C_3 + C_5) Z_0$	$\Gamma_{d5} = \Gamma_{d3} + \dfrac{4 Q_E k_{12}^2 k_{34}^2}{\omega_0 k_{23}^2 k_{45}^2}$; $k_{45} = \dfrac{\Delta\omega}{\omega_0 (g_4 g_5)^{1/2}}$
n = 6	$\Gamma_{d6} = \dfrac{4 (g_2 + g_4 + g_6)}{g_0 \Delta\omega}$	$\Gamma_{d6} = \dfrac{4 (L_2 + L_4 + L_6)}{Z_0}$	$\Gamma_{d6} = \Gamma_{d4} + \dfrac{4 k_{45}^2 k_{23}^2}{\omega_0 Q_E k_{12}^2 k_{34}^2 k_{56}^2}$; $k_{56} = \dfrac{\Delta\omega}{\omega_0 (g_5 g_6)^{1/2}}$

$\Delta\omega = \omega_2 - \omega_1$

Table 6.5 Group Delay Responses and Coupling Values

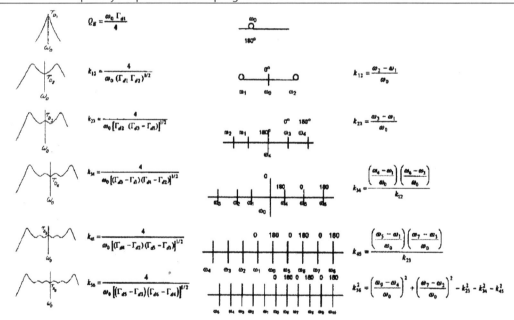

2. Adjust resonator 1 and the input coupling to obtain the specified reflection group delay (18.5 ns in the present case).
3. Tune resonator 2 and the coupling between resonator 1 and resonator 2 to obtain a symmetrical reflection group-delay response about the center frequency f_0 (2.30 GHz in the present case). It may be necessary to readjust resonator 1 if the coupling is large enough to detune resonator 1.

6.5 Dielectric Resonator Filters

Table 6.6 Computed Reflection Group Delays and Coupling Values

Resonator	Q_E or k_{ij}	Γ_d (ns)	Measured Coupling (Frequency Crossing)
1	66.8	18.5	
1, 2	.01135	32.2	.00112
1, 2, 3	.00771	58.5	.00772
1, 2, 3, 4	.00726	68.5	.00724
1, 2, 3, 4, 5	.00771	93.9	.00775
1, 2, 3, 4, 5, 6	.01135		
6	66.8		

4. Progress through the filter, tuning each resonator in turn and maintaining symmetry of reflection group-delay response by retuning the previous resonators if necessary.
5. When the last resonator is reached, observe the amplitude response of the filter and tune the last resonator and final interresonator coupling screw to obtain the desired return loss.

The only pitfall of the above method is masking of the group-delay response in the case of low-Q resonators. Finite Q has opposite effects on odd- and even-numbered resonators. In the former situation finite Q increases the measured coupling, while in the latter case it reduces the measured coupling.

6.5 Dielectric Resonator Filters

6.5.1 Introduction

In Section 6.1, we mentioned how a resonator forms the core of a microwave band-pass filter. In the preceding six sections we have discussed various types of resonators. The majority of those resonators are basically metallic cavities offering good power-handling capability and very low insertion loss or high unloaded Q-factor. A high unloaded Q-factor is the most essential factor from the standpoint of signal distortion and signal loss in a band-pass filter. Figure 6.85 shows the effect of finite Q on the frequency response of a band-pass filter.

The unloaded Q of a resonator is proportional to the stored energy to the dissipated energy in the cavity forming the resonator. Most of the electromagnetic energy in a dielectric resonator is confined within the dielectric body, when the dielectric constant ($\varepsilon_r \approx 20 - 80$) is substantially higher than that of the free space. The electromagnetic field within the resonator has the pattern of the field inside a metallic cavity of the same shape. However, unlike in a metallic cavity, it decays in space within an extremely short distance from the walls of the cavity. Consequently, the side walls of the cavity can be approximated by a magnetic wall. Figure 6.86 shows the configuration of a cylindrical dielectric resonator while Figure 6.87 shows the field pattern in the resonator. Since almost the entire field is confined to the material, the unloaded cavity factor of the resonator becomes equal to the Q-factor (1/tan δ) of the material itself. Even when the resonator is enclosed in a metallic enclosure, the enclosure walls have a negligible effect on the Q-factor of

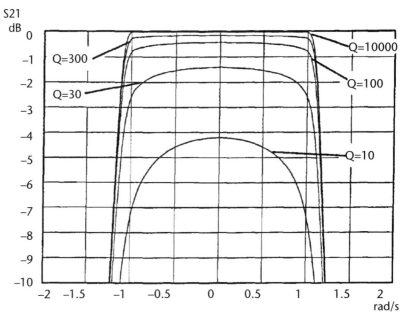

Figure 6.85 Effect of finite Q on the response of a band-pass filter. (From [51]. Courtesy of the Department of Electrical Engineering, University of Leeds.)

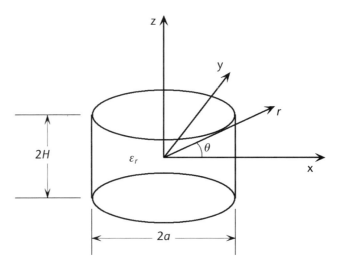

Figure 6.86 Isolated circular cylindrical resonators.

the resonator, provided the distance of the walls are more than twice the largest dimension of the resonator. Also, for the resonator mode to be completely of the dielectric resonator, the metallic enclosure should not allow any waveguide mode (i. e., the enclosure's dimensions should ensure a below cutoff or evanescent mode situation for the operating bandwidth).

Although resonance in dielectric bodies was reported first in 1939 [52], the first reported applications of dielectric resonators in microwave filters were by Harrison [53] and Cohn [54]. However, it took nearly a decade and a half for dielectric-resonator-based filter technology to become a viable option because of the

6.5 Dielectric Resonator Filters

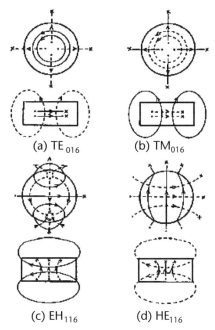

Figure 6.87 E (–) and H (-----) field plots for the pertinent resonance modes in a dielectric disk resonator.

nonavailability of suitable dielectric materials with very high temperature stability, mechanical rigidity, and low thermal expansion coefficient. The temperature stability of the dielectric resonator material is defined by a factor $TC_f = -\left[\frac{1}{2}\tau_f + \alpha_T\right]$ [55] where α_T is the linear thermal expansion coefficient and τ_f is the temperature coefficient of the dielectric constant. For microwave ceramic materials, α_T ranges between ±10 ppm/°C. It is the manufacturing process that has to take care to provide good control of the three main parameters (ε_r, Q, and TC_f) of a dielectric resonator. Appendix 6C presents typical specifications for standard dielectric resonators from TCI Ceramics.

6.5.2 Modes in a Dielectric Resonator

6.5.2.1 Modes and Mode Nomenclature

Figure 6.86 shows an isolated circular cylindrical resonator. The resonating modes of such a resonator can be classified by three distinct types: TE (TE to z), TM (TM to z), and hybrid. The fields for the TE and TM modes are axisymmetric and thus have no azimuthal variation. In contrast, the fields of hybrid modes are azimuthally dependent. The subclasses of hybrid modes are HE and EH. For HE modes, the H_z field is quite small compared to the E_z field. The other field components can be derived from the E_z only. The reverse is true for EH modes [55].

To denote the variations of the fields along the azimuthal, radial, and axial direction inside the resonator, the mode indices are added as subscripts to each family of modes. The TE, TM, HE, and EH modes are classified as $TE_{0mp+\delta}$, $TM_{0mp+\delta}$, $HE_{0mp+\delta}$, and $EH_{0mp+\delta}$ modes, respectively. The first index denotes the azimuthal variation of the fields. The azimuthal variation is of the form $\cos(n\phi)$ or $\sin(n\phi)$.

The index m ($m = 1.2.3,....$) denotes the order of the variation of the field along the radial direction and the index $p+\delta$ ($p=0,1,2,3......$) denotes the variation of the field in the z-direction.

In all filter applications, a dielectric resonator is placed in a metallic enclosure. As a result, the hybrid mode nomenclature, in some cases, becomes inconsistent. However, it is not an important issue, since most modern-day designs are based on full 3-D electromagnetic solvers.

6.5.2.2 Pertinent modes in Dielectric Resonators (DR) filters

Out of the all the resonance modes described above, the pertinent ones in filter design are $TE_{01\delta}$, $TM_{01\delta}$, $HE_{11\delta}$, and $EH_{11\delta}$. The corresponding E and H field patterns are shown in Figure 6.87 [56].

Table 6.7 shows the list of the principal modes and some higher-order modes and their field expressions [55].

6.5.2.3 Resonant Frequencies of Isolated Circular Cylindrical Dielectric Resonators

The determination of the resonant frequency of an isolated dielectric resonator requires rigorous electromagnetic and numerical analysis [55]. However, such analyses are of limited interest to a DR filter designer because of their complexity. One is interested in computing the value of the normalized wave number $k_0 a$ for a given value of ε_r and aspect ratio (H/a) of the resonator where $k_0 = 2\pi f_r/c$ denotes the free-space wave number corresponding to the resonant frequency and c is the velocity light in free space. If the value of $\varepsilon_r \gg 100$, the value of the normalized wave number varies with ε_r as

$$k_0 a \propto \frac{1}{\sqrt{\varepsilon_r}} \tag{6.77}$$

Table 6.7 Principal Modes in Dielectric Resonators

Mode	Plane of Symmetry (z=0)	Fields inside Resonator
TE_{016}	Mag. wall	$H_z = J_0(hr)\cos(\beta z)$ $E_z = 0$
TE_{011+z}	Elect. wall	$H_s = J_0(hr)\sin(\beta z)$ $E_z = 0$
TM_{011}	Elect. wall	$E_z = J_0(hr)\cos(\beta z)$ $H_z = 0$
TM_{011+z}	Mag. wall	$E_z = J_0(hr)\sin(\beta z)$ $E_z = 0$
HE_{114}	Elect. wall	$E_z = J_1(hr)\cos(\beta z)^{\cos\phi}_{\sin\phi}$ $H_z \approx 0$
HE_{214}	Elect. wall	$E_z = J_2(hr)\cos(\beta z)^{\cos 2\phi}_{\sin 2\phi}$ $H_z \geq 0$
EH_{114}	Mag. wall	$H_z = J_1(hr)\cos(\beta z)^{\cos\phi}_{\sin\phi}$ $E_z \approx 0$
EH_{214}	Mag. wall	$H_z = J_2(hr)\cos(\beta z)^{\cos 2\phi}_{\sin 2\phi}$ $E_z \geq 0$

6.5 Dielectric Resonator Filters

for a given aspect ratio (H/a) of the resonator. Mongia and Bhartia [55] presented closed-form equations for the resonant frequencies of the four most important modes $TE_{01\delta}$, $TM_{01\delta}$, $HE_{11\delta}$, and $EH_{11\delta}$. The equations were based numerical analyses results and use a modified version of (6.77) as

$$k_0 a \propto \frac{1}{\sqrt{\varepsilon_r + X}} \quad (6.78)$$

where the value of X is important for low dielectric constant materials used in dielectric-resonator-based antennas. According to [55], it has value between 1 and 2. But for filter applications, where in most cases the dielectric resonator is enclosed within a metallic enclosure, we suggest a value of −1 for X. Table 6.8 shows the closed-form equations for the resonance frequencies pertaining to various modes.

6.5.2.4 Design of Dielectric resonator Filters

Dielectric resonator filters are mostly of the band-pass type. However, dielectric resonator band-stop filters are also used in many applications [56]. As well, dielectric resonator filters are either monomode (single-mode) or multimode (more than one mode). In a multimode filter, the resonators support multiple resonances in the same single physical structure. This is similar to a metallic waveguide resonant cavity supporting degenerate resonances. This multimode supporting capability, extremely high Q, and compact structure gives dielectric resonator band-pass filters many advantages over waveguide cavity filters.

Among all multimode filters dual-mode filters are the most common in many applications. Because a dual-mode filter is the best compromise between the monomode and other multimode filters in terms of mode excitation, tuning, and fabrication, filters comprising such a resonator are most common. However, monomode as well as triple-mode filters are also used. Monomode filters are easiest to fabricate and tune. In almost all filters the resonator is enclosed within a metallic enclosure to avoid radiation losses. The dimensions of the enclosure should be

Table 6.8* Equations for Calculating Resonance Frequencies

Mode	Resonant Frequency	Range of Validity
$HE_{11\delta}$	$f_r = \frac{6.324c}{2\pi a \sqrt{\varepsilon_r - 1}} \left[0.02 \left(\frac{a}{2H}\right)^2 + 0.36 \left(\frac{a}{2H}\right) + 0.27 \right]$	$0.4 \leq a/H \leq 6.0$
$TE_{01\delta}$	$f_r = \frac{2.327c}{2\pi a \sqrt{\varepsilon_r - 1}} \left[-0.00898 \left(\frac{a}{H}\right)^2 + 0.2123 \left(\frac{a}{H}\right) + 1.00 \right]$	$0.33 \leq a/H \leq 5.0$
$TE_{011+\delta}$	$f_r = \frac{2.208c}{2\pi a \sqrt{\varepsilon_r - 1}} \left[-0.00273 \left(\frac{a}{H}\right)^2 + 0.7013 \left(\frac{a}{H}\right) + 1.00 \right]$	$0.33 \leq a/H \leq 5.0$
$TM_{01\delta}$	$f_r = \frac{c}{2\pi a \sqrt{\varepsilon_r - 1}} \sqrt{\left[\left(\frac{\pi a}{2H}\right)^2 + 14.6689\right]}$	$0.33 \leq a/H \leq 5.0$

*See Figure 6.86.

such that without the dielectric block it is a piece of waveguide below cutoff or an evanescent mode waveguide. Under no condition should the structure behave like a dielectric-loaded cavity resonator. The resonances take place only within the dielectric block. The coupling between two resonators is brought about by physical proximity, irises, or probes. The polarity of the coupling is achieved by the orientation and location of an iris. Figure 6.88 shows the two commonly used orientations of dielectric resonators in a below-cutoff waveguide.

In Figure 6.88(a) the resonators are excited in the $TE_{01\delta}$ mode. In Figure 6.88(b) the resonators are excited in the $TM_{01\delta}$ mode. The regular waveguide mode (TE_{10}) is launched by coaxial probes or loops. The resonators may be resonated in other hybrid modes. A five-pole monomode Chebyshev filter is designed with waveguide input and output based on following specifications:

1. Center frequency = 10.65 GHz
2. Passband ripple = 0.10 (−16 dB return loss)
3. Filter order 5
4. Resonator $\varepsilon_r = 36$
5. Waveguide interface is WR75 ($a = 0.75$" $b = 0.375$")

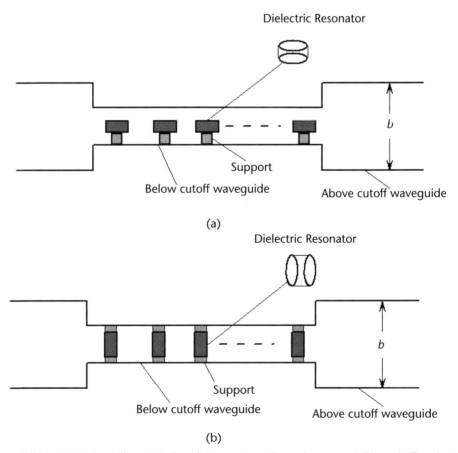

Figure 6.88 (a) Horizontally oriented puck in a rectangular enclosure, and (b) vertically oriented puck in a circular cylindrical enclosure.

6. Resonance mode is $TE_{01\delta}$

The design method is described in Section 6.2.6 for a microstrip band-pass filter. Figure 6.89(a) shows the simulation setup for the resonance frequency determination using a simulator.

The resonance frequency can be adjusted by altering the cross-sectional dimensions of the below-cutoff waveguide or by a tuning plate above the dielectric puck. Figure 6.89(b) shows the arrangement for simulation of coupling coefficient. The coupling can be adjusted by varying the distance S between the resonators or iris opening. Figure 6.89(c) shows the response of a two-resonator circuit in Figure 6.89(b). The coupling coefficient is determined by using (6.31). Once the coupling coefficients for various values of S have been obtained, a design curve like the one shown in Figure 6.35 is drawn. The required coupling coefficients for the design are obtained using the method of Section 6.2.6. They can also be obtained by multiplying the corresponding elements of the coupling matrix by the fractional ripple bandwidth. Figure 6.90 shows the coupling curve obtained via numerical simulation using WASP-NET software. Although an FEM-based software could be used, WASP-NET is very fast because it combines the mode-matching FEM and BEM methods.

Using the coupling curve in Figure 6.90, the interresonator separations are obtained and the filter is tuned using optimization by WASP-NET[3] software. The two outermost couplings are realized by irises. Figure 6.91 shows the 3-D view of the filter and the computed frequency response.

Dielectric resonator filters can also be realized using the $TM_{01\delta}$ mode. Figure 6.92 shows the configuration of a $TM_{01\delta}$ band-pass filter with coaxial line input and output [57]. The design methodology for the filter is the same as the one followed in the previous example. Figure 6.93 shows the configuration of the dielectric resonator in a below-cutoff metallic enclosure. The dielectric puck of dielectric constant ε_{rd} diameter c_1 is aligned uniaxially in the enclosure and is supported by a spacer annular support of low dielectric constant, which has a negligible effect on the resonance (i.e., $\varepsilon_b \ll \varepsilon_{rd}$).

Figure 6.94 shows a coupled pair of resonators separated by a distance $2h$. The designer must ensure that $(\pi f_0 c_2)/c < 2.405$ in order to ensure that the circular enclosure is evanescent. The structure can be analyzed using a 3-D simulator such as HFSS or WASP-NET. Once again, for every simulation for a specific value of h, a two-peak resonance curve is obtained and the coupling coefficient is calculated using (6.31). The coupling analysis can also be done using a one port simulation. The configuration for a one-port analysis is shown if Figure 6.95.

In this method, the short- and open-circuit walls are created exactly at the center between the resonators at a distance h from either resonator and the resonance frequency of S_{11} is noted. If the resonance frequencies corresponding to the short- and open-circuit conditions are f_{sh} and f_{op}, respectively, then the TM mode resonance is given by [57]

$$K = \frac{f_{op}^2 - f_{sh}^2}{f_{op}^2 + f_{sh}^2} \qquad (6.79)$$

---

3. See Chapter 5 [24].

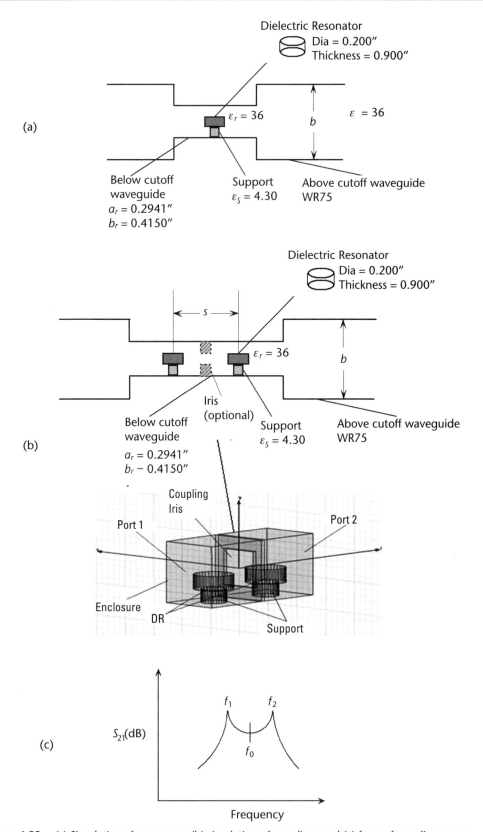

Figure 6.89 (a) Simulation of resonance, (b) simulation of coupling, and (c) form of coupling curve.

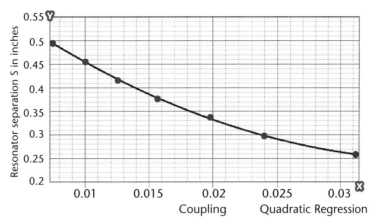

Figure 6.90 The coupling curve.

Figure 6.96(a) shows the simulated values of the resonant frequencies and Figure 6.96(b) shows the computed coupling factors for DR radius = 7.4 mm, DR thickness = 8.6 mm, cavity radius = 10.8 mm, $\varepsilon_{rd} = 82$, and $\varepsilon_b = 2.1$.

A three-pole band-pass filter was designed for 1.25% fractional bandwidth centered around 3.2-GHz and 20-dB passband return loss. Figure 6.97 shows the designed filter. Figures 6.98 and 6.99 shows the computed and measured frequency responses of the filter. Note that the center frequency of the measured filter is shifted by 60 MHz upward. This is due to the inaccuracy of the dielectric constant of the resonator.

In the above example, one notices that the difference between a sequentially coupled circular cavity waveguide filter and the above filter is that in the above filter, the dielectric puck plays the entire role of a resonator. Therefore, as in a circular waveguide cavity filter, degenerate multimodes can be coupled in the same dielectric puck in more or less the same way that is achieved in a waveguide circular or rectangular cavity. Figure 6.100 shows the similarity between a dual-mode circular cavity filter and a dual-mode dielectric resonator filter [58,59].

Figure 6.101(a) shows the practical configuration of a dual-mode dielectric resonator six-pole filter [60]. In this design the degenerate modes are two orthogonal $EH_{11\delta}$ modes. The design methodology, based on full 3-D EM simulation, is the same as in the sequentially coupled DR filter [57] described above. Figure 6.101(b) shows the response of a four-pole dual-mode filter.

The main purpose of a cross-coupled filter is to generate a higher skirt selectivity in the immediate vicinity of cutoff frequencies than in a Chebyshev filter. However, cross couplings are also used for group delay equalization. In filter design, the art of simultaneous achievement of skirt selectivity and flat-group delay, using various permutations and combinations of negative and positive cross couplings, is itself a vast subject and it is beyond the scope of this text. Also, one should keep in mind that whether the DR filter is a waveguide or coaxial line type depends only on the type of the input/output interfaces.

It is not always true that only cross coupling can generate pole extraction for higher skirt selectivity. Another way is to cascade a pole extracting band-stop section with the band-pass filter. A very unique way to generate controllable transmission zero in $TE_{01\delta}$ mode DR filter is proposed and demonstrated by Ouyang and

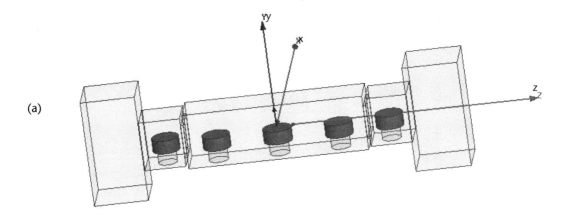

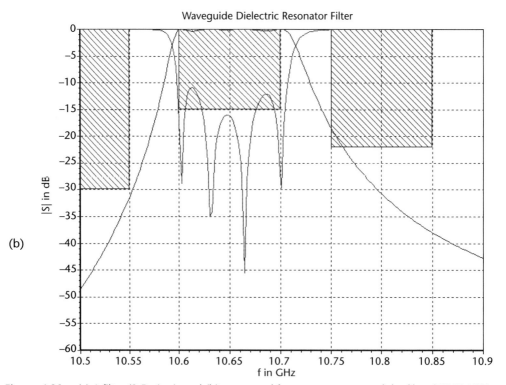

Figure 6.91 (a) A filter (3-D view), and (b) computed frequency response of the filter (WASP-NET analysis and optimization).

Wang [61]. Consider the field distribution in a circular dielectric puck in Figure 6.102(a).

The field resembles that of an axial magnetic dipole. Such resonators are often excited by a probe-type feeding structure. The magnetic field in the iris between the first resonator and the second resonator in a filter points vertically, either upward or downward. Therefore, the problem of the first cavity can be treated in such a

6.5 Dielectric Resonator Filters

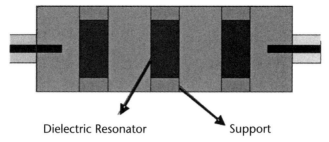

Figure 6.92 Configuration of dielectric resonator filter supporting the TM01δ mode [57]. (Reproduced courtesy of the Electromagnetics Academy.)

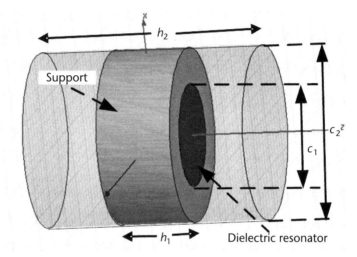

Figure 6.93 Configuration of a single resonator in a below-cutoff metallic enclosure [57]. (Reproduced courtesy of the Electromagnetics Academy.)

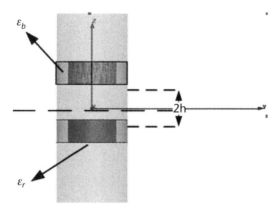

Figure 6.94 Coupled pair of resonators [57]. (Reproduced courtesy of the Electromagnetics Academy.)

way that the second cavity is replaced by a waveport, as shown in Figure 6.102(b) and (c).

Then, the structure is analyzed using HFSS. Figure 6.103 shows the computed results. From Figure 6.103, we note that there is a transmission zero f_z near

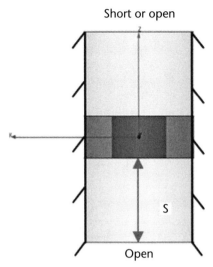

Figure 6.95 Configuration for a one-port simulation [57]. (Reproduced courtesy of the Electromagnetics Academy.)

the main resonance frequency. The location of the transmission zero follows the following rules:

1. When $\theta < 180°$, f_z is located on the left side of the resonance and its shifts leftward with increasing θ;
2. When $\theta < 180°$, f_z is located on the right side of the resonance and its shifts leftward with increasing θ;
3. When $\theta = 180°$, f_z does not exist.

The mode operation occurring for a feeding angle of 270° can be described by EM simulation of the electric field distribution shown in Figure 6.104(a).

The input or the output excites the resonant $TE_{01\delta}$ mode at the first or the last resonator, respectively. The iris is coupled to the probe by the evanescent TE_{10} mode. When the feed angle is 180° the E-field assumes an orthogonal position, as shown in Figure 6.104(b). Since there is no horizontal part of the E-field through the iris, the nonadjacent resonator is not excited and consequently no transmission zero results. In higher-order $TE_{01\delta}$ DR filters, f_z is shifted by the coupling from source/load to the nonadjacent resonator. Figure 6.105 summarizes the location of f_z for different values of the shadowed zone, while the field distribution in the resonator is still in $TE_{01\delta}$ mode. Therefore, it is possible to achieve the range for f_z between 1.7 to 2.1 GHz centered around 1.864 GHz. Figure 6.106 shows three possible configurations of the filter. The resulting triplet topology is shown in Figure 6.107.

In Figure 6.106(a), the input and the output feeding angles are at 180°; therefore, no coupling takes place from the source and the load to the adjacent resonators. Consequently, a transmission zero is created near the passband. In Figure 6.106(b), the output feeding angle is reduced to 90°, so negative coupling for a triplet section has been realized with the inline configuration. The triplet section,

6.5 Dielectric Resonator Filters

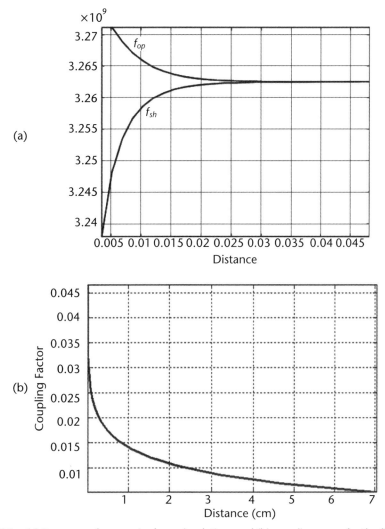

Figure 6.96 (a) Resonance frequencies from simulation, and (b) coupling curve for the filter design [57]. (Reproduced courtesy of the Electromagnetics Academy.)

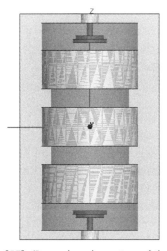

Figure 6.97 The designed filter [57]. (Reproduced courtesy of the Electromagnetics Academy.)

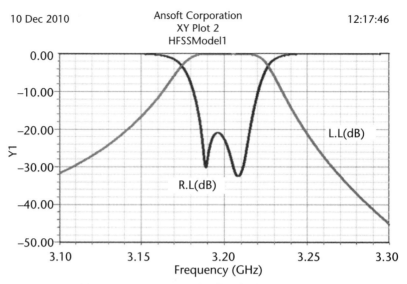

Figure 6.98 Computed frequency response of the filter [57]. (Reproduced courtesy of the Electromagnetics Academy.)

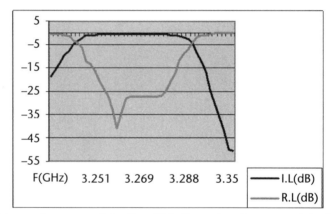

Figure 6.99 Measured response of the filter [57]. (Reproduced courtesy of the Electromagnetics Academy.)

starting from the load side to the last two resonators, introduces a transmission zero near the passband. In Figure 6.106(c), the feed angles are 90° and 270° at the input and output, respectively. The negative coupling from the input to the first two resonators generates a transmission zero below the passband. The positive coupling from the output to the last two resonators generate a transmission zero above the passband. The three situations are summarized in Figure 6.108.

A four-pole band-pass filter using $TE_{01\delta}$ was designed for a 20-MHz ripple bandwidth centered around 1.85 GHz and 0.036 dB passband ripple. The stopband zeros were located at 1.785 and 1.930 GHz. The coupling matrix (see Figure 6.108) is

6.5 Dielectric Resonator Filters

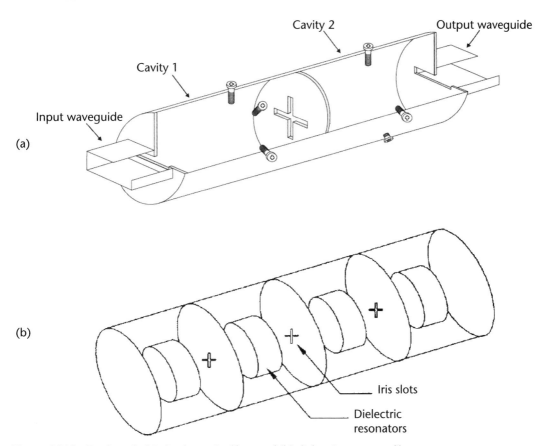

Figure 6.100 Dual-mode (a) circular cavity filter, and (b) dielectric resonator filter.

$$M = \begin{bmatrix} 0 & 1.06154 & -0.1481 & 0 & 0 & 0 \\ 1.06154 & 0.1938 & 0.93 & 0 & 0 & 0 \\ -0.1481 & 0.93 & -0.1237 & 0.7254 & 0 & 0 \\ 0 & 0 & 0.7254 & 0.1176 & 0.93 & 0.1342 \\ 0 & 0 & 0 & 0.93 & -0.1681 & 1.06154 \\ 0 & 0 & 0 & 0.1342 & 1.0654 & 0 \end{bmatrix} \quad (6.80)$$

Figure 6.109(a) shows the response of the above coupling matrix.

By interchanging the feeding angles between the input and output, the coupling matrix can be changed and one can transfer both transmission zeros on the same side of the passband. Figure 6.109(b) shows the measured frequency response of the filter.

The subjects of cross-coupled and dielectric resonator filters are of tremendous importance in modern microwave communication engineering. An excellent account of cross-coupled filter is available in the text by Cameron, Kudsia, and Mansour [62], while an excellent account of dielectric resonator filters is in Ian Hunter's book [63].

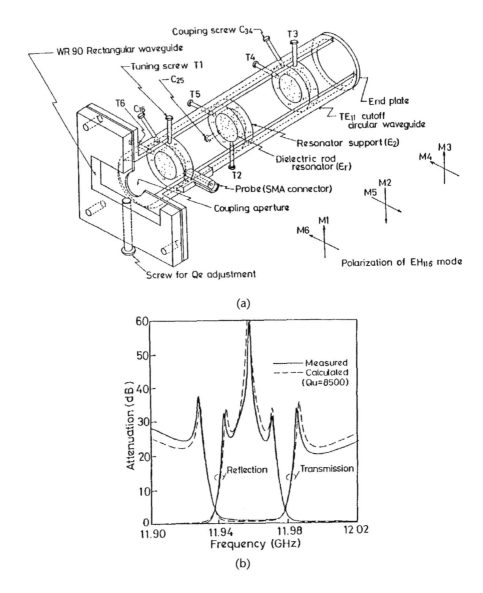

Figure 6.101 (a) Dual-mode dielectric resonator filter in a circular enclosure [60]. (Reprinted with permission from the IEEE.) (b) Frequency response of a four-pole dual-mode DR filter.

6.5 Dielectric Resonator Filters

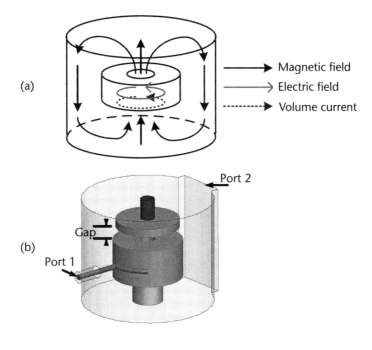

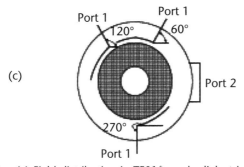

Figure 6.102 (a) Field distribution in TE01δ mode dielectric resonator figure, (b) TE01δ mode dielectric resonator excited by a probe and a wave port, and (c) top view of the dielectric resonator fed by a probe and terminated by a wave port [61]. (Reproduced courtesy of the Electromagnetics Academy.)

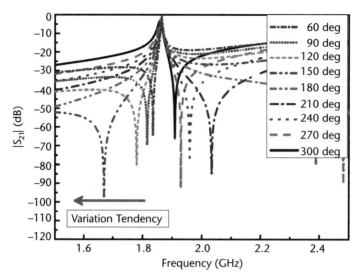

Figure 6.103 Response of the structure shown in Figure 6.101(b) for different feeding angles [61]. (Reproduced courtesy of the Electromagnetics Academy.)

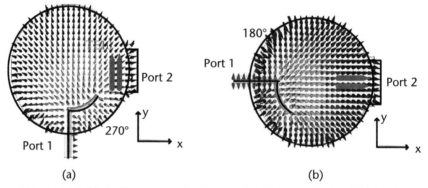

Figure 6.104 (a) E-field distribution for a feeding angle of 270°, and (b) E-field distribution for a feeding angle of 180° [61]. (Reproduced courtesy of the Electromagnetics Academy.)

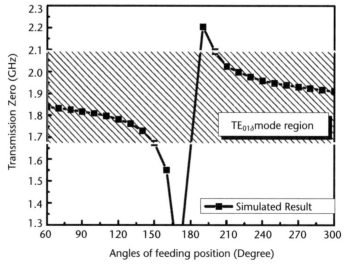

Figure 6.105 Variation of f_z with θ [61]. (Reproduced courtesy of the Electromagnetics Academy.)

6.5 Dielectric Resonator Filters

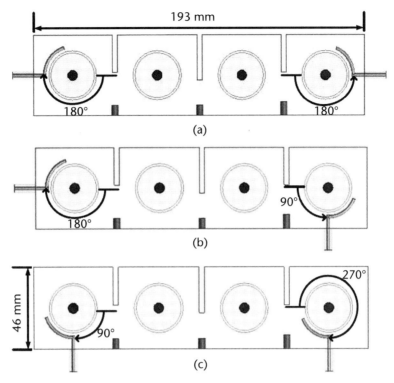

Figure 6.106 Top view of three 4-pole dielectric resonator filters [61]. (Reproduced courtesy of the Electromagnetics Academy.)

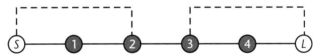

Figure 6.107 Topology of the source and load multiresonator couplings inline configuration filter [61]. (Reproduced courtesy of the Electromagnetics Academy.)

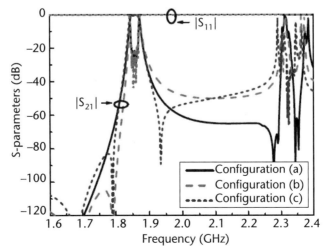

Figure 6.108 Simulation results for the three 4-pole dielectric resonator filters [61]. (Reproduced courtesy of the Electromagnetics Academy.)

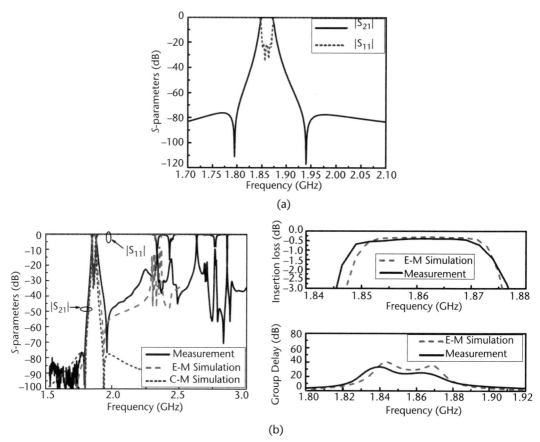

Figure 6.109 (a) Response of the coupling matrix of the filter, and (b) measured frequency response of the filter [61]. (Reproduced courtesy of the Electromagnetics Academy.)

References

[1] Collin, R. E. *Foundations for Microwave Engineering,* New York: McGraw Hill, 1966, pp. 403–433.

[2] Ansoft Designer, Ansoft Corporation, Bethlehem, PA 2005.

[3] Kajfez, D., *Notes on Microwave Circuits,* Oxford, MS: Kajfez Consulting, 1986, pp. 109–273.

[4] Matthaei, G., L. Young, and E. M. T. Jones, *Microwave Filters, Impedance Matching Networks, and Coupling Structures,* Dedham, MA: Artech House, 1980.

[5] Hunter, I., *Theory and Design of Microwave Filters,* London: The Institution of Electrical Engineers, 2001.

[6] Altschuler, H. M., and A. A. Oliner, "Discontinuities in the Center Conductor of Symmetric Strip Transmission Line," *IRE Trans., PGMTT,* Vol. MTT-8, May 1960, pp. 328–339.

[7] WAVECON, Filter design software, Escondido, CA, 1994.

[8] Matthaei, G., L. Young, and E.M.T. Jones, *Microwave Filters, Impedance Matching Networks and Coupling Structures,* Dedham, MA: Artech House, 1980, pp. 441–442.

[9] Cohn, S. B., "Parallel-Coupled Transmission Line Resonator Filters," *IRE Trans., PGMTT,* Vol. MTT-6, April 1958, pp. 223–231.

[10] Makimoto, M., and S. Yamashita, "Bandpass Filters Using Parallel-Coupled Stripline Stepped Impedance Resonators," *IEEE MTT-S Int. Symp. Digest,* 1980, pp. 141–143.

[11] Crystal, E. G., and S. Franknel, "Hairpin-Line and Hybrid Hairpin-Line /Half-Wave Parallel-Coupled Line Filters," *IEEE Trans. Microw. Theory Techn.*, Vol. MTT-20, November 1972, , pp. 719–728.

[12] Wong, J. S., "Microstrip Tapped Line Filter Design," *IEEE Trans. Microw. Theory Techn.*, Vol. MTT-27, January 1979, pp. 44–50.

[13] Dell-Imagine, R. A., "A Parallel-Coupled Microstrip Filter Design Procedure," *IEEE MTT-S Int. Symp. Digest,* June 1970, pp. 26–32.

[14] Matthaei, G., "Interdigital Band-Pass Filters," *IRE Trans. Microw. Theory Techn.*, Vol. MTT-10, No. 6, November 1962, pp. 479–491.

[15] Crystal, E. G., "Tapped-Line Coupled Transmission Line with Applications to Interdigital and Combline Filters," *IEEE Trans. Microw. Theory Techn.*, Vol. MTT-23, December 1975, pp. 1007–1012.

[16] Matthaei, G., L. Young, and E. M. T. Jones, *Microwave Filters, Impedance Matching Networks and Coupling Structures,* Dedham, MA: Artech House, 1980, pp. 614–633.

[17] Caspi, S., and J. Adelman, "Design of Combline and Interdigital Filters with Tapped Line Input," *IEEE Trans. Microw. Theory Techn.*, Vol. MTT-36, April 1988, p. 759.

[18] Atia, A. E., A. E. Williams, and R. W. Newcomb, "Narrow-band Multiple Couple Cavity Synthesis," *IEEE Trans. CAS-21,* September 1974, pp. 649–655.

[19] EM3DS-Full-Wave Electromagnetic Simulation Software, MEMRESEARCH, www.mem-research.com, Italy, 2006.

[20] Hong, J. -S., and M. J. Lancaster, *Microstrip Filters for RF/Microwave Applications,* Wiley Series in Microwave and Optical Engineering, New York: John Wiley and Sons, 2001.

[21] Dishal, M., "Alignment and Adjustment of Synchronously Tuned Multiple Resonant Circuit Filters," *Proc. IRE.,* Vol. 30, November 1951, pp. 1448–1455.

[22] Matthaei, G., "Combline Bandpass Filters of Narrow and Moderate Bandwidth," *Microw. J.,* August 1963, pp. 82–91.

[23] Rhodes, J. D., *Theory of Electrical Filters,* London: John Wiley and Sons, 1976.

[24] WAVEFIL, Waveguide component design software, Polar Waves Consulting, Saskatoon, Saskatchewan, Canada, 2000.

[25] Kirilenko, A. A., L. Rud, V. Tkachenko, and D. Kulik, "Evanescent-Mode Bandpass Filter Based on Waveguide Sections and Inductive Strips," *IEEE MTT-S Int. Microw. Symp.,* June 2001, pp. 1317–1320.

[26] Kurzrok, R. M., "General Four-Resonator Filters at Microwave Frequencies," *IEEE Trans. Microw. Theory Techn.*, Vol. MTT-14, June 1966, pp. 295–296.

[27] Thomas, J. B., "Cross-Coupling in Coaxial Cavity Filters: A Tutorial Overview," *IEEE Trans. Microw. Theory Techn.*, Special Issue on RF and Microwave Tutorials, 2003.

[28] Pfitzenmaier, G., "Synthesis of Narrow-Band Canonic Microwave Bandpass Filters Exhibiting Linear Phase and Transmission Zeros," *IEEE Trans. Microw. Theory Techn.*, Vol. MTT-30, No. 9, September 1982, pp. 1300–1311.

[29] Cameron, R., "General Coupling Matrix Synthesis Methods for Chebyshev Filtering Functions," *IEEE Trans. Microw. Theory Techn.*, Vol. MTT-47, No. 4, April 1999, pp. 433–442.

[30] Amari, S., "Synthesis of Cross-Coupled Resonator Filters Using an Analytical Gradient Based Optimization Technique," *IEEE Trans. Microw. Theory Techn.*, Vol. MTT-48, September 2000, pp. 1559–1564.

[31] Atia, W. A., K. A. Zaki, and A. E. Atia, "Synthesis of General Topology Multiple Coupled Resonator Filters by Optimization," *IEEE MTT-S Digest,* Vol. 2, 1998, pp. 821–824.

[32] Mayer, B., and M. Vogel, *Rapid Elliptic-Filter Design,* Design Tutorial, online, Ansoft Corporation, 2003.

[33] Montejo-Garai, J. R. and J. Zpata, "Full-Wave Design and Realization of Multicoupled Dual-Mode Circular Waveguide Filters," *IEEE Trans. Microw. Theory Techn.*, Vol. MTT-43, No. 6, June 1995, pp. 1290–1297.

[34] Liang, J. F., X. P. Liang, K. A. Zaki, and A. E. Atia, "Dual-Mode Dielectric or Air-Filled Rectangular Waveguide Filters," *IEEE Trans. Microw. Theory Techn.*, Vol. MTT-42, No. 7, July 1994, pp. 1330–1336.

[35] Atia, A. E., and A. E. Williams, "Measurement of Intercavity Couplings," *IEEE Trans. Microw. Theory Techn.*, Vol. MTT-23, No. 6, June 1975, pp. 519–522.

[36] N. A. McDonald, "Measurement of Intercavity Couplings," *IEEE Trans. Microw. Theory Techn.*, Vol. MTT-24, No. 6, p. 162, March, 1976.

[37] Levy, R., "The Relationship between Dual Mode Cavity Cross-Coupling and Waveguide Polarizers," *IEEE Trans. Microw. Theory Techn.*, Vol. MTT-43, No. 11, November 1995, pp. 2614–2620.

[38] Accatino, L., G. Bertin, and M. Mongiardo, "Elliptic Cavity Resonators for Dual-Mode Narrow-Band Filters," *IEEE Trans. Microw. Theory Techn.*, Vol. MTT-45, No. 12, December 1997, pp. 2393–2401.

[39] Ofli, E., "Analysis and Design of Microwave and Millimeter-Wave Filters and Diplexer," M.Sc. thesis, Swiss Federal Institute of Zurich, 2004.

[40] Kochbach, J., and K. Folgero, "Design Procedure for Waveguide Filters with Cross-Couplings," *IEEE MTT-S Int. Microw. Symp.*, June 2002, pp. 1449–1452.

[41] Hong, J. S. and M. J. Lancaster, *Microstrip Filters for RF/Microwave Applications*, New York: John Wiley Interscience, 2001.

[42] Iliev, I. G., and M. V. Nedelchev, "CAD of Cross-Coupled Miniaturized Hairpin Bandpass Filters," *Microw. Rev.*, Bulgaria, December 2002, pp. 49–52.

[43] Nedelchev, M. V., and I. G. Iliev, "Accurate Design of Triplet Square Open-Loop Resonator Filters," *Microtalasna Revija*, November 2006, pp. 36–40.

[44] Levy, R., "Filters with Single Transmission Zeros at Real or Imaginary Frequencies," *IEEE Trans. Microw. Theory Techn.*, Vol. MTT-34, No. 4, April 1976, pp. 172–181.

[45] Levy, R., "Direct Synthesis of Cascaded Quadruplet (CQ) Filters," *IEEE Trans. Microw. Theory Techn.*, Vol. MTT-43, No. 12, December 1995, pp. 172–181.

[46] Sen, O. A., Y. Sen, and N. Yildirim, "Synthesis of Cascaded Quadruplet Filters Involving Complex Transmission Zeros," *IEEE MTT-S 200 Symp. Digest*, Boston, June 2000, pp. 1177–1180.

[47] Yildirim, N., O. A. Sen, and Y. Sen, "Synthesis of Cascaded N-Tuplets," *Microw. Rev.*, December 2001.

[48] FILPRO, http://www.eee.metu.edu.tr/~nyil/filpro.html.

[49] Ness, J. B., "A Unified Approach to the Design, Measurement, and Tuning of Coupled-Resonator Filters," *IEEE Trans. Microw. Theory Techn.*, Vol. 46, No. 4, April 1998, pp. 343–352.

[50] Benhamed, N., and M. Feham, "Rigorous Analytical Expressions for Electromagnetic Parameters of Transmission Lines: Coupled Sliced Coaxial Cables," *Microw. J.*, Vol. 44, No. 11, November 2001, pp. 130–138.

[51] Walker, V. E. G., "Dielectric Resonators and Filters for Cellular Base Stations," Ph.D. thesis, Department of Electrical Engineering, University of Leeds, December 2003.

[52] Richtmeyer, R. D., "Dielectric Resonators," *J. Appl. Phys.*, Vol. 10, June 1939, p. 391.

[53] Harrison, W., "A Miniature High-Q Bandpass Filter Employing Dielectric Resonators," *IEEE Trans. Microw. Theory Techn.*, Vol. 16, No. 4, April 1968, pp. 210–218.

[54] Cohn, S. B., "Microwave Bandpass Filters Containing High-Q Dielectric Resonators," *IEEE Trans. Microw. Theory Techn.*, Vol. 16, No. 4, April 1968, pp. 815–824.

[55] Mongia, R. K., and P. Bhartia, ""Dielectric Resonator Antenna- A Review," *Int. J. MIMICAE*, Vol. 4, No. 3, 1994, pp. 230–247.

[56] Kobayashi, Y. and C. Inoue, " Bandpass and Bandstop Filters Using Dominant $TM_{01\delta}$ Mode Dielectric Rod Resonators," *IEEE MTT Symp. Digest*, Vol. 2, 1997, pp. 793–796.

[57] Salimnijad, R., and M. R. Ghafaurifard, "A Novel and Accurate Method for Designing Dielectric Resonator Filter," *PIER B*, Vol. 8, 2008, pp. 293–306.

[58] Vincente, M. B., "Design of Microwave Filters and Multiplexers in Waveguide Technology Using Distributed Models," Ph.D dissertation, Department of Communications, Universidad Politechnica de Valencia, December 2014.

[59] Kwok, R. S., and S. J. Fiedziuszko, "Advanced Filter Technology in Communications Satellite Systems," *Proc. ICASS,* Shanghai, 1996, pp. 155.

[60] Kobayashi, Y., and K. Kubo, "Canonical Bandpass Filters Using Dual-Mode Dielectric Resonators," *IEEE MTT Symp. Digest,* 1987, pp. 134–140.

[61] Ouyang, X., and Bo-Y. Wang, "Inline $TE_{01\delta}$ Mode Dielectric Resonator Filters with Controllable Transmission Zeros for Wireless Base Stations," *Prog. Electromagn. Res. B Pier B,* Vol. 38, 2013, pp. 101–110.

[62] Cameron, R., C. Kudsia, and R. R. Mansour, *Microwave Filters for Communication Systems,* Hoboken, NJ: John Wiley and Sons, 2007.

[63] Hunter, I., *Theory and Design of Microwave Filters,* London: Institution of Electrical Engineers, 2001.

Appendix 6A Slot Coupled Coaxial Combline Filter Design[1]

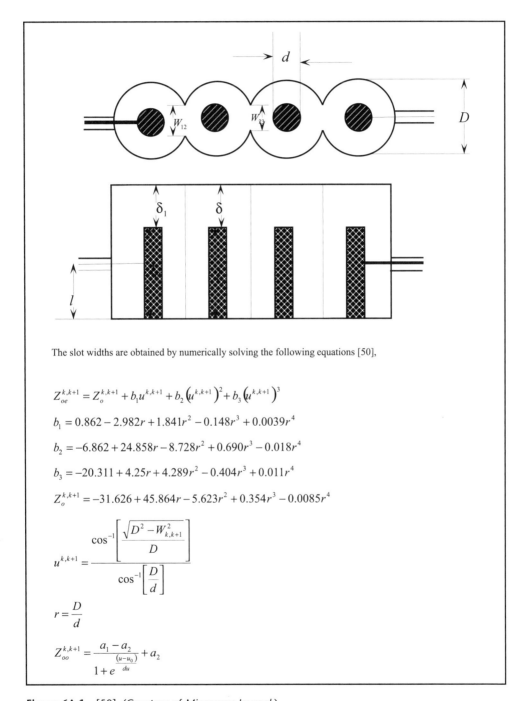

The slot widths are obtained by numerically solving the following equations [50],

$$Z_{oe}^{k,k+1} = Z_o^{k,k+1} + b_1 u^{k,k+1} + b_2 \left(u^{k,k+1}\right)^2 + b_3 \left(u^{k,k+1}\right)^3$$

$$b_1 = 0.862 - 2.982r + 1.841r^2 - 0.148r^3 + 0.0039r^4$$

$$b_2 = -6.862 + 24.858r - 8.728r^2 + 0.690r^3 - 0.018r^4$$

$$b_3 = -20.311 + 4.25r + 4.289r^2 - 0.404r^3 + 0.011r^4$$

$$Z_o^{k,k+1} = -31.626 + 45.864r - 5.623r^2 + 0.354r^3 - 0.0085r^4$$

$$u^{k,k+1} = \frac{\cos^{-1}\left[\frac{\sqrt{D^2 - W_{k,k+1}^2}}{D}\right]}{\cos^{-1}\left[\frac{D}{d}\right]}$$

$$r = \frac{D}{d}$$

$$Z_{oo}^{k,k+1} = \frac{a_1 - a_2}{1 + e^{\frac{(u-u_0)}{du}}} + a_2$$

Figure 6A.1 [50]. (Courtesy of *Microwave Journal*.)

---

1. *From* [50].

$$a_1 = -37.78 + 50.089r - 6.398r^2 + 0.417r^3 - 0.01r^4$$

$$a_2 = 310.831 - 366.622r + 47.908r^2 - 3.263r^3 + 0.083r^4$$

$$u_0 = 1.407 - 0.017r + 0.0024r^2 - 1.507 \times 10^{-4} r^3 + 3.554 \times 10^{-6} r^4$$

$$du = 0.153 + 0.0045r - 11 \times 10^{-4} r^2 + 8.483 \times 10^{-5} r^3 - 2.21 \times 10^{-6} r^4$$

The even and the odd mode impedances are given by

$$Z_{oe}^{12} = \frac{Z_R}{1 - y_{12} Z_R} = Z_{oe}^{N-1,N}$$

$$Z_{oo}^{12} = \frac{Z_R}{1 + y_{12} Z_R} = Z_{oo}^{N-1,N}$$

For $k=2$ to $N-2$

$$Z_{oe}^{k,k+1} = \frac{1}{\dfrac{2}{Z_R} - \dfrac{1}{Z_{oe}^{k-1,k}} - y_{k+1,k} - y_{k-1,k}}$$

$$Z_{oo}^{k,k+1} = \frac{1}{2 y_{k,k+1} + \dfrac{1}{Z_{oe}^{k,k+1}}}$$

The values of $y_{k,k+1}$ are obtained from equation (6.38) with $Y_r = 1/Z_R$. The gap (δ) between the tip of a resonator and the top wall is obtained by using equation in Figure 6.42b with h replaced by δ and C replaced by C_{Sk} respectively.

Figure 6A.2 [50]. (Courtesy of *Microwave Journal*.)

Appendix 6B A Step-By-Step Procedure for Waveguide Folded and Elliptic Filter Design

Filter Specifications:

Center frequency: $f_0 = 6.84$ GHz

Bandwidth: $\Delta f = 0.320$ GHz

Filter order: $N = = 12$

Passband return loss: RL = 24 dB

Waveguide dimensions: $a = 34.89$ mm, $b = 8.45$ mm

Locations of transmission zeros: $f_{zL} = 6.38$ GHz, $f_{zU} = 7.38$ GHz

Step 1: Designing the Basic Filter

Design a 12-pole waveguide Chebyshev filter using a suitable filter wizard. One can use WASP-NET or MICIAN or any other software that has optimization capability.

Figure 6B.1 shows the designed filter and the optimized frequency response.

Step 2: Bringing the Nonadjacent Resonators to Be Cross Coupled Close to Each Other

Fold the filter at the center iris (iris 7) as shown in Figure 6B.2(a) and optimize the frequency response. Figure 6B.2(b) shows the computed frequency response.

Step 3: Generating the Cross Coupling

Cut a square hole in the common wall between resonator 3 and 10, as shown in Figure 6B.3(a). Optimize the frequency response. Figure 6B.3(b) shows the optimized frequency response.

In the above example, E-plane folding with a negative cross coupling is used. A positive cross coupling results in a loss of stopband rejection, no transmission zeros. and a flatter group delay. Figure 6B.4(a) shows a WR137 cross-coupled filter with H-plane folding and positive cross coupling. All couplings, including the cross coupling, is inductive. Figure 6B.4(b) shows the frequency response of the filter.

The above method can be used with many other types of resonators like coaxial, dielectric resonators, and planar transmission lines.

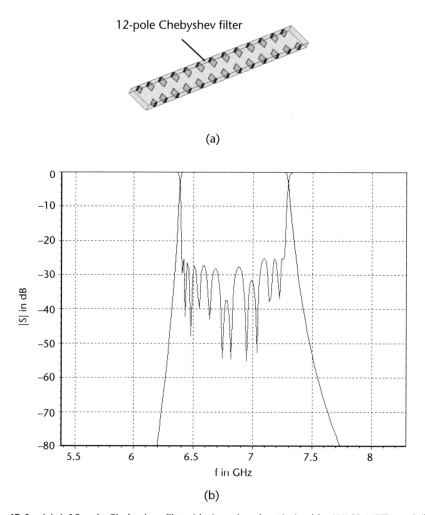

Figure 6B.1 (a) A 12-pole Chebyshev filter (designed and optimized by WASP-NET), and (b) the optimized frequency response of the filter in (a). [Readers may contact Protap Pramanick at protap-pramanick2@comcast.net for the CAD files (STL, SAT, and STEP). The filters were optimized using WASP-NET software.]

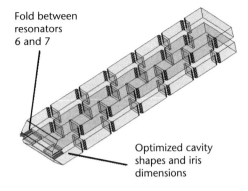

(a)

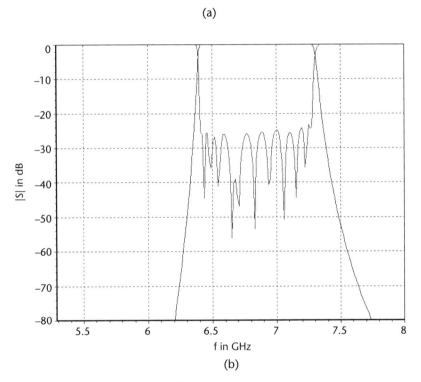

(b)

Figure 6B.2 (a) The folded Chebyshev filter, and (b) the optimized frequency response of the folded filter.

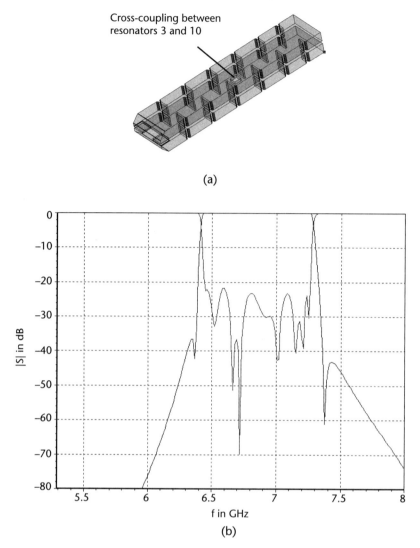

Figure 6B.3 (a) The cross-coupled filter, and (b) the optimized and computed frequency response of the 12-pole elliptic filter.

Appendix 6B A Step-By-Step Procedure for Waveguide Folded and Elliptic Filter Design   379

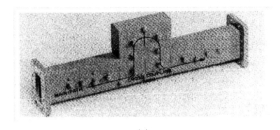

(a)

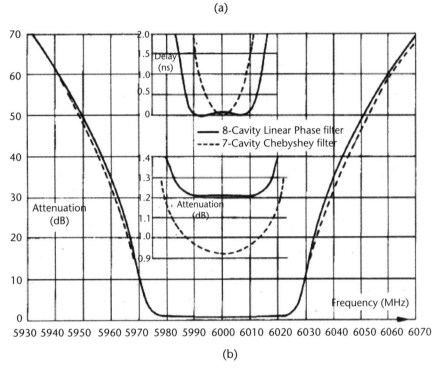

(b)

Figure 6B.4(a) Eight--cavity waveguide linear phase filter, and (b) frequency response of filter shown in (a). (From [64]. Courtesy of Dr. Ralph Levy.)

Appendix 6C Design of Dielectric Resonator Filters

DIELECTRIC RESONATORS

General Overview
TCI Ceramics, Inc. offers a broad range of high dielectric constant resonator materials that are temperature compensated with high Q. Dielectric resonators are used in microwave oscillator and filter applications where miniaturization, temperature stability and low loss are required.

Typical Applications
- Cellular Base Station Filters and Combiners
- PCS/PCN Base Station Filters and Combiners
- Direct Broadcast Satellite LNB
- TV Receive Only Satellite LNB
- Radar Detectors
- LMDS / MMDS Wireless Cable TV

Part numbering guide

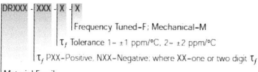

DRXXX - XXX - X - X

Frequency Tuned-F; Mechanical-M

τ_f Tolerance 1- ±1 ppm/°C, 2- ±2 ppm/°C

τ_f PXX-Positive, NXX-Negative; where XX-one or two digit τ_f

Material Family

Example: DR36-P9-2F

Material Family DR36; τ_f - +9; τ_f Tolerance - ± 2ppm/°C; Frequency Tuned
List product dimensions as description under part numbers.

Note: TCI Ceramics, Inc. can work with you to determine the correct frequency in your device, or the part may be ordered to mechanical dimensions-tolerances.

Standard Cofiguration - Disks/Cylinders/Assemblies

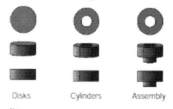

Disks   Cylinders   Assembly

Note:
[1] Assemblies are non standard

TCI Ceramics, a Division of National Magnetics Group Inc., 1210 Win Drive, Bethlehem, PA 18017 USA
Tel: 610 867 7600 • Fax: 610 867 0200 • Email: sales@tciceramics.com

Appendix 6C Design of Dielectric Resonator Filters

Standard Diameters Availability - Disks/Cylinders/Assemblies

Disks/Cylinders/Assemblies Standard Outside Diameter						
2.475	1.400	.840	.505	.305	.190	.120
2.275	1.300	.785	.470	.285	.180	.112
2.100	1.210	.730	.435	.265	.170	.104
1.940	1.125	.675	.405	.245	.160	.096
1.790	1.045	.630	.375	.230	.150	.089
1.650	.975	.585	.350	.215	.140	.082
1.520	.905	.545	.325	.200	.130	.076

Units (inches)

Cylinders/Assemblies	
Outside Diameter	Std. I.D.
.785 < 1.400	.162
.585 < .784	.122
.245 < .585	.083

Units (inches)

Note:
[1] When determining the dielectric resonator size, use the nearest standard diameter and adjust the thickness for frequency. Calculate resonator dimensions as discussed under the section "Product Selection" on page 22.

Dielectric Supports - Material Characteristics

Material Type MS–4

Composition .... Cordierite

Dielectric Constant (ε) .... 4.3 ± 5%

Dielectric Loss (tan δ) .... ≤ 0.0002

Temperature Coefficient of Resonant
Frequency (τ_f) (ppm/°C, 25° to 60°C) .... +55

Volume Resistivity (ohm · cm) @ 20°C .... 10^{14}

Thermal Expansion (ppm/°C) 2.4

Thermal Conductivity (cal/(sec)(cm)(°C)) x 10^3 .... 7.0

Dielectric Supports - Available Configurations

Dielectric supports can be used with both cylinders or assembly type resonators to improve coupling and temperature stability while reducing phase noise and cavity losses. Supports can be supplied in either disk or cylinder form. Please consult the factory for details. Support material specifications are shown above. Consult factory for alternative support materials and sizes.

Note:
[1] Length of support available in two sizes per diameter range.

Materials Overview

Features
- High ε'
- High Q
- Temperature Compensated
- Excellent Frequency Stability

Benefits
- Reduced Size
- Reduced Weight
- Low Loss
- Close Channel Spacing

DR30 This product offers the user the ultimate in a high Q resonator along with a very linear temperature coefficient. It is an excellent choice for the most demanding circuit requirements, especially in high frequency applications.

- Extremely high Q (> 10,000 @ 10 GHz)
- Very linear τ_f
- Excellent high frequency performance
- Typical uses include satellite communications, military ECM, high frequency filters and DRO's

DR36 This product offers excellent overall performance at a very affordable price for all applications from 800 MHz to 20 GHz. Q is extremely high in the cellular & PCS frequency bands (800MHz-2.0GHz) and the temperature coefficient linearity is excellent over a wide temperature range.

- High ε' (36) for circuit miniaturization
- Excellent τ_f linearity
- Very high Q
  (> 35,000 @ 850 MHz; > 10,000 @ 4 GHz)
- Typical applications include GSM, PCS, TVRO, DBS, DRO's, microwave filters

DR45 This material offers a higher dielectric constant for further circuit miniaturization while maintaining an excellent Q value, making it ideal as an alternative to DR36 for cellular and PCS applications when small size is important. It performs well over a wide frequency range.

- Very high ε' (45) for greater circuit miniaturization
- Very high Q
  (> 35,000 @ 850 MHz; > 10,000 @ 4 GHz)
- Typical applications include GSM, PCS, TVRO, DBS, DRO's, microwave filters
- Wide range of τ_f available

DR80 Our highest dielectric constant standard resonator material offers the benefit of greatly enhanced circuit miniaturization while maintaining a good Q. Available in a wide range of temperature coefficients and sizes.

- Extremely high ε' (80) for a very high degree of circuit miniaturization
- Wide range of τ_f available
- High Q (>3000 @ 3 GHz)
- Typical applications include filters, duplexers, TVRO, and cellular radio

Non Standard, Available Upon Request

DR38 This material is made available for applications requiring a direct replacement for products with a dielectric constant of 38 and positive τ_f values. Except for members in the τ_f - 6-8 ppm/°C range, these materials are not as linear as the τ_f for DR36.

- Broad range of positive τ_f, +4 to +14
- High Q (> 10,000 @ 4 GHz)
- Typical applications include DBS, DRO's and microwave filters

TCI Ceramics, a Division of National Magnetics Group Inc., 1210 Win Drive, Bethlehem, PA 18017 USA
Tel: 610 867 7600 • Fax: 610 867 0200 • Email: sales@tciceramics.com

Appendix 6C Design of Dielectric Resonator Filters

Material Specification Summary

Material Family	Resonators			
	DR30	DR36	DR45	DR80
Dielectric Constant (ε) Nominal	30±1	36±1	45±1	80±2
Temperature Coefficient of Resonant Frequency (τ_f) (ppm/°C, 25° to 60°C)	+4 to -2	+9 to -3	+6 to -3	+9 to -3
Insulation Resistance (ohm-cm)	>10^{14}	>10^{13}	>10^{13}	>10^{12}
Thermal Expansion (ppm/°C) Nominal	10.0	9.0	6.0	8.5
Thermal Conductivity (cal/cm^2/cm/sec/°C) x 10^3 Nominal	6	5	5	5

Notes:
[1] Other temperature compensated dielectrics available upon request (i.e., ε' -10, 20, 36, 38, 90, 120). Consult factory for details
[2] All material densities are > 95% of theoretical and, therefore, moisture absorption is nil.
[3] Customized properties are available upon request.

Material Code	Dielectric Constant	τ_f	Unloaded Q	Diameter Range Available	Recommended Frequency
DR30 – P4	31.0 ± 1	+4	≥ 10,000 @ 10GHz	.076" to 1.500"	1.5 – 20.0 GHz
DR30 – P2	30.5 ± 1	+2	"	"	"
DR30 – 0	30.0 ± 1	0	"	"	"
DR30 – N2	29.5 ± 1	-2	"	"	"
DR36 – P9	36.5 ± 1	+9	≥ 10,000 @ 4GHz 35,000 @ 850 MHz	.160" to 3.500"	0.8 – 13.5 GHz
DR36 – P6	36.0 ± 1	+6		"	"
DR36 – P3	35.7 ± 1	+3	"	"	"
DR36 – 0	35.5 ± 1	0	"	"	"
DR36 – N3	35.0 ± 1	-3	"	"	"
DR45 – P9	46.5 ± 1	+9	≥ 10,000 @ 4GHz 35,000 @ 850 MHz	.160" to 3.500"	0.4 – 13.5 GHz
DR45 – P6	45.5 ± 1	+6		"	"
DR45 – P3	45.0 ± 1	+3	"	"	"
DR45 – 0	45.0 ± 1	0	"	"	"
DR45 – N3	45.0 ± 1	-3	"	"	"
DR80 – P9	80.0 ± 2	+9	≥ 3,000 @ 3GHz	.405" to 1.125"	0.4 – 4.0 GHz
DR80 – P6	80.0 ± 2	+6	"	"	"
DR80 – P3	80.0 ± 2	+3	"	"	"
DR80 – 0	80.0 ± 2	0	"	"	"
DR80 – N3	80.0 ± 2	-3	"	"	"

Notes:
[1] All τ_f specifications are measured between 25° to 60°C using a TCI Ceramics, Inc. standard cavity.
[2] All materials are available in ±1ppm/°C and ±2 ppm/°C tolerance, except DR80 which is available only in ±2ppm/°C tolerance.

TCI Ceramics, a Division of National Magnetics Group Inc., 1210 Win Drive, Bethlehem, PA 18017 USA
Tel: 610 867 7600 • Fax: 610 867 0200 • Email: sales@tciceramics.com

CHAPTER 7

Design of Multiplexers

7.1 Definition of A Multiplexer

By definition, a multiplexer is a device for separating the signals of different frequencies present within the same system and directing them into subsidiary systems, or vice versa [1]. Multiplexers permit the transmission of a number of signals from one station to another without the introduction of appreciable cross-talk. They therefore constitute one of the main applications of filters. Unlike a duplexer, which is an electronic switch in the time domain, a multiplexer can be defined as a switch in the frequency domain. We briefly touched upon the issue of multiplexing in Chapter 1 when describing the significance of filters in microwave communication. Figure 7.1 shows the block diagram of a triplexer, which is a slightly different version of the diplexer shown in Figure 1.4 in Chapter 1. It uses three separate band-pass filters. The first one separates the 3.0–3.7 GHz band, the second one separates the 3.7–4.4 GHz band, and the third filter separates the 4.4–5.0 GHz band. At first glance, it may appear that the task of implementing the multiplexer or triplexer is simply completed by connecting the filters at a common point that will act as the input port of the resulting four-port device. However, in reality, it is not quite so simple. The art of multiplexer design depends on how the filters are connected together. Forming a simple common junction will result in undesired loading among the filters. This will give rise to interactions among the channels and seriously degrade the multiplexer's performance. There are various traditional approaches towards multiplexer design [1], which have been followed over many decades. Modern computer-aided techniques have made many such approaches obsolete. Although using the filters in a multiplexer can be combined effectively by using brute force or systematic CAD optimization techniques, the basic multiplexer configurations are still very useful. Such configurations are

Common junction with susceptance annulling network;
Cascaded directional filters;
Channel filters separated by isolators;
Manifold multiplexer.

In the following sections, we will briefly describe these approaches.

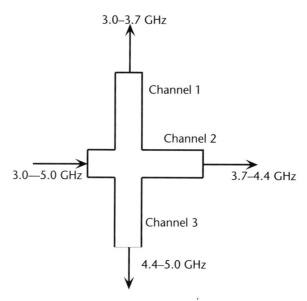

Figure 7.1 A three-channel multiplexer.

7.2 Common Junction Multiplexer with Susceptance Annulling Network

For a contiguous multiplexer wherein the adjacent channels have attenuation characteristics that typically cross over at their 3-dB points, each channel filter is constructed from a singly terminated low-pass prototype and a shunt susceptance annulling network at the common junction, as schematically shown in Figure 7.2. The susceptance annulling network helps provide a nearly constant total input admittance, Y_{TN}. By design, Y_{TN} is nearly real and approximates the generator output conductance, G_B, across the entire operating band of the multiplexer. The use of singly terminated low-pass prototypes for the band-pass filters ensure that the real part of the input admittance of each filter has the same frequency response as the transmission characteristic. A dual of the above multiplexer is one that is series-connected, in which all channel filters are connected in series and a reactance annulling network is connected in series with the channel filters and the generator [1].

As mentioned earlier, each channel filter is designed from the corresponding singly terminated low-pass prototype. A singly terminated low-pass prototype is synthesized under the assumption that it is driven by a voltage source of zero series impedance, as shown in Figure 7.3. The power delivered to the load connected to the filter is given by

$$P = |E_S|^2 \operatorname{Re} Y_k' \tag{7.1}$$

which implies that the real part of the input admittance of the filter has the same frequency response as that of the transfer function of the filter. Figure 7.4 shows typical $\operatorname{Re} Y_k'$ characteristics for low-pass Chebyshev prototype filters designed to be driven by zero impedance generators. The input admittances Y_i of the band-pass

7.2 Common Junction Multiplexer with Susceptance Annulling Network

channel filters in Figure 7.2 are the band-pass mappings of such low-pass prototype admittance characteristics.

In Figure 7.3, a low-pass prototype filter parameter g''_{n+1} is defined. Ideally, by definition, the parameter is equal to infinity. However, as shown in Figure 7.4, the value of g''_{n+1} or $1/g''_{n+1}$ should be the geometric mean between the values of $\mathrm{Re}\,Y'_k$ at the top and bottom of the passband ripple. This admittance level for the low-pass prototype is analogous to the driving source impedance G_B in Figure 7.2.

The above approach to multiplexer design is basically applicable to contiguous multiplexers where any two adjacent channels have their crossover frequency points at –3dB. However, the approach is applicable to noncontiguous multiplexers as well. Rhodes and Levy developed a more accurate theory and analytical equations for common junction multiplexer designs [2].

While the above approach is considerably analytical, a direct approach can be based on computer optimization. Let us consider a real-world triplexer design in order to establish the strength of modern CAD method for common junction multiplexer designs. The electrical specifications for the triplexer are as follows:

Channel 1

Passband 154 MHz – 174 MHz

Passband return loss: > –18 dB

Rejection between 400 MHz and 512 MHz: >20 dB

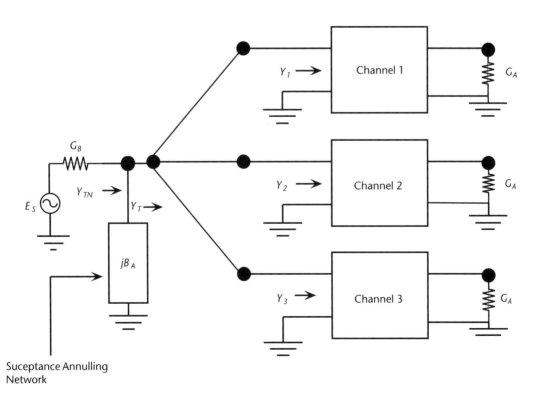

Figure 7.2 Shunt connected multiplexer (triplexer) with susceptance annulling network.

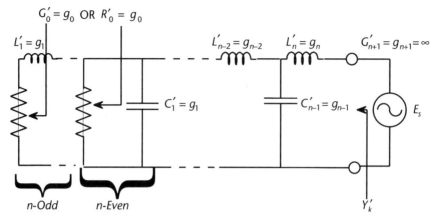

Figure 7.3 A singly terminated low-pass filter.

Channel 2
Passband 400 MHz – 512 MHz
Passband return loss: > –18 dB
Rejection between 154 MHz and 174 MHz: >20 dB
Rejection between 760 MHz and 870 MHz: >20 dB

Channel 3
Passband 760 MHz – 870 MHz
Passband return loss: > –18 dB
Rejection between 400 MHz and 512 MHz: >20 dB

In the first step of designing the triplexer, we design three doubly terminated band-pass filters that meet the channel requirements and the rejection levels within the adjacent channels. Note that the filters are simple doubly terminated ones. Figure 7.5(a) shows that the triplexer has been formed by creating a common junction at the input ports of the individual filters. Figure 7.5(b) and Table 7.1 show the component values after the frequency response was optimized using a circuit simulator (Ansoft Designer [3]). Figure 7.6 shows the frequency response of the triplexer before and after optimization.

7.3 Cascaded Directional Filter

A directional filter is a four-port device, as shown in Figure 7.7(a), with a theoretical insertion loss characteristic, as shown in Figure 7.7(b). It is assumed that the ports of the network are terminated in their characteristic impedances. The transmission between ports 1 and 4 has a band-pass response, whereas the transmission between ports 1 and 2 has a complementary band-stop response. Ideally, no power emerges at port 3. The performance is obtained in an analogous way no matter which port is used as the input port.

7.3 Cascaded Directional Filter

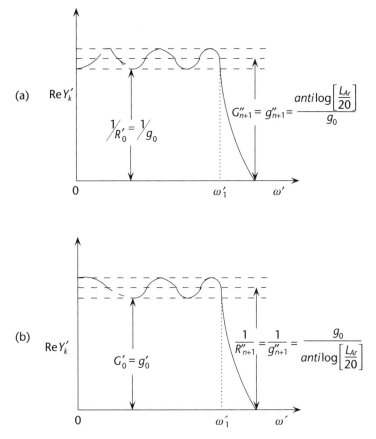

Figure 7.4 Real part of the input admittance for Chebyshev filters of the type in Figure 7.3.

The midband insertion loss, between the ports having the band-pass frequency response, of an actual directional filter employing resonators with a finite unloaded Q factor is the same as that of a two-port band-pass filter having the same frequency response and using the resonators with same unloaded Q factor.

A directional filter can either be a single band-pass filter with dual-coupler type or a dual band-pass filter with dual hybrid-coupler type. Figure 7.8 shows the waveguide and planar transmission line versions of the first type of directional filter. The waveguide version (Figure 7.8(a)) [4] uses two TE_{10} mode supporting rectangular waveguides connected to a circularly polarized TE_{11} mode-based coupled circular cavity band-pass filter. At the center frequency of the passband of the band-pass filter, an incident TE_{10} mode signal on port 1 excites circularly polarized TE_{11} mode in the first cavity of the circular cavity band-pass filter. The circularly polarized TE_{11} mode signal in the last (nth) cavity launches a TE_{10} mode signal in the second rectangular waveguide. No power is transmitted to port 3. The frequency response between port 1 and port 4 is the same as that of the band-pass filter.

A complementary band reject response is obtained between port 1 and port 2. The available bandwidth of a directional filter depends on how the coupling slots are implemented. The band-pass filter can be designed using a filter design wizard and CAD optimization procedure [5]. It should be noted that the length of each circular cavity is nearly half a wavelength. Also, the coupling aperture between the

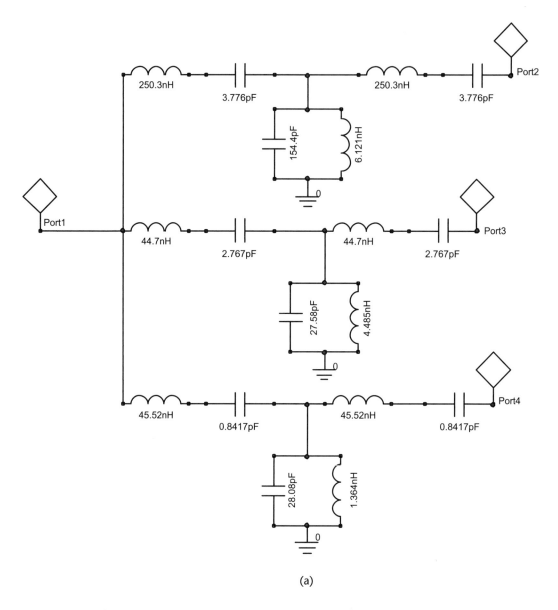

(a)

Figure 7.5 Triplexer network before and after compensation and optimization. (Courtesy of Sandeep Palreddy.)

end cavity and the rectangular waveguide uses offset apertures of special shapes in order to launch pure circularly polarized waves in the cylindrical cavities. However, offset simple rectangular or circular shaped irises can be used for coupling to rectangular input/output waveguides.

The planar form of the directional filter (also called "ring type"), shown in Figure 7.8(b) [1], often needs tuning screws to minimize the effects of the reverse wave that are caused by the right angle bend discontinuities, coupled sections, and step discontinuities. However, if adequate fabrication accuracy that matches optimum design as per modern CAD techniques is guaranteed, no tuning screw will be required after production. Figure 7.8(b) also presents the design equations for the

7.3 Cascaded Directional Filter

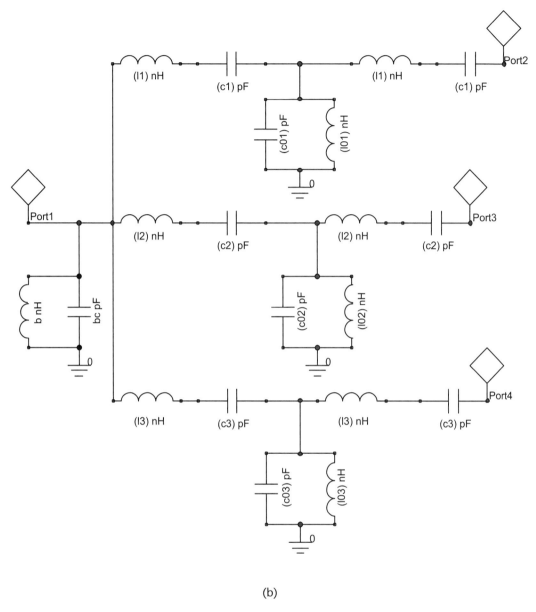

(b)

Figure 7.5 (continued)

filter. Once the even- and odd-mode characteristic impedances of the coupled sections are known, the corresponding line widths and the gaps can be obtained from (3.136)–(3.141) in Chapter 3 (depending on the type of transmission line used). It should be mentioned here that the ring type directional filters can also be realized using waveguides.

Figure 7.9 shows the schematic diagram of a hybrid-coupled directional filter. Such a directional filter can be realized using any type of transmission line. It consists of two identical band-pass filters and two identical 90°, −3 dB hybrid couplers [6]. Figure 7.10 shows the waveguide and planar circuit versions of a typical directional filter. In either structure, each filter can be separately tuned without affecting

Table 7.1 Component Values

Name	Include	Nominal Value
l1	✓	97.5695
l2	✓	61.7481
l3	✓	28.6409
c1	✓	8.26091
c2	✓	1.99395
c3	✓	1.31816
c01	✓	15.2162
c02	✓	29.8598
c03	✓	17.9341
l01	✓	32.2863
l02	✓	4.09344
l03	✓	2.16701
b	✓	9894.92
bc	✓	0.355643

the other filter's frequency response. Also, the dimensional tolerances of the hybrids are not so critical. In the case of waveguide, the branch-guide hybrid can be replaced by a short slot Riblet hybrid [7]. In case of a planar circuit, the branch-guide hybrid can be replaced by 3 dB interdigital couplers [8]. Also, the band-pass filters can be replaced by any other suitable form of planar band-pass filter. Figure 7.11 shows a block diagram of a cascaded directional filter-based multiplexer. Usually, the loading effects on any filter from the adjacent ones are minimal. However, using modern CAD techniques, such effects can always be minimized.

In a hybrid-coupled directional filter approach to multiplexing, one half of the input power passes through each filter. This is extremely advantageous for high-power applications because the filter specifications can be more relaxed. Also, due to minimal interactions between adjacent filters, a modular approach can be used towards realization of the overall multiplexer and additional channel filters can be added easily at any time.

Multiplexers based on conventional directional filter are reasonably compact. However, those consisting of hybrid-coupled directional filters are bulky and heavy for obvious reasons. Since symmetry is a key requirement for maintaining phase balance between different paths, the multiplexer must be fabricated with tight tolerances in order to minimize phase deviation.

7.4 Circulator Based Multiplexer

A circulator is a passive multiport device used in modern RF and microwave equipment. By using circulators the stability, performance, and reliability of a system can be improved and often enabling better and cheaper solutions than other alternatives. In addition, in certain applications, the use of circulators is a must. It is defined as a device with three or more ports, where power is transferred from one port to another in a prescribed order. That means for a three-port circulator (shown in Figure 7.12 [9]), power entering port 1 leaves port 2, port 3 is decoupled; power

7.4 Circulator Based Multiplexer

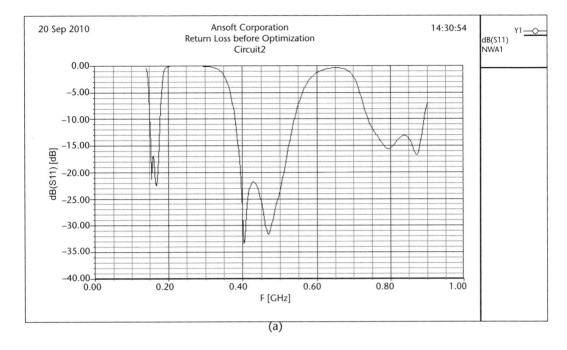

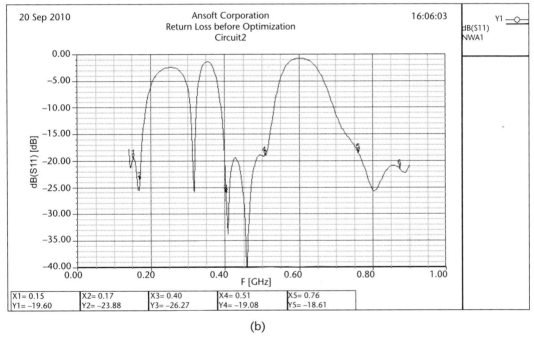

Figure 7.6 Input return loss of the triplexer before the addition of susceptance annulling network and optimization. (Courtesy of Sandeep Palreddy.)

entering port 2 leaves port 3, port 1 is decoupled; and power entering port 3 leaves port 1, port 2 is decoupled.

The operation of a circulator is based on the unique properties of ferrite materials. The reader is referred to [10] for a complete understanding of circulators. The junction circulator is the most common circulator. It is realized in waveguide and

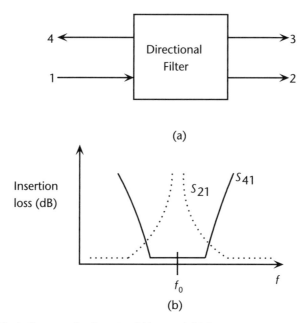

Figure 7.7 (a) Block diagram of a directional filter, and (b) frequency response of a directional filter.

stripline forms. The stripline form of a junction usually has coaxial line interfaces. Figure 7.13 shows the circuit symbol for a three-port junction circulator. Figure 7.14 shows the schematic diagram of a circulator-coupled multiplexer. In terms of amenability to modular integration and assembly, the unidirectional property of a circulator provides the same advantages as the directional filter or hybrid-coupled directional filter approach. However, isolators are inherently lossy.

Consequently, the multiplexer has higher insertion loss. While the first channel suffers the combined insertion loss from one circulator and the filter, the last channel suffers the insertion loss from all of the circulators and the filter. Figure 7.15 shows a practically realized isolator-based, space-qualified multiplexer [11]. Figure 7.16 shows how an extra circulator can be used to put the transmitter and receiver multiplexers on the same antenna [12]. The designer must bear in mind that the transmitter and the receiver multiplexers need to fulfill stringent requirements for intermodulation.

7.5 Manifold Multiplexer

In a manifold multiplexer, each filter channel is attached to a manifold transmission line using T-junctions. Figure 7.17 shows the schematic diagram of a manifold multiplexer. The direct connection of all filters to a single manifold via several T-junctions makes the performance of each channel highly dependent on the performance of the rest of the channels [13]. This makes the entire system very complex. Therefore, a heavy amount of design optimization is required in order to achieve the optimum performance of a multiplexer. Also, unlike the directional filter-based approach and the circulator-based approach, the addition of more channels to an already optimized manifold multiplexer requires a redesign. However, such

7.5 Manifold Multiplexer

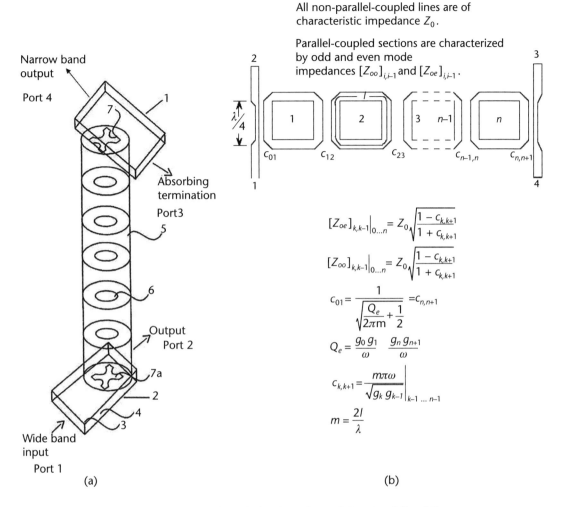

Figure 7.8 (a) Waveguide directional filter [4]. (b) Planar form of directional filter [1].

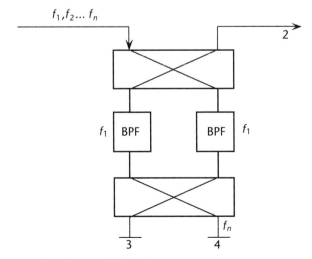

Figure 7.9 Hybrid-coupled directional filter schematic.

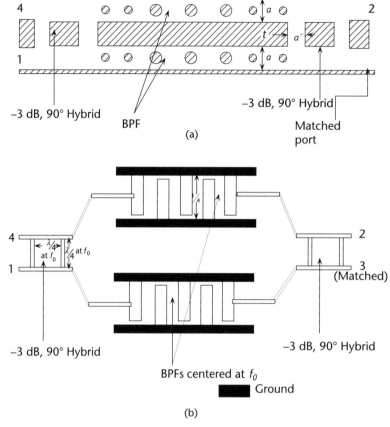

Figure 7.10 (a) Waveguide version of a hybrid-coupled directional filter. (b) Planar circuit version of a hybrid-coupled directional filter.

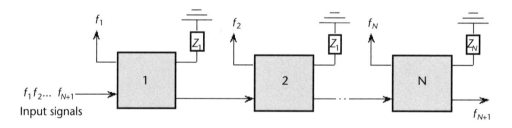

Figure 7.11 Multiplexer based on cascaded directional filter approach.

shortcomings outweigh the advantages of manifold multiplexers, especially in satellite communication systems. Such multiplexers have the smallest weight because they are devoid of extra filters and hybrids as in directional filters or circulators, containing ferrites and magnets, in a circulator coupled multiplexer. In addition, manifold multiplexers offer the lowest insertion loss. That is extremely important in satellite systems where the available dc power is limited.

Over the past several decades, manifold multiplexers were designed using various semianalytical and experimental methods. An excellent account of such methods have been recounted by Matthaei, Young, and Jones [1], various papers [14],

7.5 Manifold Multiplexer

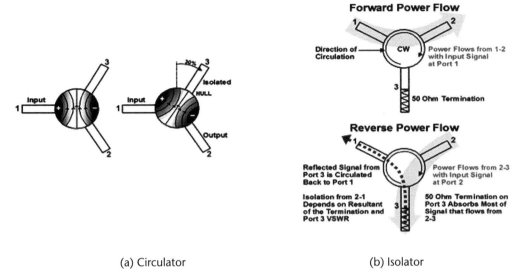

(a) Circulator (b) Isolator

Figure 7.12 Power flow in circulators. (Courtesy of Douglas F. Carlberg, M2Global, San Antonio, TX.)

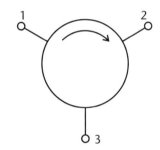

Figure 7.13 Symbol for a three-port circulator.

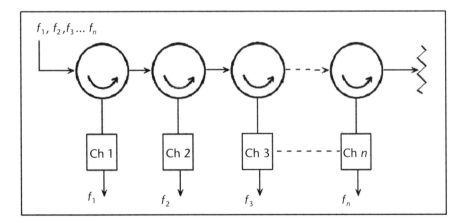

Figure 7.14 Circulator-coupled multiplexer.

and a review article by Cameron and Yu [15]. Today, the most common and effective method for the design and optimization of a manifold multiplexer is based

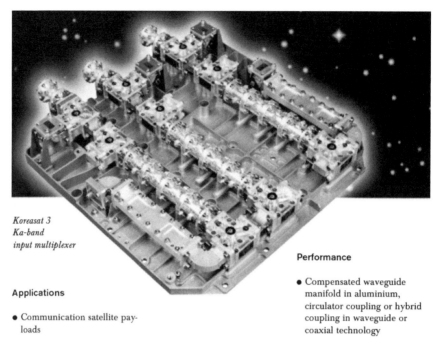

Koreasat 3 Ka-band input multiplexer

Applications

- Communication satellite payloads

Performance

- Compensated waveguide manifold in aluminium, circulator coupling or hybrid coupling in waveguide or coaxial technology

Figure 7.15 Three-channel circulator-based space qualified multiplexer. (From [11].)

on computer optimization [16–18]. While most of the earlier methods started with singly terminated channel filters, modern methods start with doubly terminated filters.

The methods have three basic steps:

1. Designs of channel filters as doubly terminated filters;
2. Determination of locations of a filter from the corresponding T-junction on the manifold and the spacings between two successive T-junctions;
3. Full electromagnetic analysis-based optimization of the scattering matrix of the entire structure.

As in computer aided optimization, the initial guess of the T-junction spacings and the locations of the channel filters play a very significant role in the convergence of the optimization procedure. Morini, Rozzi, and Morelli [19] presented very useful analytical method and closed-form equations for the initial guess of the T-junction spacings and filter locations. The method is described in the following sections.

7.5.1 Diplexer Design

Let S be the scattering matrix of a symmetrical T-junction J shown in Figure 7.18.

At a given frequency, f, it is always possible to minimize the reflection coefficient of the two-port junction formed by terminating one of the ports of the reciprocal and lossless three-port (say, port 2) by a load jX, provided that the load is positioned at a distance of

7.5 Manifold Multiplexer

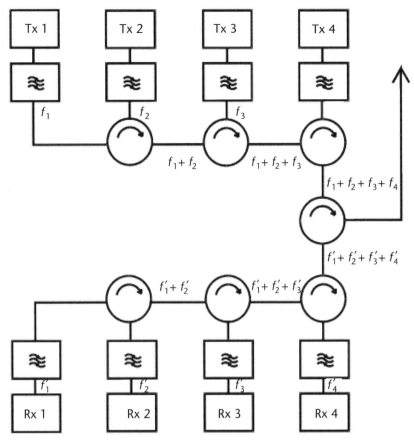

Figure 7.16 Circulator-based transmitter and receiver multiplexers on the same antenna. (From [12].)

$$l(f) = \frac{\psi - \varphi}{2\beta} \qquad (7.2)$$

where $e^{j\psi}$ is the reflection phase of the reactive load, β is the propagation constant of the feed lines, and

$$\phi = 2\tan^{-1}\frac{-b + \sqrt{a^2 + b^2 - c^2}}{c - a} \qquad (7.3)$$

The real quantities a, b, and c are given by the following expressions:

$$a = \frac{a_{33}}{a_{11}}\left(1 + a_{22}^2\right)\sin(\phi_s - \phi_{11} - \phi_{33}) - a_{22}\left\{1 + \left(\frac{a_{33}}{a_{11}}\right)^2\right\}\sin\phi_{22} \qquad (7.4)$$

$$b = \frac{a_{33}}{a_{11}}\left(1 + a_{22}^2\right)\sin(\phi_s - \phi_{11} - \phi_{33}) - a_{22}\left\{1 + \left(\frac{a_{33}}{a_{11}}\right)^2\right\}\cos\phi_{22} \qquad (7.5)$$

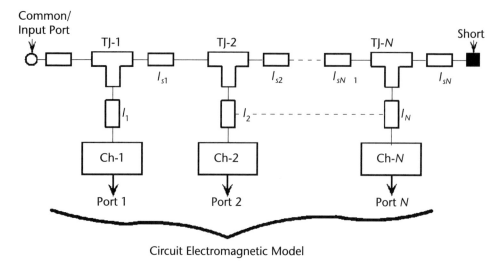

Figure 7.17 Schematic of a manifold multiplexer. (From [13].)

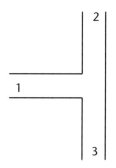

Figure 7.18 Transverse section of T-junction.

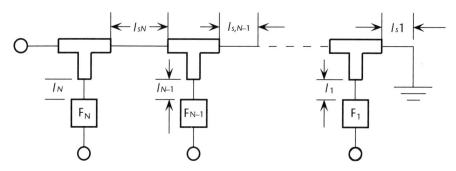

Figure 7.19 Layout of a T-manifold multiplexer.

$$c = \frac{a_{33}}{a_{11}} \sin(\phi_{11} + \phi_{22} + \phi_{33} - \phi_s) \qquad (7.6)$$

7.5 Manifold Multiplexer

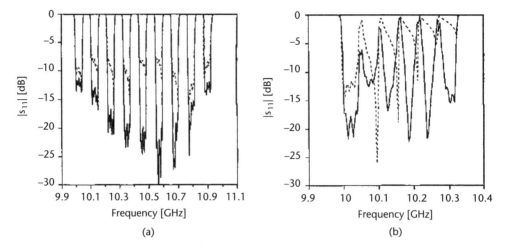

Figure 7.20 (a) The common port reflection response of an unoptimized, nine-channel, noncontiguous multiplexer, employing septum filters and E-plane T-junction manifold. The continuous lines are according to the present method. The dashed lines are obtained using the method described in [20]. (Reprinted with permission from the IEEE.) (b) The common port reflection response of an unoptimized, six-channel, contiguous multiplexer, employing septum filters and E-plane T-junction manifold. The continuous lines are according to the present method. The dashed lines are obtained using the method described in [20]. (Reprinted with permission from the IEEE.)

where $S_{ii} = a_{ii} e^{j\varphi_{ii}}$ are the diagonal terms of the scattering matrix of the three-port junction and $e^{j\varphi_s}$ is the determinant of the matrix computed at frequency, f. The corresponding value of the minimum reflection coefficient is

$$\rho_{min} = a_{11} \left| \frac{1 - \frac{a_{33}}{a_{11}} e^{j(\phi + \phi_s - \phi_{11} - \phi_{33})}}{1 - a_{22} e^{j(\phi + \phi_{22})}} \right| \quad (7.7)$$

Note that the minimum reflection coefficient ρ_{min} is zero only when $a_{11} = a_{33}$. The next step is to realize a diplexer using such a T-junction and two given filters F_1 and F_2. The approach to the design is based on the fact that the out-of-band input impedance of a filter acts as an almost reactive load. Equation (7.2) gives the locations l_2 and l_2 of the filters F_1 and F_2, respectively, with the phase ψ appearing in the equation chosen as $\angle S_{11}^{F_1}(f_2)$ and $\angle S_{11}^{F_2}(f_1)$, respectively. Since the required T-junction scattering matrix conditions are fulfilled at the center frequencies of the two filters, the diplexing effects of the junction degrades near the band edges of the filters. However, the frequency response remains acceptable for all practical purposes.

7.5.2 Multiplexer Design

Multiplexer design is an extension of the diplexer design approach. Let us consider an N channel multiplexer consisting of N band-pass filters F_i ($i = 1, 2, ..., N$) centered at fi such that $f_i < f_i + 1$. The filters are connected to a cascade of N identical

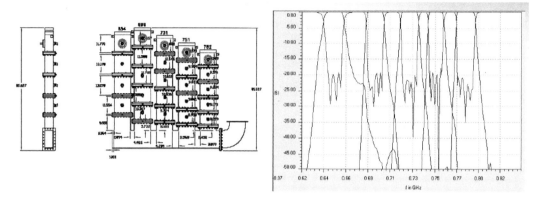

Figure 7.21 Five-channel multiplexer for TV broadcasting and the optimized frequency response (dimensions are in inches).

T-junctions of scattering matrix S^T, as shown in Figure 7.19. The distances l_k of the filters from the junction and the separations l_{sk} of the T-junctions are calculated as follows:

- l_k is computed so that a matched two-port is obtained between ports 3 and 2 of the kth junction at frequency f^* (the arithmetic mean of the center frequencies of all the channels), when port 1 is closed on F_k. In that case, $e^{j\psi} = S_{11}^{F_k}(f^*)$.
- l_{sk} is the distance between port 3 of the kth junction and the reactive load jX that minimizes the reflection of the resulting two-port at frequency f_k for $k > 1$, jX is the input impedance seen to the right of port 2 of $(k-1)$th junction calculated at frequency f_k; for $k = 1$, $jX = 0$, being the input impedance of a short circuit.

The above approach disregards multimode interactions among the involved T-junction in manifold multiplexer. Consequently, the designer has to carry out a multimode analysis-based optimization of the entire multiplexer with the initial values obtained by that method.

Figure 7.20 shows computed responses of a noncontiguous and contiguous multiplexer designed using this method [20]. In contiguous multiplexers the interactions among the constituent filters are more pronounced, and overall optimization is a must in all cases. Figure 7.21 shows a fabricated 5-channel waveguide multiplexer for TV broadcasting.

References

[1] Matthaei, G. L. Young, and E. M.T. Jones, *Microwave Filters, Impedance Matching Networks and Coupling Structures,* Norwood, MA: Artech House Inc, 1980.

[2] Rhodes, J. D., and R. Levy, "Generalized Multiplexer Theory," *IEEE Trans. Microw. Theory Techn.,* Vol. 27, No. 2, February 1979, pp. 99–111.

[3] Ansoft Designer, Ansoft Corporation, Pittsburgh, PA, 2008.

[4] Broad, G. J., N. A. McDonald, and C Williams, Waveguide Directional Filters, U.S. Patent 2005/0231301 A1, Alcatel Paris.

[5] WASP-NET Filter Wizard, MIG, Bremen, Germany, 2010.

[6] Matthaei, G., L. Young, and E. M. T. Jones, *Microwave Filters, Impedance Matching Networks and Coupling Structures,* Norwood, MA: Artech House, 1980, pp. 967–968.

[7] Riblet, H. J., "Short-Slot Hybrid Junction," *IRE Proc.,* Vol. 40, 1952, pp. 180–184.

[8] Pressure, A., "Interdigiteted Microstrip Coupler Design," *IEEE Trans. Microw. Theory Techn.,* Vol. 26, No. 10, October 1978, pp. 801–805.

[9] Phillips Application Notes, Circulators and Isolators, Unique Passive Devices, AN98035, March 23, 1998.

[10] Helszajn, J., *Waveguide Junction Circulators,* New York: John Wiley and Sons, 2009.

[11] Passive Components for Space Applications, BOSCH, Bosch Telecom GmbH, Germany.

[12] http://www.m2global.com/esources.circulators-isolators.

[13] Zhang, Y., Modeling and Design of Microwave and Millimeter Wave Filters and Multiplexers, Ph.D. thesis, University of Maryland, College Park, MD, 2006.

[14] Crystal, E. G., and G. L. Matthaei, "A Technique for the Design of Multiplexers Having Contiguous Channels," *IEEE Trans. Microw. Theory Techn.,* January 1964, Vol. 10, No. 1.

[15] Cameron, R. J., and M. Yu, "Design of Manifold Coupled Multiplexers," *IEEE Microwave Magazine,* October 2007, pp. 46–59.

[16] Guglielmi, M., "Simple CAD Procedure for Microwave Filters and Multiplexers," *IEEE Trans. Microw. Theory Techn.,* Vol. 42, No. 6, June 1994, pp. 1347–1352.

[17] Rosenberg, U., D. Wolk, and H. Zeh, "High Performance Output Multiplexers for Ku-Band Satellites."

[18] 13 AIAA International Communication Satellites Systems Conference, Los Angeles, pp. 747–752, March 1990.

[19] Morini, A., T. Rozzi, and M. Morelli, "New Formula for the Initial Design in the Optimization of T-Junction Manifold Multiplexers," *IEEE MTT-S International Microwave Symposium Digest,* 1997, pp. 1025–1028.

[20] Uher, J., J. Bornemann, and U. Rosenberg, *Waveguide Components for Antenna Feed Systems,* Norwood, MA: Artech House, 1993.

About the Authors

Dr. Protap Pramanick received his B.Tech in Electronics and Communication Engineering from IIT (Kharagpur) and PhD in microwave engineering from IIT [Kanpur], India, in 1982. In 1984, he joined Canadian Marconi Company in Ottawa, Canada, as a senior design specialist in the Avionics Division. In 1986, he moved to COMDEV in Cambridge, Ontario. He served as associate professor from 1992 to 2000 in electrical engineering at the University of Saskatchewan, Saskatoon, and then as a filter design engineer at K & L Microwaves in Salisbury. He currently works at Microwave Engineering Corp. in Andover, MA, with design responsibility for the company's filter and antenna products. He has acted as an independent consultant and filter and passive microwave component designer for Motorola, Ericsson, K&L Microwave, M2Global Incs., and others.

Dr. Pramanick co-edited: E-*Plane Integrated Circuits* [Artech House, 1987] and contributed chapters to *Topics in Millimeter Wave Technology* [Academic Press, 1988], *Microwave Solid State Circuit Design* [Wiley, 1995 and 2006], and *The Handbook of Microwave Engineering* [Academic Press, 1995] and *Encyclopedia of Electrical Engineering* [Wiley, 2002]. He holds several patents on filters for cellular radios.

Dr. Prakash Bhartia graduated with a B. Tech. (Hons.) from IIT, Bombay, and a Ph.D. from the University of Manitoba, Winnipeg. Over a 25-year career with the Department of National Defense in Canada, Dr. Bhartia has held four director-level and two director general level positions. Dr. Bhartia has published extensively with more than 200 publications. He also has five patents and nine books to his credit. He was appointed to the Order of Canada in 2002, and is a Fellow of the Royal Society of Canada, an IEEE Fellow, and a Fellow of The Engineering Institute of Canada, The Canadian Academy of Engineer, and the Institute of Electronic and Telecommunication Engineers. Dr. Bhartia received the IEEE McNaughton Gold Medal for his contributions to engineering. He is currently the executive vice president of Natel Engineering in Chartsworth, CA.

Index

A

ABCD matrix
   of cascaded network, 35
   cascaded UEs, 210
   in computing network properties, 36–37
   defined, 34
   impedance inverter, 187
   multiport network representation, 35
   normalized, 37–40
   ridged waveguide low-pass filter, 221
   slot, 221
   two-port network, 36
   unnormalized in common networks, 38–39
   usefulness of, 35
Admittance inverter
   physical realization of, 186–89
   transmission line, 186
Admittance inverter-coupled Butterworth filter, 182
Admittance matrix
   defined, 17–18
   obtaining, 18
   of two-port networks, 20
Attenuation
   Chebyshev low-pass filters, 163
   in circular waveguides, 112–16
   in coupled striplines, 78–79
   peak, inside stop band, 252
   in rectangular waveguides, 113
Attenuation constant
   striplines, 66–67
   transmission lines, 41
Attenuation poles, 10

B

Band-pass filters
   based on coupling matrix, 295–303
   capacitively coupled, 274, 275–344

coaxial cavity, 303, 304–5
combline, 303–10
coupled resonators, 266
cross-coupled design with dual-mode resonators, 330–34
cross-coupled design with independently controlled transmission zeros, 345–46
cross-coupled design with planar transmission lines, 338–44
cross-coupled resonator design, 319–30
dielectric resonator, 349–68
Dishal's method of design, 303
edge parallel-coupled, 277–78
evanescent-mode waveguide design, 314–19
evolution from low-pass prototype, 269
extracted pole, 273
four-pole, 362
gap-coupled transmission line, 275–77
hairpin-line, 284–89
input/output coupling capacitors, 265
interdigital, 289–92
interdigital, capacitively loaded, 292–95
K-inverter-coupled prototype, 274
multiple-coupled resonator, 295–97
narrowband lumped element inverter, 274
prototype frequency response, 265
realizable response, 3
realization of, 271
theory of, 265–75
transformation of low-pass filter to, 267, 268, 270, 271
waveguide design, 310–14
See also Filters
Band-stop filters
   capacitor stub-coupled narrowband, 248
   defined, 243
   design, 243–63
   dielectric resonator, 353
   effects of dissipation loss in, 252

407

Band-stop filters (continued)
    equivalent circuit, 252
    frequency response, 244, 251
    inverter-coupled prototype, 247–48
    K-inverters, 271
    low-pass filter transformation, 243–51
    peak attenuation, 252
    prototype element values, 245
    prototype networks, 245
    reactance slope parameter, 246, 249–50
    realizable response, 3
    realization of resonant circuits, 249
    reflection coefficient, 252
    resonant electrical length, 251
    response for finite conductivity, 253
    slope parameters, 246
    stripline design, 250, 251
    stripline layout, 256
    stripline response, 256
    stub impedance, 251
    stub-loaded, 253–58
    susceptance slope parameter, 250
    waveguide, 257, 258–63
    See also Filters
Belevitch matrix, 148–49
Bend discontinuity, 73–74
Broadside-coupled striplines
    defined, 79
    even mode impedance, 80
    illustrated, 79
    loose coupling, 81, 82
    offset, 81–83
    slot, 83–85
    suspended, 102–5
    synthesis of, 80–81
    tight coupling, 81–83
    See also Striplines
Butterworth filters
    admittance inverter-coupled, 182
    approximation, 150–53
    Cauer network realization, 159
    in complex frequency, 151
    cutoff frequency, 157
    distributed elements, 215, 216
    first Cauer form, 156
    frequency response, 150–51, 157
    general solution for response, 157–61

Nth-order inverter-coupled, 181–82
    poles of, 151
    power reflection coefficient, 157
    required order of, 160
    second-order, time delay response, 161
    third-order, 156
    time delay of, 161
    See also Low-pass filters

C

Capacitively coupled band-pass filters
    based on coupling matrix, 295–303
    capacitively loaded interdigital, 292–95
    coaxial cavity, 303, 304–5
    combline, 303–10
    cross-coupled design with dual-mode
        resonators, 330–34
    cross-coupled design with planar
        transmission lines, 338–44
    cross-coupled resonator design, 319–30
    distributed transmission line form of,
        275–344
    edge parallel-coupled, 277–84
    evanescent-mode waveguide design, 314–19
    gap-coupled transmission line, 275–77
    hairpin-line, 384–89
    interdigital, 289–92
    schematic diagram, 274
    waveguide design, 310–14
    See also Band-pass filters
Capacitively loaded interdigital filters
    advantage of, 292
    design approach, 292–95
    equivalent circuit, 294
    illustrated, 294
    realization of, 295
    See also Interdigital filters
Cascaded directional filter
    block diagram, 394
    defined, 388
    dual band-pass filter, 389
    frequency response, 394
    hybrid-coupled, 391, 392, 395, 396
    performance, 389
    planar form, 390
    ring type, 390

single band-pass filter, 389
waveguide, 395
Cascaded UE low-pass filters, 219, 220
Cauer network realization, 159, 160
Cauer synthesis, 153–57
  Cauer I, 152, 154–55
  Cauer II, 152
  with continued fraction expansion, 158
Causal networks, 8–9
Chain matrix. *See* ABCD matrix
Channel combiners, 4
Characteristic impedance
  circular waveguides, 112
  coaxial lines, 47
  coupled microstrip line, 95–96
  definition of, 17
  open circuit transmission line, 208
  rectangular coaxial lines, 60
  shielded microstrip line, 89, 91
  striplines, 62–64
  suspended microstrip line, 96–97
  transmission lines, 41, 42
  transmission line theory, 15
Chebyshev filters
  12-pole, 376
  approximation, 161–68
  attenuation, 163
  cutoff frequency, 163, 209
  defined, 162
  distributed elements, 215, 216
  folded, 377
  fourth-order, 163
  frequency response, 164–65
  generalized, 175–78
  input admittance, 389
  locus of poles, 166
  lumped element, 209
  Nth-order inverter-coupled, 182–83
  parameters, 164
  prototype, 197, 230
  required order of, 164–67
  response factors, 164
  third-order, time delay response, 168
  time delay of, 168
  transfer function, 166
  See also Low-pass filters
Chebyshev minimax theorem, 161

Chebyshev polynomials
  defined, 161–62
  higher-order, 162
  important properties of, 162
Circular disk capacitor, 234–37
Circular hole discontinuity, 71
Circular waveguides
  analytical equations, 113
  attenuation in, 112–16
  characteristic impedance, 112
  cutoff frequency, 110
  cutoff wavelengths, 110
  discontinuities, 120–22
  frequency dependence of Q-factor in, 114
  illustrated, 106
  mode sequence, 110
  power-handling capability, 119–20
  TE and TM mode fields, 111
Circulator based multiplexers, 392–94
  illustrated, 397
  receiver, 399
  transmitter, 399
  See also Multiplexers
Circulators
  defined, 392
  operation of, 393
  power flow in, 397
  three-channel, 398
  three-port, symbol for, 397
Coaxial cavity band-pass filters, 303, 304–5
Coaxial line high-pass filters, 236
Coaxial lines
  cascaded UE low-pass filters, 220
  characteristic impedance, 47
  characteristics, 48
  conductor loss, 52
  conductor Q, 52
  cross section, 47
  discontinuities, 54–59
  discontinuity illustration, 58
  field pattern of TEM mode, 49
  formulas for derivatives, 51
  gap discontinuity, 58, 59
  general equation for attenuation, 50–52
  loss calculation, 51
  maximum power-handling capability of, 52–54, 55

Coaxial lines (continued)
  order of cutoff frequencies, 50
  overall attenuation, 52
  Q-factor of, 46–50
  rectangular, 59–61
  standard, attenuation in, 53
  step discontinuity, 56, 58
  T-junctions, 58, 59
  transverse field patterns, 49
  See also Transmission lines
Combline filters
  compensating capacitance, 305
  defined, 303
  design equations, 303–8
  design example, 308–10
  dimensions, 310
  equivalent network, 306
  frequency response, 309
  practical schematic, 307
  Q-factor, 309
  quarter-wavelength resonator, 308
  schematic, 306
  size, 306
  slot-coupled coaxial-resonator-based, 309, 373–74
  specifications, 308
  square-rod, 308–10
  stopband, 306
  terminating capacitors, 305
  See also Band-pass filters
Commensurate line filters
  analytical formulas, 213
  defined, 213
  line length and impedance effects, 212–33
  transfer function, 213
Complex frequency, 5–6
Corrugated waveguide harmonic reject filter
  frequency response, 229
  longitudinal slots, 228–29
  use of, 228
Coupled microstrip line
  characteristic impedance, 95–96
  cross section, 92
  defined, 92
  frequency dependence, 93–95
  full-wave analysis programs, 96

  static odd-mode effective dielectric constant, 93
  See also Microstrip lines
Coupled resonator filters
  circuit elements, 347
  computed reflection group delays, 349
  reflection group delay responses, 347, 348
  reflection group delays, 348
  unified approach to tuning, 346–49
Coupled slablines
  analysis equations, 86
  defined, 85
  illustrated, 85
  optimization scheme, 86–87
Coupled striplines
  analysis curves, 77
  attenuation in, 78–79
  broadside, 79–85
  edge, 74–77
  synthesis equations, 77–78
  See also Striplines
Coupling coefficient, 31
Coupling matrix, 128
Cross-coupled filters
  hairpin-line, 339
  with independently controlled transmission zeros, 345–46
  with planar transmission lines, 338–44
  purpose of, 357
Cross-coupled low-pass prototype
  conventional method comparison, 193
  cost function, 191
  coupling coefficients, 191
  coupling values, 190
  defined, 189
  derivative-based optimization, 191–92
  generalized network, 191
  group delay, 192
  topology matrix, 191
  two-port scattering parameters, 191
Cross-coupled planar filters
  admittance slope parameter, 339–41
  coupled modified hairpin resonators, 340
  coupling coefficient, 341, 342
  coupling matrix, 343
  defined, 338–39
  design method, 339

design steps for, 343–44
frequency responses, 344
mixed coupled modified hairpin resonator configurations, 340
modified hairpin-line, 339
tapped input/output locations, 343
tap-point location, 341
Cross-coupled resonator filters
advantages of, 320
basic resonator orientation, 328
Chebyshev, 321
circuit arrangement, 329
computed frequency response, 325, 326, 330
coupling matrix, 320, 322, 323, 324, 327
coupling topology, 325, 326
coupling values, 327
curve fitting, 329
design, 319–30
design with dual-mode resonators, 330–34
dual-resonator orientation, 328
E-plane folded, 337
equivalent network, 326–27
frequency response, 323, 331
generalized network, 320
general multicoupled network, 325
H-plane folded, 336
interresonator couplings, 328
iris opening and, 328
location of irises, 329
normalized frequency, 320
physical structure, 322
quintuplet-coupling configurations, 324
scattering parameters, 321
specifications, 326–27
See also Band-pass filters
Current
backward, 16
defined, 14
equivalent, 13–17
Cutoff frequency
Butterworth filters, 157
Chebyshev filters, 163, 209
circular waveguides, 110
coaxial lines, 49, 50
defined, 1

electrical length of transmission line at, 215
elliptic filters, 169
error, 218
fundamental mode, 16, 137
higher-order mode, 60
high-pass filters, 234, 240
illustrated, 2
lowest-order TE mode, 91
propagation constant and, 109
quasi-distributed element TEM filters, 230
striplines and, 62
waveguide high-pass filter, 242
waveguides, 129, 228
Cutoff wavelength
circular waveguides, 110
fundamental mode, 137
striplines, 63

D

Darlington's method
driving-point impedance, 201
realization, 203
realized network, 204
steps, 200
synthesis, 201–6
z-parameters, 203
Derating factor, 119
Dielectric resonator filters
computed frequency response, 362
configurations, 359–60
coupled pair of resonators, 359
coupling curve, 355, 357
coupling matrix, 368
designed filter, 361
design of, 353–68, 380–83
dual-mode, 353, 357, 363, 364
effect of finite Q, 350
E-field distribution, 366
equations for resonance frequencies, 353
feeding angles, 363
field distribution, 365
four-pole, 362, 367
horizontally oriented puck, 354
introduction to, 349–51
isolated circular cylindrical resonators, 350

Dielectric resonator filters (continued)
  measured response, 362
  in microwave communication engineering, 363
  modes and mode nomenclature, 351–52
  multiresonator couplings, 367
  one-port simulation configuration, 360
  principle modes, 352
  realization of, 355
  resonance frequencies from simulation, 361
  resonance frequency adjustment, 355
  resonant frequencies, 352–53
  resonator in below-cutoff metallic enclosure, 359
  resonator modes, 351–68
  simulation of resonance, 356
  types of, 353
  vertically oriented puck, 354
  See also Band-pass filters
Diplexers
  block diagram, 4
  defined, 2
  design, 398–401
  See also Multiplexers
Directional couplers
  insertion loss, 31
  primary line, 30
  reflection meter, 32
  scattering matrix, 32
  schematic diagram, 30
Discontinuities
  bend discontinuity, 73–74
  circular hole discontinuity, 71
  coaxial lines, 54–59
  E-plane, 135–37
  gap discontinuity, 70–71
  H-plane, 133–35, 138
  inhomogeneous transmission lines, 103–5
  iris, 131
  microstrip and suspended stripline, 103–5
  open-end discontinuity, 71–72
  slot, 222
  step discontinuity, 58, 69–70, 123
  striplines, 67–74
  three-dimensional, 137–41
  T-junction discontinuity, 72–73
  in waveguides, 120–22

Distributed circuits
  band-stop filter design, 243–63
  commensurate line filters, 212–33
  cutoff frequency error, 218
  design example, 217–19
  lumped elements and, 207–10
  quasi-distributed element TEM filters, 230–33
  theory of, 207–63
  TLE/UE and Kuroda identity, 210–12
Distributed elements
  categories, 214
  equivalence of lumped elements, 207–10
  input impedance, 207
Diversity, 31
Driving-point impedances
  Darlington synthesis, 201
  impedance inverter, 179
  realizable, 10–11
Dual-mode circular cavity filter, 357, 363
Dual-mode cross-coupled filters
  circular cavity, 331
  coupling coefficient, 332–34
  design dimensions, 334, 335
  equivalent circuit, 332
  K-inverter values, 331–32
  McDonald's method, 334
  operation of, 330
  optimization-based iteration scheme, 332, 333
  tilted iris, 332
  See also Cross-coupled resonator filters
Dual-mode dielectric resonator filters
  in applications, 353
  in circular enclosure, 364
  dual-mode circular cavity filter and, 357
  illustrated, 363
Duplexers, 119

E

Edge-coupled striplines, 74–77
Edge-coupled suspended microstrip lines, 100–102
Edge parallel-coupled band-pass filters
  defined, 277
  design equations, 283–84

drawback, 284
within enclosure, 282
equivalent network, 280
first coupled pair of lines, 277
illustrated, 280
input file for, 281–82
intermediate coupled pair of lines, 278
outermost coupled pair of lines, 279
planar, 280
stepped impedance, 283
See also Band-pass filters
Electrical discontinuity, 44–45
Electrical filters, 11
Elliptic filters
12-pole, 378
approximation, 168–75
cutoff frequency, 169
design specifications, 172
discrimination parameter, 169–70
frequency response, 170, 175
input impedance, 174
input reflection coefficient, 174
inverter-coupled prototype for, 184
normalized transducer power gain, 173
normalized transfer function, 169
order of, 172–75
prototype, 197, 230
synthesis of, 169
synthesized, 174
theory of approximation, 169–72
transfer function, 169
See also Low-pass filters
Elliptic integral function, 171
E-plane folded cross-coupled resonator filters, 337
E-plane iris-coupled stub-loaded band-stop filter, 261
E-plane waveguide discontinuities, 135–37
Equal ripple filter, 215
Equivalent current, 13–17
Equivalent voltage, 13–17
Error function, 194
Evanescent-mode waveguide band-pass filters
advantages of, 315–16
computed frequency response, 318, 319
conventional, 319
defined, 314
double-ridge, 317
equivalent circuit, 318
interposed inductive strips, 319
single-mode approach, 318
single-ridge, 317
stopband width, 318
See also Band-pass filters
Extracted pole low-pass filters
circuit configuration, 198
frequency response, 198
network of, 273
response, 272

F

Filters
cascaded directional, 388–92
combline, 303–10
commensurate line, 212–33
corrugated waveguide harmonic reject, 228, 229
cutoff frequency, 1
defined, 1
electrical, 11
fabrication of, xi
noncommensurate line, 213
quasi-distributed element TEM, 230–33
types of, 1
waffle-iron, 229
See also Band-pass filters; Band-stop filters; High-pass filters; Low-pass filters
Finite element modal expansion method
E-plane waveguide discontinuities, 135–37
H-plane waveguide discontinuities, 133–35
First Cauer form, 152, 154–55
FMINCON, 196
Folded resonator cross-coupled filters
defined, 334
design steps for, 334–37
E-plane, 337
frequency response, 338
H-plane, 336
See also Cross-coupled resonator filters
Four-pole band-pass filters, 362
Frequency scaling, 230–31

G

Gain-bandwidth integral restriction, 4
Gap-coupled transmission line band-pass filters, 275–77, 278
Gap discontinuities, 58, 59, 70–71
Generalized Chebyshev low-pass filters
    conservation of power, 176
    degrees of freedom, 175
    frequency response, 179
    operation theory, 175
    recursive procedure, 177–78
Generators, 147
Group delay
    coupled resonator filters, 347–49
    cross-coupled low-pass prototype, 192

H

Hairpin-line filter
    cross-coupled, 339
    design equations, 285–86
    illustrated, 284
    input and output connections, 286
    interdigital filter, 289
    normalized capacitance matrix, 286
    resonator degeneration, 290
    tapered, 288
    tap-point location, 341
    WAVECON input file, 287–88
    See also Band-pass filters
Heat transfer, power equation for, 118
Helmholtz equations, 108, 135, 136
High-pass filters, 240
    characteristic of, 2
    coaxial line, 236
    cutoff frequency, 234
    design, 233–43
    homogeneous distributed, 238
    Levy's procedure for, 237–41
    low-pass transformation, 233–34, 235
    quasi-lumped element, 234–37
    realizable response, 3
    waveguide design, 242–43
Hollow metallic waveguides
    cutoff wavenumber, 109
    defined, 105
    electric and magnetic field components, 109
    illustrated, 106
    TE and TM mode fields, 107, 111
    See also Waveguides
Homogeneous distributed high-pass filters, 238
H-plane folded cross-coupled resonator filters, 336
H-plane waveguide discontinuities, 133–35
Hurwitz polynomial, 157, 201
Hybrid-coupled directional filter, 391, 392, 395, 396

I

Impedance inverters
    ABCD matrix, 187
    absorption of negative elements, 188
    compensated transmission line, 187
    concept of, 178–89
    coupled elliptic low-pass filter, 184
    driving-point impedance, 179
    extraction of, 180
    extraction of second, 181
    input impedance, 179
    mixture with UEs, 219–29
    Nth-order, 181–82
    physical realization of, 186–89
    schematic diagram, 179
    terminated by load, 179
    values, 223, 227
    waveguide iris discontinuity, 189
Impedance matching networks, 4–5
Impedance matrix, 19
Impedances
    driving-point, 10–11, 179, 201
    input, 174, 179, 201, 207, 208
    See also Characteristic impedance
Impedance scaling, 230
Inhomogeneous coupled line section, 238, 239
Inhomogeneous transmission lines
    broadside-coupled suspended striplines, 102–3
    coupled microstrip line, 92–96
    defined, 87
    discontinuities, 103–5

edge-coupled suspended microstrip lines, 100–102
illustrated, 87
shielded microstrip line, 88–92
shielded suspended microstrip line, 98–100
suspended microstrip line, 96–98
See also Transmission lines
Input admittance, 182
Input impedance
distributed elements, 207
elliptic low-pass filter, 174
impedance inverter, 179
one-port network, 201
open circuit transmission line, 208
Insertion loss
directional couplers, 31
method, 145–48
scattering matrix and, 28–30
Interdigital filters
capacitively loaded, 292–95
configurations, 291
defined, 289
equivalent circuit, 292
microstrip, 297–303
slabline, WAVECON input file, 293–94
tap point location, 291
See also Band-pass filters
Iris analysis, 130–32
Iris-coupled waveguide band-pass filters, 312

J

Jacobian sine elliptic function, 171

K

K-inverters, 187, 221, 226, 271
Kuroda's identity, 211–12

L

Ladder networks, 159, 167
realization of, 178
synthesizing from input admittance, 182
LaTourrette, Peter, 237
Levy's procedure, 237–41
equivalent circuit, 241
inhomogeneous coupled line section, 238, 239
step-by-step, 240
Linear time-invariant networks, 6–8
Lossless nonreciprocal networks, 27
Low-pass filters
band-stop filter transformation, 243–51
as basic building block, 145
Belevitch matrix, 148–49
Butterworth, 150–61
Cauer synthesis, 153–57
Chebyshev, 161–68, 175–78
design, 145–97
distributed prototype ridged waveguide, 222
elliptic, 168–75
extracted pole, 198
general response, 234
to high-pass transformation, 233–34, 235
ideal and distributed response comparison, 214
impedance inverter, 178–89
insertion loss method of design, 145–48
midseries configuration, 204, 205
overview, 145
realizable response, 3
ridged waveguide, 221–23
singly terminated, 388
stripline stepped impedance, 217
tapered corrugated, 224, 225–27
transformation to band-pass filter, 267, 270, 271
Low-pass prototypes
band-pass filter evolution from, 269
Chebyshev filter, 230
Chebyshev response, 197
comparison, 196
with cross-coupled networks, 189–93
elliptic filter, 230
elliptic response, 197
frequency responses, 197
midshunt, 205, 206
with mixture of UEs and impedance inverters, 219–29
with optimization and MLS, 193–98
UE, 217
Lumped element Chebyshev low-pass filters, 209
Lumped element inverters, 187, 188

M

Manifold multiplexers
  defined, 394
  design of, 396
  diplexer design, 398–401
  five-channel, for broadcast TV, 402
  minimum reflection coefficient, 401
  multiplexer design, 401–2
  optimization, 397–98
  port reflection response, 401
  schematic, 400
  shortcomings and advantages, 396
  T, layout, 400
  T-junction, 400
  See also Multiplexers
MATLAB toolbox, 196
Maximum power-handling capability
  of circular waveguides, 119–20
  of coaxial lines, 52–54, 55
  of waveguides, 116–19
McDonald's method, 334
Method of least squares (MLS)
  for adaptive filters, 193
  cascaded connection of shunt and series units, 196
  error function, 194
  features, 193
  flowchart, 195, 196
  in low-pass prototypes, 193–98
  optimization, 196
Microstrip interdigital filter
  arrangements of coupled pair of resonators, 299
  computed frequency response, 302
  conclusions, 302–3
  coupling as function of resonator spacing, 300
  final layout, 302
  illustrated, 298
  low-pass prototype parameters, 298–300
  resonator coupling coefficient, 298
  resonator spacings, 298–300
  specifications, 297
  structure for Q determination, 301
  tap-point locations, 300–303
Microstrip lines
  coupled, 92–96
  edge-coupled suspended, 100–102
  shielded, 88–92
  shielded suspended, 98–100
  suspended, 96–98
Microwave filters. *See* Filters
Microwave network theory
  ABCD matrix, 34–40
  equivalent voltage and current, 13–17
  impedance and admittance matrices, 17–19
  introduction to, 13
  measurement of scattering matrix, 30–34
  scattering matrix, 19–30
Microwave transmission lines. *See* Transmission lines
Midseries network, 204, 205
Midshunt low-pass prototype, 205, 206
Mixed UE inverter, 221, 222
Mode matching method
  analysis of an iris, 130–32
  coupling matrix, 128
  offset step junction, 132, 133
  TE and TM mode cutoff frequencies and, 128
  waveguide discontinuities and, 124–32
Multiple-coupled resonator filter
  coupling coefficient, 297
  generalized network, 296
  impedance matrix, 296
  network equation, 295
  scattering parameters, 296–97
Multiplexers
  cascaded directional filter, 388–92
  circulator based, 392–94
  configurations, 385
  contiguous, 386
  defined, 2, 385
  design of, 385–402
  directional filter approach, 396
  manifold, 394–402
  shunt connected, 387
  with susceptance annulling network, 386–88
Multiport networks
  ABCD matrix, 35
  passive components, 11
  scattering matrix of, 21–22
  termination by different impedances, 22

Multiresonator couplings inline configuration filter, 367

N

Network analyzers, 34
Network characteristics, determination of, 13
Noncommensurate line filters, 213
Nonsinusoidal voltages, 6–8
Normalized ABCD matrix, 37–40
N-port network, 17
Nth-order inverter-coupled Butterworth low-pass filter, 181–82
Nth-order inverter-coupled Chebyshev filters, 182–83
Numerical electromagnetic (EM) analysis, 57

O

Offset step junction, 132, 133
One-port networks
    illustrated, 21
    input impedance, 201
    parameters, 9
Open circuit transmission line
    characteristic impedance, 208
    input impedance, 208
Open-end discontinuity, 71–72
Optimization
    dual-mode cross-coupled filters, 332–33
    in low-pass prototypes, 193–98
    manifold multiplexers, 397–98
    method of least squares (MLS), 196
Organization, this book, xi–xiii

P

Paley-Wiener criterion, 2, 5
Parallel-coupled lines
    attenuation in, 78–79
    broadside-coupled striplines, 79–85
    coupled-slab lines, 85–87
    defined, 74
    edge-coupled striplines, 74–77
    synthesis equations, 77–78
    See also Transmission lines
Passive networks, 8
Peak power handling capability, 41

Planar edge parallel-coupled band-pass filters, 280
Propagation constant
    striplines, 65
    transmission lines, 41, 42
    waveguides, 109

Q

Quadruplexers, 2
Quasi-distributed element TEM filters
    defined, 230
    frequency scaling, 230–31
    impedance scaling, 230
    shunt capacitances, 231
    WAVECON analysis and design file, 232
    WAVECON layout and frequency response, 233
Quasi-lumped element high-pass filters
    circular disk capacitor, 234–37
    coaxial line, 236
    defined, 234
    impedance and admittance inverters, 236

R

Reciprocal networks
    defined, 8
    two-port, 35
Rectangular coaxial lines
    characteristic impedance, 60
    cross section, 59
    higher-order modes in, 60
    square, 60–61
    use of, 59
    See also Coaxial lines
Rectangular waveguides
    analytical equations, 113
    asymmetric step discontinuity, 123
    attenuation in, 113
    discontinuities, 120–22
    EH-plane step junction, 124
    E-plane multiport, 135
    frequency dependence of Q-factor in, 114
    high-pass filters, 243
    H-plane iris discontinuity, 187
    H-plane multiport, 134
    illustrated, 106

Rectangular waveguides (continued)
  maximum power-handling capability, 116
  mode sequence in, 107
  TE and TM mode fields, 107
  three-dimensional discontinuities, 137–41
  See also Waveguides
Reference planes shift
  illustrated, 25
  scattering matrix transformation due to, 24–27
Reflection coefficient
  band-stop filters, 252
  defined, 21
  elliptic low-pass filter, 174
  of one-port device, 33
  open-end discontinuity, 72
  ridged waveguide low-pass filter, 223
Reflection meters, 32
RF filters. *See* Filters
Ridged waveguide low-pass filter
  ABCD matrix, 221
  distributed prototype, 222
  frequency response, 224
  impedance inverter values, 223
  reflection coefficient, 223
  slot discontinuity, 222
  Ridge waveguides, 137–40

S

Scalar network analyzers, 34
Scattering matrix, 19–30
  capacitive iris, 227
  of common networks, 26
  defined, 19–24
  directional couplers, 32
  elements, 27
  equivalent T-network and, 187
  insertion loss concept and, 28–30
  of loss-less passive two-port network, 148–49
  of lossless three-port network, 28
  measurement of, 30–34
  of multiport networks, 21–22
  reference plane shift and, 24–27
  T-junctions, 73
  transformation of, 24–27

usefulness of, 28
Second Cauer form, 152, 156
Series stub-loaded band-stop filters, 257
Shielded microstrip line
  analysis equations, 92
  characteristic impedance, 89, 91
  defined, 88
  effective dielectric constant, 88
  frequency dispersion, 90
  homogeneous equivalent, 88
  suspended, 98–100
  See also Microstrip lines
Shunt inductor, 148
Shunt series resonant circuits, 231, 232
Single-conductor closed transmission lines
  circular waveguides, 112–16
  discontinuities, 120
  finite element modal expansion method, 133–37
  hollow metallic waveguides, 105–12
  maximum power-handling capability, 116
  mode matching method, 124–32
  three-dimensional discontinuity, 137–41
  waveguide asymmetric H-plane step, 122–24
  waveguide discontinuity analysis, 132–33
  See also Transmission lines
Sinusoidal voltage, 6–8
Slot-coupled coaxial-resonator-based combline filter, 309, 373–74
Slot discontinuity, 222
Square coaxial line, 60–61
Step discontinuity, 58, 69–70, 123
Stepped impedance edge parallel-coupled band-pass filters, 283
Stripline low-pass filter, 232
Striplines, 103–4
  attenuation constant, 66–67
  balanced, 61, 62–64
  basic configuration, 61–62
  bend discontinuity, 73–74
  broadside-coupled suspended, 102–3
  characteristic impedance, 62–64
  circular hole discontinuity, 71
  cutoff wavelength, 63
  defined, 61
  discontinuities, 67–74
  edge-coupled, 74–77

field configuration, 62
gap discontinuity, 70–71
low-pass cascaded UE filter, 218
modes in, 62
open-end discontinuity, 71–72
parallel plate equivalent, 69
power-handling capability, 67
propagation constant, 65
step discontinuity, 69–70
stepped impedance low-pass filter, 217
synthesis of, 65–66
T-junction discontinuity, 72–73
unbalanced, 64–65
unloaded Q-factor, 66–67
See also Transmission lines
Stub-loaded band-stop filters
bandwidth parameter, 254
defined, 253
design parameters, 253
E-plane iris-coupled, 261
illustrated, 253
series, 257
shunt, 255
WAVECON input file, 255
WAVECON output file, 256
wide stopband shunt, 254
Suspended microstrip line
characteristic impedance, 96–97
defined, 96
discontinuities, 103–4
open, 97
shielded, 98–100
See also Microstrip lines
Suspended striplines, broadside-coupled, 102–3

T

Tapered corrugated low-pass filters
defined, 224
equivalent network, 225
illustrated, 224
impedances of distributed elements, 225–26
performance parameters, 224–25
specifications and dimensions, 228
tapering profiles, 227
Tap-point locations, 300–303

Three-dimensional discontinuities, in rectangular waveguide, 137–41
Time delay
Butterworth filter, 161
Chebyshev filter, 168
T-junction discontinuity, 72–73
T-junctions, 58, 59, 400
Transfer function
Chebyshev low-pass filters, 166
commensurate line filters, 213
elliptic low-pass filters, 169
Transmission line elements (TLEs)
defined, 210
electrical lengths, 213
Transmission line equivalent circuit, 209
Transmission lines, 41–141
characteristic impedance, 15
with electrical discontinuity, 44–45
equations, 41–44
frequency dependence of attenuation, 141
illustrated, 14
inhomogeneous, 87–105
introduction to, 41
parallel-coupled, 74–87
parameters, 41, 43
planar, 338–44
single-conductor closed, 105–41
strip, 61–74
theory, 14
two-conductor, 45–59
with voltage standing wave, 44
Transmission zeros, 10
Triplexers, 2
Two-conductor transmission lines
coaxial line, 46–59
conductors of equal diameter, 45–46
cross section, 46
general equation for attenuation, 50–52
loss calculation, 51
See also Transmission lines
Two-port networks
ABCD matrix representation, 36
admittance matrix, 20
circuit representation, 28
doubly terminated, 201
with generator, 147

Two-port networks (continued)
    illustrated, 27, 194
    inserted, 29
    parameters, 11
    power loss ratio, 193
    reciprocal, 35
    in tandem, 36
    terminated, 146

U

Unbalanced striplines, 64–65
Unidirectional scalar network analyzer, 33
Unit elements (UEs)
    cascaded to series short-circuited transmission line, 210
    defined, 210
    Kuroda's identity using, 211–12
    low-pass filter cascaded with, 212
    low-pass prototype, 217
    mixture with impedance inverters, 219–29
    shunt open-circuited transmission line cascaded to, 211
Unnormalized ABCD matrix, 38–39

V

Voltage
    backward, 16
    defined, 14
    equivalent, 13–17
Voltage standing wave, 44

W

Waffle-iron filter, 229
WAVEFIL, 224
Waveguide band-pass filters
    computed frequency response, 315
    design, 310–14
    equivalent circuit, 312
    equivalent network, 314
    evanescent-mode, 314–19
    have-wavelength-long cavities, 310
    input file, 315
    iris-coupled, 312
    K-inverter value, 313, 314
    output file, 315
    procedure, 311–13
    typical interface, 316
    waveguide length, 314
Waveguide band-stop filters
    characteristic admittance, 259
    computed response, 262
    configurations, 257
    design example, 259–63
    design steps, 258–63
    E-plane iris-coupled, 261
    equation for, 258
    guided wavelength, 260
    magnetic polarizabilities, 260
    optimized response, 263
    ridged waveguide resonator elements, 258
    susceptance slope parameters, 259
    synthesis of, 259
    See also Band-stop filters
Waveguide directional filter, 395, 396
Waveguide discontinuities, 120–22
    analysis, 132–33
    asymmetric step, 123
    E-plane discontinuities, 135–37
    finite element modal expansion method, 133–37
    H-plane discontinuities, 133–35
    mode matching method and, 124–32
    step discontinuity, 121
    three-dimensional, 137–41
Waveguide folded and elliptic filter
    12-pole Chebyshev filter, 376
    basic design, 375
    cross-coupled filter, 378
    cross-coupling generation, 375
    design, 375–79
    eight-cavity linear phase filter, 379
    folded Chebyshev filter, 377
    nonadjacent resonators, 375
    specifications, 375
Waveguide high-pass filter design, 242–43
Waveguides
    asymmetric H-plane step, 122–24
    average power rating of, 117
    cutoff frequency, 129, 228
    cutoff wavenumber, 109
    dielectric loss in, 115–16

double ridge, 138
electric and magnetic field components, 109
generalized step junction, 129
hollow metallic, 105–12
iris analysis, 130–32
maximum power-handling capability of, 116–19
position for heat transfer relations, 116
propagation constant, 109
ridge, 137–40
TE and TM mode fields, 107, 111
See also Circular waveguides; Rectangular waveguides

Artech House Microwave Library

Behavioral Modeling and Linearization of RF Power Amplifiers, John Wood

Chipless RFID Reader Architecture, Nemai Chandra Karmakar, Prasanna Kalansuriya, Randika Koswatta, and Rubayet E-Azim

Control Components Using Si, GaAs, and GaN Technologies, Inder J. Bahl

Design of Linear RF Outphasing Power Amplifiers, Xuejun Zhang, Lawrence E. Larson, and Peter M. Asbeck

Design Methodology for RF CMOS Phase Locked Loops, Carlos Quemada, Guillermo Bistué, and Iñigo Adin

Design of CMOS Operational Amplifiers, Rasoul Dehghani

Design of RF and Microwave Amplifiers and Oscillators, Second Edition, Pieter L. D. Abrie

Digital Filter Design Solutions, Jolyon M. De Freitas

Discrete Oscillator Design Linear, Nonlinear, Transient, and Noise Domains, Randall W. Rhea

Distortion in RF Power Amplifiers, Joel Vuolevi and Timo Rahkonen

Distributed Power Amplifiers for RF and Microwave Communications, Narendra Kumar and Andrei Grebennikov

Electronics for Microwave Backhaul, Vittorio Camarchia, Roberto Quaglia, and Marco Pirola, editors

EMPLAN: Electromagnetic Analysis of Printed Structures in Planarly Layered Media, Software and User's Manual, Noyan Kinayman and M. I. Aksun

An Engineer's Guide to Automated Testing of High-Speed Interfaces, Second Edition, José Moreira and Hubert Werkmann

Envelope Tracking Power Amplifiers for Wireless Communications, Zhancang Wang

Essentials of RF and Microwave Grounding, Eric Holzman

FAST: Fast Amplifier Synthesis Tool—Software and User's Guide, Dale D. Henkes

Feedforward Linear Power Amplifiers, Nick Pothecary

Filter Synthesis Using Genesys S/Filter, Randall W. Rhea

Foundations of Oscillator Circuit Design, Guillermo Gonzalez

Frequency Synthesizers: Concept to Product, Alexander Chenakin

Fundamentals of Nonlinear Behavioral Modeling for RF and Microwave Design, John Wood and David E. Root, editors

Generalized Filter Design by Computer Optimization, Djuradj Budimir

Handbook of Dielectric and Thermal Properties of Materials at Microwave Frequencies, Vyacheslav V. Komarov

Handbook of RF, Microwave, and Millimeter-Wave Components, Leonid A. Belov, Sergey M. Smolskiy, and Victor N. Kochemasov

High-Linearity RF Amplifier Design, Peter B. Kenington

High-Speed Circuit Board Signal Integrity, Stephen C. Thierauf

Integrated Microwave Front-Ends with Avionics Applications, Leo G. Maloratsky

Intermodulation Distortion in Microwave and Wireless Circuits, José Carlos Pedro and Nuno Borges Carvalho

Introduction to Modeling HBTs, Matthias Rudolph

Introduction to RF Design Using EM Simulators, Hiroaki Kogure, Yoshie Kogure, and James C. Rautio

Introduction to RF and Microwave Passive Components, Richard Wallace and Krister Andreasson

Klystrons, Traveling Wave Tubes, Magnetrons, Crossed-Field Amplifiers, and Gyrotrons, A. S. Gilmour, Jr.

Lumped Elements for RF and Microwave Circuits, Inder Bahl

Lumped Element Quadrature Hybrids, David Andrews

Microstrip Lines and Slotlines, Third Edition, Ramesh Garg, Inder Bahl, and Maurizio Bozzi

Microwave Circuit Modeling Using Electromagnetic Field Simulation, Daniel G. Swanson, Jr. and Wolfgang J. R. Hoefer

Microwave Component Mechanics, Harri Eskelinen and Pekka Eskelinen

Microwave Differential Circuit Design Using Mixed-Mode S-Parameters, William R. Eisenstadt, Robert Stengel, and Bruce M. Thompson

Microwave Engineers' Handbook, Two Volumes, Theodore Saad, editor

Microwave Filters, Impedance-Matching Networks, and Coupling Structures, George L. Matthaei, Leo Young, and E. M. T. Jones

Microwave Materials and Fabrication Techniques, Second Edition, Thomas S. Laverghetta

Microwave Materials for Wireless Applications, David B. Cruickshank

Microwave Mixer Technology and Applications, Bert Henderson and Edmar Camargo

Microwave Mixers, Second Edition, Stephen A. Maas

Microwave Network Design Using the Scattering Matrix, Janusz A. Dobrowolski

Microwave Radio Transmission Design Guide, Second Edition, Trevor Manning

Microwave and RF Semiconductor Control Device Modeling, Robert H. Caverly

Microwave Transmission Line Circuits, William T. Joines, W. Devereux Palmer, and Jennifer T. Bernhard

Microwaves and Wireless Simplified, Third Edition, Thomas S. Laverghetta

Modern Microwave Circuits, Noyan Kinayman and M. I. Aksun

Modern Microwave Measurements and Techniques, Second Edition, Thomas S. Laverghetta

Modern RF and Microwave Filter Design, Protap Pramanick and Prakash Bhartia

Neural Networks for RF and Microwave Design, Q. J. Zhang and K. C. Gupta

Noise in Linear and Nonlinear Circuits, Stephen A. Maas

Nonlinear Microwave and RF Circuits, Second Edition, Stephen A. Maas

On-Wafer Microwave Measurements and De-Embedding, Errikos Lourandakis

Passive RF Component Technology: Materials, Techniques, and Applications, Guoan Wang and Bo Pan, editors

Practical Analog and Digital Filter Design, Les Thede

Practical Microstrip Design and Applications, Günter Kompa

Practical Microwave Circuits, Stephen Maas

Practical RF Circuit Design for Modern Wireless Systems, Volume I: Passive Circuits and Systems, Les Besser and Rowan Gilmore

Practical RF Circuit Design for Modern Wireless Systems, Volume II: Active Circuits and Systems, Rowan Gilmore and Les Besser

Production Testing of RF and System-on-a-Chip Devices for Wireless Communications, Keith B. Schaub and Joe Kelly

Q Factor Measurements Using MATLAB®, Darko Kajfez

QMATCH: Lumped-Element Impedance Matching, Software and User's Guide, Pieter L. D. Abrie

Radio Frequency Integrated Circuit Design, Second Edition, John W. M. Rogers and Calvin Plett

RF Bulk Acoustic Wave Filters for Communications, Ken-ya Hashimoto

RF Design Guide: Systems, Circuits, and Equations, Peter Vizmuller

RF Linear Accelerators for Medical and Industrial Applications, Samy Hanna

RF Measurements of Die and Packages, Scott A. Wartenberg

The RF and Microwave Circuit Design Handbook, Stephen A. Maas

RF and Microwave Coupled-Line Circuits, Rajesh Mongia, Inder Bahl, and Prakash Bhartia

RF and Microwave Oscillator Design, Michal Odyniec, editor

RF Power Amplifiers for Wireless Communications, Second Edition, Steve C. Cripps

RF Systems, Components, and Circuits Handbook, Ferril A. Losee

Scattering Parameters in RF and Microwave Circuit Analysis and Design, Janusz A. Dobrowolski

The Six-Port Technique with Microwave and Wireless Applications, Fadhel M. Ghannouchi and Abbas Mohammadi

Solid-State Microwave High-Power Amplifiers, Franco Sechi and Marina Bujatti

Stability Analysis of Nonlinear Microwave Circuits, Almudena Suárez and Raymond Quéré

Substrate Noise Coupling in Analog/RF Circuits, Stephane Bronckers, Geert Van der Plas, Gerd Vandersteen, and Yves Rolain

System-in-Package RF Design and Applications, Michael P. Gaynor

Terahertz Metrology, Mira Naftaly, editor

TRAVIS 2.0: Transmission Line Visualization Software and User's Guide, Version 2.0, Robert G. Kaires and Barton T. Hickman

Understanding Microwave Heating Cavities, Tse V. Chow Ting Chan and Howard C. Reader

Understanding Quartz Crystals and Oscillators, Ramón M. Cerda

For further information on these and other Artech House titles, including previously considered out-of-print books now available through our In-Print-Forever® (IPF®) program, contact:

Artech House Publishers
685 Canton Street
Norwood, MA 02062
Phone: 781-769-9750
Fax: 781-769-6334
e-mail: artech@artechhouse.com

Artech House Books
16 Sussex Street
London SW1V 4RW UK
Phone: +44 (0)20 7596 8750
Fax: +44 (0)20 7630 0166
e-mail: artech-uk@artechhouse.com

Find us on the World Wide Web at: www.artechhouse.com